Geophysical Monograph Series

Including

IUGG Volumes

Maurice Ewing Volumes
Mineral Physics Volumes

GEOPHYSICAL MONOGRAPH SERIES

Geophysical Monograph 72
IUGG Volume 12

Dynamics of Earth's Deep Interior and Earth Rotation

J.-L. Le Mouël
D.E. Smylie
T. Herring

Editors

American Geophysical Union

International Union of Geodesy and Geophysics

Published under the aegis of the AGU Books Board

Library of Congress Cataloging-in-Publication Data

Dynamics of earth's deep interior and earth rotation / J.-L. Le Mouël, D.E. Smylie,
 T. Herring, editors.
      p.   cm. — (Geophysical monograph : 72)
ISBN 0-87590-463-7
1. Earth—Core.   2. Earth—Mantle.   3. Geodynamics.   I. Le Mouël,
 J. L.   II. Smylie, D. E.   III. Herring, T. (Thomas)   IV. Series.
QE509.D87   1993
551.1'12—dc20
                                                          93-7784
                                                          CIP

ISSN 0065-8448
ISBN 087590-463-7

CONTENTS

PREFACE

The study of the Earth's deep interior is the object of a spectacular development due both to new techniques of observation (including very long baseline interferometry and superconducting gravimeters) and to progress in theory spurred by new computing capability. Stimulated by the international SEDI group, founded in 1986, geophysicists from different disciplines—Earth dynamicists, seismologists, geomagneticians, mineral physicists--began to cooperate and integrate more fully one another's work. SEDI meetings favor and promote those close contacts and cooperation. Great efforts will still be needed before all the disciplinary divisions dissolve--if they ever do--but things are clearly improving, as shown by this AGU monograph. We think indeed that this volume is a good, although incomplete, illustration of the situation as described above and that it is a benchmark in the exciting story of the progress in knowledge of the deep interior of our planet.

More than half the papers in this volume concern the dynamics of the core and rotation of the Earth. Interest in this topic has, of course, been renewed and enhanced by the introduction of high-sensitivity superconducting gravimeters and new techniques for measuring Earth rotation, especially VLBI. One of the first results coming from these new data was the 12.9-hour peak in the oldest series, the Brussels one. A new analysis of the series confirms the existence of the peak, but it seems to be due to a regional effect not yet taken into account in the tidal residues rather than being the gravimetric signal of a core mode. In fact, a comparison of the existing records makes it clear that a stacking of signals coming from a worldwide array of stations, over a common period of several years, will be necessary to detect geodynamical signals from the core safely. But right now, existing gravity data from the IDA array can be used to determine the period of the nearly diurnal free wobble and the corresponding quality factor Q. These values can be compared with those derived from VLBI nutation measurements; a significant discrepancy appears in the values of Q, which could be due to an inaccurate ocean loading correction in tidal studies. Other comparisons of VLBI and gravity data, theoretically equivalent, can be performed when searching for short-period components of the polar motion; encouraging agreement has been observed, although the results are close to the limit of resolution. At the same time, while data acquisition and analysis are improving, theoretical studies of the dynamics of the core are being vigorously developed. More and

more interest is being paid to the motions of the solid inner core, the torques which act on it, and the effect of these motions on Earth rotation. Much attention is also devoted to the computation of the core modes and of their period (theory leading observation and guiding it). Different models of Earth are considered, rotating or not rotating, and different techniques of computation, either numerical or analytical, are used, leading to somewhat different results. The triplet of periods of the translational modes computed for a rotating Earth, in the subseismic approximation, seems to be compatible with peaks identified in the signal of the superconducting gravimeters. This exciting suggestion probably needs more work, both observational and theoretical. The tidal forcing of core modes is further examined, including the now well-established enhancement of the gravimetric factor due to existence of the nearly diurnal free wobble. Another point of interest is the temporal variation of the degree 2 zonal term, J2, of the terrestrial gravity potential; generally attributed to relaxation from the removal of external loads, such as that due to deglaciation. It could also result from internal loads located, for example, at the core-mantle boundary (CMB). Other papers are devoted to the core flow, which could balance the decade variations of the length of the day. It seems that particular zonal motions which extend through the whole core but can be determined from the flow at the CMB derived from geomagnetic secular variation data have the right temporal behavior and order of magnitude to conserve the angular momentum of the core-mantle system. The required coupling mechanism between the core and the mantle could be topographic, due to interaction between the flow at the top of the core and a bumpy CMB. This mechanism, and particularly the computation of the flow at the top of the core, probably still needs further work.

A topic which has attracted much attention and is progressing rapidly is seismic imaging of the aspherical structure of the mantle, the core-mantle boundary, and the uppermost layers of the core. Of particular interest is its relation to convection in the Earth. For example, a flow model in the viscous mantle can be derived from the computed density heterogeneities, and the topography of the CMB induced by this flow can be inferred. This flow-induced topography appears to be in fair agreement with previous models of topography derived from PcP and PKP travel times. Lateral velocity heterogeneities in the outermost 200 km of the core are now investigated using SmKS travel times; a dominant degree 1 pattern appears.

Many important problems related to the structure and composition of the mantle core are still unsolved. Is mantle convection two-layered or whole? What is the nature of the D" layer and the temperature at the base of the mantle? Seismology, numerical simulations, and mineral physics data are the tools to approach these problems. Diffraction of P and S waves around the CMB reveals heterogeneities in the velocities in the D" layer of a few percent. When estimating the seismic velocity changes that could be due to changes in temperature or composition, it turns out that the latter has a predominant influence. Numerical modeling predicts chaotic behavior for the mixing rate of the upper and lower mantle, although this depends strongly on the Clapeyron slope of the phase transformation at 670 km. If this is the case, then the mantle is likely to be layered or intermittently layered.

Resolution of the issues raised in this volume will clearly depend on the interdisciplinary collaborations that SEDI has been so successful in stimulating.

J. L. Le Mouël
D. E. Smylie
T. A. Herring
Editors

FOREWORD

The scientific work of the International Union of Geodesy and Geophysics (IUGG) is primarily carried out through its seven associations: IAG (briefly, Geodesy), IASPEI (Seismology), IAVCEI (Volcanology), IAGA (Geomagnetism), IAMAP (Meteorology), IAPSO (Oceanography), and IAHS (Hydrology). The work of these associations is documented in various ways.

Dynamics of Earth's Deep Interior and Earth Rotation is one of a group of volumes published jointly by IUGG and AGU that are based on work presented at the Inter-Association Symposia as part of the IUGG General Assembly held in Vienna, Austria, in August 1991. Each symposium was organized by several of IUGG's member associations and comprised topics of interdisciplinary relevance. The subject areas of the symposia were chosen such that they would be of wide interest. Also, the speakers were selected accordingly, and in many cases, invited papers of review character were solicited. The series of symposia were designed to give a picture of contemporary geophysical activity, results, and problems to scientists having a general interest in geodesy and geophysics.

In view of the importance of these interdisciplinary symposia, IUGG is grateful to AGU for having put its unique resources in geophysical publishing expertise and experience at the disposal of IUGG. This ensures accurate editorial work, including the use of peer reviewing. So the reader can expect to find expertly published scientific material of general interest and general relevance.

Helmut Moritz
President, IUGG

Core Dynamics and Surface Gravity Changes

JACQUES HINDERER[1] AND DAVID CROSSLEY

Geophysics Laboratory, McGill University, 3450 University Street,
Montreal, Quebec, Canada H3A 2A7

The surface gravity changes caused by planetary-scale dynamical phenomena taking place in the Earth's core are investigated. We first show that specific boundary conditions related to the core flow (in normal or transverse traction, in gravitational potential and gradient) have to be considered at the different interfaces (inner core-outer core, outer core-mantle) and induce an elastogravitational deformation of the solid parts. The resulting gravity changes can then be expressed in terms of gravimetric factors which are a combination of generalized Love numbers. Several problems dealing with large-scale oscillatory flows in the fluid outer core are addressed: excitation level caused by earthquakes, anelastic damping, dependence of the surface gravity estimate on some classical approximations for the core flow. Recent work in observational detection is also reviewed. We finally show how the wobbling motions of the inner and outer cores lead to a double resonance in the tidal gravimetric factor, which means that a very accurate determination of tidal gravity changes is able to provide useful information on some physical or geometrical parameters of the Earth's deep interior.

Introduction

The spectrum of gravity disturbances that can be observed at the Earth's surface with high-precision superconducting gravimeters ranges from low-frequency variations like those induced by polar motion (annual and Chandler wobbles) to high-frequency changes of seismic origin (e.g. normal modes). The largest contributions are of lunisolar tidal origin and a non-negligible part comes from the loading of the atmosphere. We investigate here gravity changes related to long-period oscillatory motions in the Earth's core having typical periods from a few hours to one day; this includes oscillations in the liquid core driven by buoyancy forces, as well as wobbles of the inner and outer cores. For this we consider the elasto-gravitational deformation with the help of a Love number formalism which has been developed elsewhere [Hinderer and Legros, 1989; Hinderer et al., 1991a]. In particular, we

---

[1] Permanently at Institut de Physique du Globe, 5, rue René Descartes, 67084 Strasbourg Cedex, France.

Dynamics of Earth's Deep Interior and Earth Rotation
Geophysical Monograph 72, IUGG Volume 12
Published in 1993 by the International Union of Geodesy and Geophysics and the American Geophysical Union.

give the surface gravity amplitude associated with possible seismic excitation of core modes and the resultant damping times due to the anelastic behaviour of the Earth. The influence of several classical approximations for the core flow on surface gravity is estimated. Recent attempts to detect core modes effects in gravity observations are discussed, the ultimate goal being to determine the stability of the density stratification in the fluid outer core. Special attention is paid to the double resonance effect caused by the inner and outer core wobbles which arises in the gravimetric factor of diurnal tides.

Theoretical Background

We briefly review the equations and boundary conditions which are needed for expressing the elasto-gravitational static deformation of an Earth model (hydrostatically prestressed and spherically symmetric) with a liquid outer core where dynamical process occur; more details can be found in Hinderer et al. [1991a]. The linearized set of equations for a spheroidal mode of deformation can be written as a first-order linear differential system:

$$\dot{y}_i(r) = c_{ij}(r)y_j \qquad i,j = 1,...,6 \quad (1)$$

where $c_{ij}(r)$ are algebraic functions of radius r and the elastic shear modulus, compressibility, and density of the Earth model. Equations (1) reduce to a set of six equations for the six unknowns y_i (first introduced by Alterman et al., [1959]) related to the displacement $\mathbf{u}$, the traction $\mathbf{T}$ and the mass redistribution potential ϕ:

$$\mathbf{u} = \sum_n [y_{1n}(r)Y_n(\theta,\lambda)\frac{\mathbf{r}}{r} + ry_{3n}(r)\nabla Y_n(\theta,\lambda)]$$

$$\mathbf{T} = \sum_n [y_{2n}(r)Y_n(\theta,\lambda)\frac{\mathbf{r}}{r} + ry_{4n}(r)\nabla Y_n(\theta,\lambda)]$$

$$\hspace{5cm}(2)$$

$$V + \phi = \sum_n [y_{5n}(r)Y_n(\theta,\lambda)]$$

$$y_{6n}(r) = \frac{dy_{5n}(r)}{dr} - 4\pi G\rho(r)y_{1n}(r)$$

In equation (2), $\rho(r)$ is the unperturbed density and V(r) any external gravitational potential (e.g. tidal); G is the gravitational constant, $\mathbf{r}$ the radius vector and $Y_n(\theta,\lambda)$ a surface spherical harmonic function of degree n (colatitude θ; longitude λ).

In equations (1) and (2), $y_1(r)$ is the radial displacement, $y_2(r)$ the normal stress, $y_3(r)$ the transverse displacement, $y_4(r)$ the transverse stress, $y_5(r)$ the gravitational potential and $y_6(r)$ the gravitational flux density.

Toroidal modes of deformation are of no interest here because, for a spherical Earth model, they do not perturb the gravity [e.g., Merriam, 1985]; they do however couple spheroidal modes in a rotating reference frame [Crossley, 1975a].

Outer Core Dynamics

The dynamical behaviour of the liquid outer core depends in the general case on rotational, magnetic, viscous and buoyancy forces. These forces give rise to the following source terms (or specific boundary conditions) at the core-mantle and inner core boundaries:

at ICB :	at CMB :
Y_n^{22}	Y_n^{12}
Y_n^{24}	Y_n^{14}
Y_n^{25}	Y_n^{15}
Y_n^{26}	Y_n^{16}

We followed the notation introduced by Crossley et al. [1991] where the source terms appearing because of the fluid dynamics (e.g., oscillations in the fluid core) are denoted $Y^{\mathcal{ST}}$, where the index $\mathcal{S}$ stands for the spatial location (1 = core-mantle boundary CMB, 2 = inner core boundary ICB) and the index $\mathcal{T}$ for the type (2 = normal stress, 4 = transverse stress, 5 = gravitational potential, 6 = gravitational flux density).

Elastogravitational Solution

Owing to these source terms located at the two different solid-fluid boundaries, the inner core and the mantle will be elastically deformed. When the different continuity conditions in traction or potential at the ICB and CMB are taken into account (see, e.g., Dahlen, [1974]), as well as at the Earth's center and surface, the solid elastogravitational deformation is then entirely given by the radially dependent solutions y_i.

It is known that Love numbers are a convenient and useful description of the response of the solid parts of the Earth to dynamics in the fluid regions [Hinderer and Legros, 1989]. It means that the solutions y_i in the solid parts (equation (1)) can be assumed proportional to any source term with the help of a specific Love number.

For instance, the radial displacement and the change in gravitational potential at the surface (r = a) become:

$$y_1(a) = (h_0^{12})_n \frac{Y_n^{12}}{\rho g} + (h_0^{14})_n \frac{Y_n^{14}}{\rho g} + (h_0^{15})_n \frac{Y_n^{15}}{g}$$
$$+ (h_0^{16})_n \frac{a Y_n^{16}}{g} + (h_0^{22})_n \frac{Y_n^{22}}{\rho g} + (h_0^{24})_n \frac{Y_n^{24}}{\rho g}$$
$$+ (h_0^{25})_n \frac{Y_n^{25}}{g} + (h_0^{26})_n \frac{a Y_n^{26}}{g}$$

$$\text{(3)}$$

$$y_5(a) = (k_0^{12})_n \frac{Y_n^{12}}{\rho} + (k_0^{14})_n \frac{Y_n^{14}}{\rho} + (k_0^{15})_n Y_n^{15}$$
$$+ (k_0^{16})_n a Y_n^{16} + (k_0^{22})_n \frac{Y_n^{22}}{\rho} + (k_0^{24})_n \frac{Y_n^{24}}{\rho}$$
$$+ (k_0^{25})_n Y_n^{25} + (k_0^{26})_n a Y_n^{26}$$

for each spheroidal deformation of degree n (g is the surface gravity). The notation for the different kinds of Love numbers (h_0^{ST}, k_0^{ST}) is similar to the one introduced for the source terms, except there is in addition a subscript (here 0) indicating the location (outer surface) of the

elastic effect (see also Table 2 in Hinderer et al., [1991a]). Due to the requirement of non-singular solutions at the Earth's centre, there are only 3 independent triplets of Love numbers to describe the deformation of the ICB (as a function of any 3 arbitrary core variables); similarly, because of the boundary conditions at the Earth's surface, there are only 3 independent triplets of Love numbers for the CMB deformation. In the case that the core has zero rigidity and zero viscosity, the Love numbers for transverse stress (at each boundary) also vanish leaving three pairs of independent Love numbers in the inner core and mantle. Additionally the pair of Love numbers for the transverse displacement is uninteresting as this quantity is discontinuous at the fluid-solid boundaries; thus we are left with two pairs of independent Love numbers at each boundary (in terms of 2 arbitrary core variables as source terms).

From the set of equations (3), taking into account the displacement in the initial gravitational field (terms in y_1), as well as the mass redistribution potential (terms in y_5) in addition to any direct potential (if existing), the surface gravity change becomes:

$$- \frac{a}{n} \Delta g_n(a) = (\delta_0^{12})_n \frac{Y_n^{12}}{\rho} + (\delta_0^{14})_n \frac{Y_n^{14}}{\rho}$$
$$+ (\delta_0^{15})_n Y_n^{15} + (\delta_0^{16})_n a Y_n^{16} + (\delta_0^{22})_n \frac{Y_n^{22}}{\rho} \text{(4)}$$
$$+ (\delta_0^{24})_n \frac{Y_n^{24}}{\rho} + (\delta_0^{25})_n Y_n^{25} + (\delta_0^{26})_n a Y_n^{26}$$

with the help of generalized gravimetric factors which are a combination of the Love numbers appearing in equations (3) [Hinderer and Legros, 1989]:

$$(\delta_0^{ST})_n = \frac{2}{n}(h_0^{ST})_n - \frac{n+1}{n}(k_0^{ST})_n$$

Each gravimetric factor of degree n is related to a specific source term of the same degree. If a dynamic process involves source terms of different degrees, then the surface gravity changes are obtained by summing up the different harmonic contributions.

Core Oscillations

In this section, we will review several aspects concerning oscillatory flows which may exist in the Earth's fluid core: excitation, damping, dependence of surface gravity on flow approximation and, finally, observational detection.

In the liquid outer core, we have the same equations of motion as in a solid, except that Hooke's law reduces to a state equation relating the change in the fluid pressure to the elastic compressibility and to the displacement in the initial gravity field [Crossley and Rochester, 1980].

In the most general case of a core flow, there is no simple relationship between different types of boundary conditions and the induced surface gravity change requires the use of all the 8 gravimetric factors related to the various sources introduced in equation (4). Moreover, as previously said, if several spherical harmonic terms of different degree n are present in the boundary conditions (as it is indeed the case when Coriolis coupling is taken into account in the equations of motion), one has to sum up the different degrees for each gravimetric factor.

Viscosity estimates for the Earth's fluid outer core can vary by as much as 10 orders of magnitude according to the method [Lumb and Aldridge, 1991, Table 1]. If laboratory experiments give very low values close to that of water [Poirier, 1988], estimates of core viscosity from geophysical or astronomical observations like the damping of the Free Core Nutation in gravimetry [Neuberg et al., 1987] or the the phase lag of the annual retrograde nutation in Very Long Baseline Interferometry (VLBI) [Gwinn et al., 1986] suggest much larger values. However viscosity is usually neglected in the equations of motion. The role of the magnetic field at periods less than several years has

also been shown to be negligible [Crossley and Smylie, 1975]. In both cases, shear stress boundary conditions identically vanish (source terms of type 4 vanish).

The large and complex effect of the Earth's rotation on the fluid motion has prompted the consideration of further simplifications to the inviscid equations of motion. Often the full fourth-order system of fluid equations (without viscosity and magnetic field) is simplified in order to decouple the pressure perturbation from the gravity perturbations. Two major approximations are usually assumed for the Earth's liquid core:

— the Boussinesq approximation where the fluid is taken as incompressible and the gravitational disturbances are negligibly small [Crossley and Rochester, 1980]; notice that this approximation takes into account the core density stratification through a parameterised form of N^2 in the momentum equation, where N is the buoyancy frequency. This means that source terms of type 5 and 6 are ignored at the ICB and CMB, and the surface gravity depends only on the pressure conditions at these boundaries [Crossley et al., 1991];

— the subseismic (below acoustic frequencies) approximation, first introduced by Smylie and Rochester [1981], which takes into account the core compressibility but modifies the equation of state. It neglects the contribution of flow pressure to elastic compression compared to transport through the hydrostatic pressure field and represents therefore an intermediate step between the Boussinesq approximation and the full system of equations. There is a perturbation in the gravitational potential related to the fluid motion (Poisson's equation) and the surface gravity change requires the introduction of two internal load gravimetric factors [Smylie et al., 1990; Crossley et al., 1991].

Excitation

The excitation problem for long period core oscillations has been investigated for a non-rotating Earth by Crossley [1988]. This preliminary study dealing only with low degree harmonic modes has been recently extended to higher degrees (up to 20) by Crossley et al. [1991].

Using a standard seismic source model (the 1960 Chilean earthquake), these authors computed the inferred excitation of different core modes and the total resulting gravity perturbation at the Earth's surface. The results for low-degree oscillations of large spatial scale have been recomputed using an improved Earth model and are shown on Figure 1, where n and m are the degree and order of the spherical har-

monic component and where seismic overtones are denoted by a positive radial number and core undertones by a negative one. The detection limit in gravity has been taken as 1 nanogal, which is a very optimistic value even for superconducting gravimeters as will be seen later. All degree 2 overtones easily exceed this threshold of detectability, whereas it is evident that all low degree (from n= 1 to n = 10) core modes lie well below this threshold.

The Slichter mode of degree n = 1 and overtone number 1 is the primary translational motion of the inner core and is excited at just below

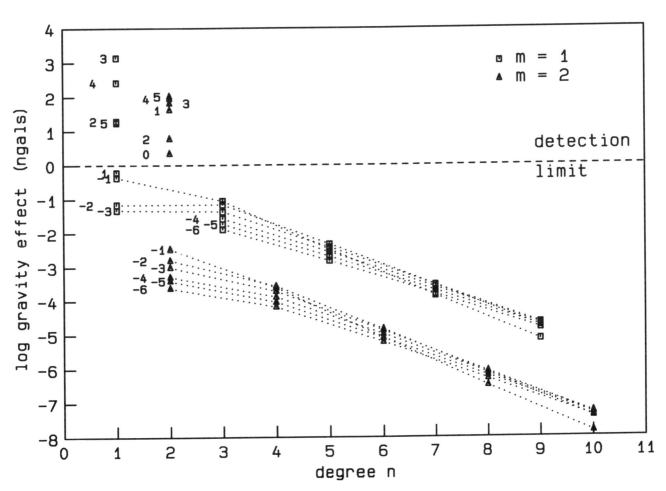

Fig. 1. Amplitude of surface gravity for different seismic overtones (positive radial number) and core undertones (negative radial number) excited by the 1960 Chilean earthquake; n is the degree and m the order.

the nanogal level. On the other hand, the core undertones of this degree (radial numbers -1, -2, -3), which also include a translational motion of the solid inner core, have amplitudes much less than a nanogal. Note in contrast to this calculation, that the triplet of Slichter modes in a rotating Earth has recently been claimed to be observed, at the 5 nanogal level, in a stacked spectrum of superconducting gravimeter records [Smylie, 1992].

Moreover there is an exponential decrease of gravity with degree n, suggesting that there should be a rapid convergence of the surface gravity when introducing, in the rotating case, a coupling chain (sum of spheroidal and torsional terms) in the expression of the core motion. This means that, even assuming a strong earthquake, there is only a weak excitation of core modes leading to surface gravity changes unlikely to be detectable. A possible alternative would be to study other possible source mechanisms such as volcanic eruptions, which have been suggested for the Slichter mode excitation [W. Zürn, 1991, private communication].

Damping

There are, in principle, three major reasons for core modes to be damped: viscosity in the fluid core, electromagnetic dissipation in the inner and outer cores and possibly at the base of the mantle, and anelasticity which can be bulk and shear in the solid parts but only bulk in the liquid core.

The very low viscosity of the liquid core, as well as electromagnetic dissipation, would lead to negligible damping with spin-down times for a 12 hr period oscillation of about 10^4 years. The influence of anelastic dissipation on core modes was studied by Crossley et al. [1991] assuming shear and compressibility dissipation in the inner core and mantle in addition to bulk

dissipation in the fluid outer core. Starting from the Preliminary Reference Earth Model (PREM) Q values extrapolated from the seismic model [Dziewonski and Anderson, 1981] to core periods (say 12-24 hr), the anelastic decay times, which are of the order of a few days for seismic modes, were found to be much larger than one year, increasing with radial number and decreasing with degree (cf. Figure 2). The only exception is the Slichter mode (of degree 1) having a damping time of about 420 days.

Notice that the dominant contribution to the anelastic damping actually comes from the poorly known Q value associated with the outer core compressibility and not from shear dissipation in the mantle. In a more recent Earth model CORE11 [Widmer et al., 1988], this anelastic Q is set to infinity in the fluid core which implies that the anelastic damping of the core modes comes only from the mantle with a consequent lengthening of the damping times in Figure 2. For such a model, viscous effects in the liquid core would become a significant factor.

Influence of Flow Approximation on Gravity

We address here the problem of the computational dependence of surface gravity change on simplifications to the model. There are essentially two major steps: the first one deals with approximations to the equations for the flow itself leading to perturbations in eigenfunction and eigenfrequency with respect to the full elastic theory; the second step concerns the use of a Love number formalism to extend the eigensolutions from the core to the solid parts in order to evaluate the gravity changes.

On one hand, it was shown by Crossley et al. [1991] that the eigenfrequencies of core modes computed for a non-rotating Earth using the full elastic theory vary only very little (about 2.5%) when introducing rigid boundaries and either a Boussinesq or a subseismic approxima-

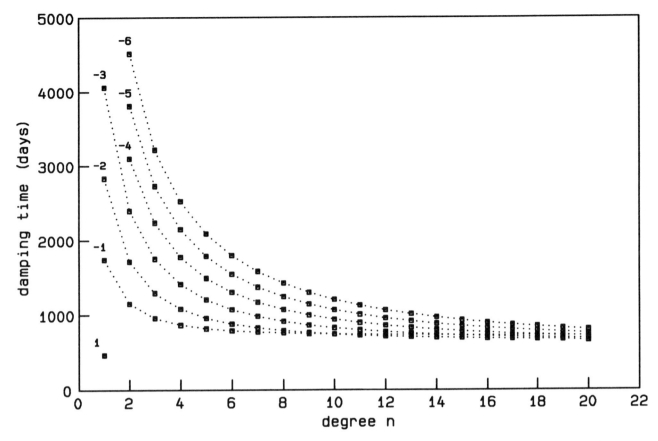

Fig. 2. Damping time of different core undertones due to seismic anelasticity; n is the degree.

tion in the core flow. Moreover the same remark applies to the eigenfunctions themselves (y_1 to y_6) except that, by definition, the Boussinesq solution ignores gravity perturbations (y_5 and y_6)) in the core. On the other hand, the subseismic solution is very close to the rigid boundary solution for all the variables, verifying the utility of the subseismic approximation for describing core flows. Note that the solutions for the core modes were computed assuming rigid boundaries, an approach that Crossley and Rochester [1980] have shown to be valid due to the large rigidity contrast across the core boundaries. Of course rotation will have a major effect in coupling these non-rotating modes into a new eigenfrequency structure (including rotational splitting) that is now just beginning

to be understood (Smylie et al., 1992b). The problem of the excitation of these modes in a rotating Earth has yet to be satisfactorily solved.

On the other hand, the viability of a Love number approach for taking into account the elasticity of the inner core and mantle and for expressing surface gravity changes can be seen from Figure 3, where the synthesized elastic values (based on internal load Love numbers as defined previously) are extremely close (to 1/1000) to the full elastic solution (without Love numbers). Because the load Love numbers are static, a small discrepancy still exists between the use of them and with the full elastic solution due to the inclusion of the inertial term in the solid parts of the Earth model in the latter formulation. Details on the computation of the internal

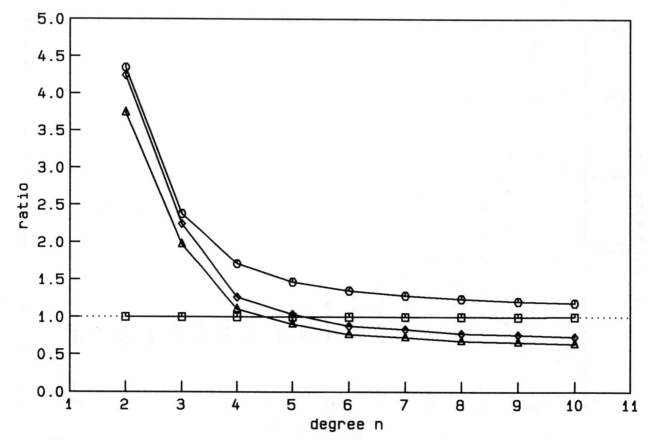

Fig. 3. Ratio of surface gravity computations using different approximations for the core flow to the full elastic case. All symbols refer to a computation using an elastic Love number approach and a perfect agreement would be 1 (dotted line); the squares are for the full equations in the core with elastic boundaries, the circles are for the full equations in the core with rigid boundaries, the triangles and diamonds refer respectively to the Boussinesq and subseismic approximations in the core with rigid boundaries.

load Love numbers relating the elastic solutions to source terms at the ICB and CMB can be found in Crossley et al. [section 7, 1991].

The influence of the two other approximations (Boussinesq and subseismic) is to overestimate the gravity change with respect to the complete elastic theory for low degrees (say n = 2-4), but there is only a weak difference for higher degrees. As indicated above, the Love number formalism for incorporating elasticity in the solid parts was applied using core eigenfunctions computed in a first step for rigid boundaries. A further iteration assuming deformable

boundaries in the computation of the core flow could partly explain the discrepancy between the subseismic ans the full solution in the surface gravity change, but further work is required to verify this point. We note that the degree n=1 case is not relevant here because with rigid core boundaries it is not possible to compute accurately the surface gravity associated with motions which involve translation of the inner core.

How the surface gravity fluctuation explicitly depends on the type of source terms for each core mode of degree n was investigated by Hinderer et al. [1991a] and is summarized in Figure 4. The curve with squares is the contribution only from pressure conditions (type 2) at the core boundaries and shows that the overall effect is close to 50 %. The curve with circles is the effect of the harmonic part of the gravitational potential, as if the mantle would be perfectly rigid. The overall effect of this part is close to 35 %. The curve with triangles shows the remaining contributions i.e. the ones coming from the total deformation of the inner core and the mantle deformation caused by changes in core densities and associated gravitational potential. In general, this part is quite small (less than 12 % for degree n larger than

7). However, the degree 2 is very peculiar in the sense that this non-harmonic part reaches about 50 % and dominates the two other contributions. One of the consequences is that the Boussinesq approximation, in conjunction with the rigid boundary assumption, is certainly incorrect for a degree 2 core mode. These calculations are currently being updated to use elastic boundaries to more accurately evaluate the effect of these approximations on surface gravity.

Detection of Core Modes in Gravity

There has been a considerable observational effort in the last decade in trying to identify

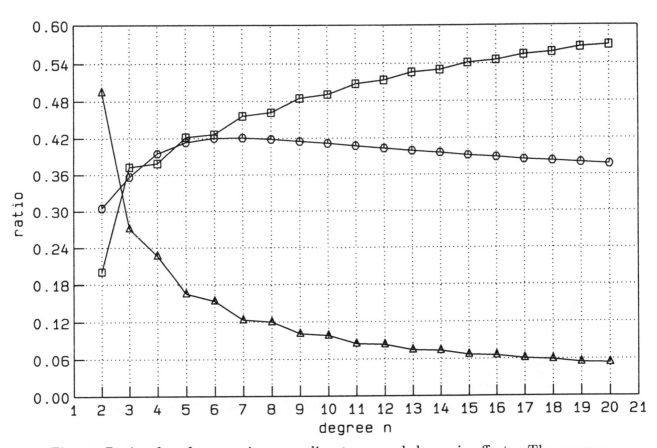

Fig. 4. Ratio of surface gravity according to several dynamic effects. The squares denote the effects of pressure at the ICB and CMB. The circles denote the harmonic part of the gravity effect from the core that would exist even if the mantle was rigid. The triangles give the effect of perturbations in core density which elastically deform the mantle. All effects are normalized by the total gravity.

gravity signals originating from the Earth's core, especially with the help of superconducting gravimeters which have sensitivities of the order of one nanogal (one part in 10^{12} of the Earth's surface gravity).

The first claim of a 15 nanogal signal at 13.9 hr period in the Brussels (Belgium) superconducting gravimeter (SCG) was made by Melchior and Ducarme [1986] and there have been related studies on the same gravity record by Melchior et al. [1988] , Aldridge and Lumb [1987] and Aldridge et al. [1988]. Zürn et al. [1987] investigated the gravity record from the German SCG in Bad Homburg, which is located only 320 km away from Brussels and found no evidence for such a signal. The Brussels record was re-examined again by Mansinha et al. [1990]

showing no statistical evidence for spectral peaks Moreover, Florsch et al. [1991] analyzed a 1.5 year record of the French SCG located near Strasbourg and concluded that there are only a few signals marginally above the ambient noise level, if one excludes the harmonic components of the solar day of atmospheric thermal origin.

A typical residual harmonic line amplitude plot obtained with a long gravity record from a superconducting gravimeter (we used a 3.4 year record (1987-1991) from the Strasbourg instrument) is shown in Figure 5. Such a plot shows the true amplitudes of harmonic signals present in the record, but misrepresents aperiodic noise components (for which a power spectral density plot is more appropriate). This unsmoothed

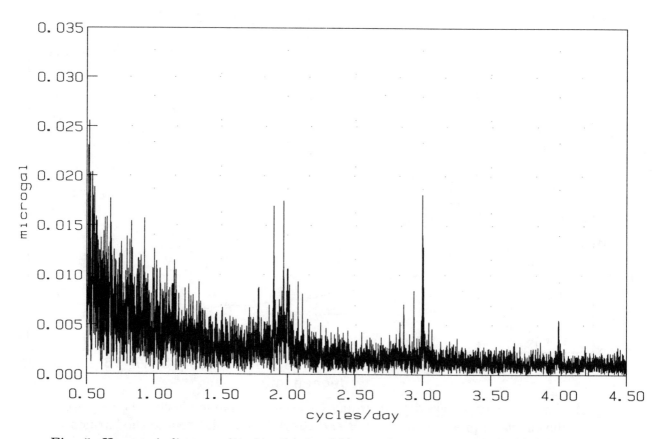

Fig. 5. Harmonic line amplitudes obtained from a 3.4 year record of the Strasbourg superconducting gravimeter after correction for tides, instrumental drift and local barometric effects.

spectrum is of course unreliable due to the aperiodic noise components, and we are currently comparing the full statistics (smoothed and unsmoothed) of this record in comparison with a new Canadian superconducting gravimeter data set. We see that the residual gravity variations (observations corrected for instrumental drift, tides and local barometric pressure) are still slightly larger in the tidal bands (1, 2, 3 cycle per day (cpd)) reaching a few nanogal in amplitude; this represents nevertheless a huge reduction (by a factor of about 10^4) with respect to the raw gravity variations (typically 100 microgal at 1 cpd). It appears clear that the nanogal level we referred to in a previous section is achievable in the subtidal band for periods below the fourth diurnal tidal band using long SCG records and an appropriate data processing. Another important remark is that, because of the decrease of the local atmospheric pressure fluctuations with frequency [see e.g. Florsch et al., 1991], the high-frequency gravity noise (say below 1 day period) also exhibits a strong decrease with frequency [Agnew and Berger, 1978]. This implies that signals of geophysical interest are more easy to detect in higher frequency bands, as would be the case for the Slichter translational modes having eigenperiods in the sub-tidal band below 6 hr.

The question of the exact values of the frequencies of these modes is of course of primary importance for any attempt to detect them in gravity data. In response to a request from one reviewer, we include some calculations of the Slichter modes for two Earth models, PREM (Dziewonski and Anderson, 1981) and 1066A (Gilbert and Dziewonski, 1975). In particular we note that it is essential to begin the integrations for such modes at the origin of the coordinate system and to correctly allow for displacements of the inner core's centre of mass. In our calculations, we use the starting solutions of Crossley [1975b] to provide the initial constants for the integration of the 4th order system in the transformed variables using a 4th

order Runge-Kutta routine with automatic step selection. We do not see the need to first expand the solutions away from the Earth's centre as a power series before continuing with the numerical integration. It can be seen (Table 1) that stable eigenperiods of 5.42065 hr and 4.59920 hr are obtained for formal error tolerances of about 10^{-6} and that approximately 100-200 steps are needed in the inner core to achieve this accuracy. It should be noted that the period for 1066A compares very favourably with that quoted by Smith [1976], who found 4.5996 hr in the non-rotating limit; however, for an alternative opinion on the location of the frequencies of the Slichter mode, see Smylie et al. [1992b].

After these investigations of gravity spectra which were relative only to individual stations, Cummins et al. [1991] performed the first stack using gravity records from the IDA network of Lacoste-Romberg spring meters; the only signal detected above the noise level was a 9 nanogal signal at 9.54 hr period, of unknown origin for the moment.

Finally, very recently, the first stack using several superconducting gravity meters was done by Smylie et al. [1992a], where each record was segmented and windowed with a 12 000 hr Parzen window using 75 % overlap [see also

TABLE 1

Integration Accuracy for $_1S_1^0, \Omega = 0$.

model	error	NI*	NO*	NT*	period(h)
PREM	10^{-5}	56	34	185	5.4208365
	10^{-6}	98	60	263	5.4206578
	10^{-7}	168	112	403	5.4206474
	10^{-8}	294	198	658	5.4206468
1066A	10^{-5}	65	36	148	4.5989396
	10^{-6}	113	59	254	4.5991992
	10^{-7}	189	106	437	4.5991977
	10^{-8}	325	196	761	4.5991991

* NI, NO, NT = number of integration steps in inner core, outer core and whole Earth.

Mansinha et al., 1990]. The resulting product of the power spectral densities of four SCG records (2 from Brussels (1982-1986, 1987-1991), 1 from Bad Homburg (1986-1988), 1 from Strasbourg (1987-1991)) in the first intertidal band (between 1 and 2 cpd) is shown on Figure 6. One can see that the average noise level fits more or less a parabolic line with a minimum around 14-15 hr. The effect of the stacking method with respect to an individual record analysis is clearly revealed by the enhancement of the two tidal lines at the right extremity. Once again, there are only a few lines above the noise level and none strictly above the 95 % confidence interval. Nevertheless there is a very recent claim by Smylie [1992] of the identification of the translational triplet of the solid inner core in the subtidal part of the stacked gravity spectral product.

Clearly this stacking method is promising and needs to be extended in the next future by adding more SCG records from outside Europe, namely Canada, Japan and China where such instruments are in operation.

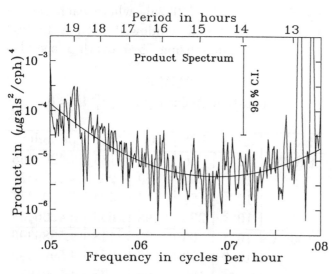

Fig. 6. Product of the power spectral densities of four European superconducting gravimeter records in the first intertidal band (between 1 and 2 cpd).

Wobble Resonances

Another fundamental use of surface gravity observations is to show the appearance of resonances in the tidal gravity changes. If there is an external tidal volume potential whose spherical harmonic components are denoted V_n, the gravity changes at the Earth's surface can be expressed with the help of classical volume Love numbers h_n, k_n:

$$\Delta g(\sigma, a) = -\delta_n \frac{n V_n}{a} \qquad (5)$$

introducing the standard gravimetric coefficient

$$\delta_n = 1 + \frac{2}{n} h_n - \frac{n+1}{n} k_n$$

The unit coefficient represents the direct effect of the potential, the term in h_n represents an elevation change through the unperturbed gravity field and the term (in k_n) expresses the mass redistribution potential due to the elastic deformation. Because the tidal potential of degree n is inversely proportional to d^{n+1}, where d is the distance between the Moon or Sun and the Earth, the major contribution to the total gravity is from the degree n = 2, but additional terms coming from the degrees 3 and 4 are also significant at the level of precision of the superconducting gravimeters [Hinderer et al., 1991b].

If there is a differential rotation of the homogeneous fluid outer core with respect to the mantle, a resulting inertial overpressure of degree n = 2 and order m = 1(linearly dependent on the rotation) occurs both at ICB and CMB [e.g. Hinderer et al., 1982]; for a stratified outer core, there are additional conditions in gravitational potential and derivative, as stated by Sasao et al. [1980]. This wobbling mode is called the nearly diurnal free wobble (NDFW) because of its eigenfrequency close to 1 cpd and the associated nutation in space the free core nutation (FCN). Moreover, the solid inner core is also able to have its own rotation motion with respect to the outer core and mantle, and this leads to complementary source terms at the ICB. This second wobbling mode, called free inner core nutation (FICN), also of nearly diurnal frequency, was computed independently by De Vries and Wahr [1991] and Mathews et al. [1991a, b], assuming gravitational and inertial coupling torques between the inner core and the rest of the Earth, as well as contributions to the inertia tensors due to the elastogravitational deformation (see also Dehant et al., [1991]).

The diurnal lunisolar tidal potential of degree 2 is able to excite these two wobbles and the rotational amplitudes can be found by solving the Euler equations for conservation of angular momentum. Knowing these amplitudes, it is then easy to derive the source terms at ICB and CMB which lead to an elastogravitational deformation of the inner core and mantle. The total tidal gravity change becomes then [e.g., Hinderer et al., 1991a]:

$$\Delta g(\sigma, a) = -[\delta_2 + \frac{\bar{\delta}_2^1 A}{\sigma - \tilde{\sigma}^{oc}} + \frac{\bar{\delta}_2^2 B}{\sigma - \tilde{\sigma}^{ic}}]\frac{2V_2}{a} \quad (6)$$

where $\tilde{\sigma}^{oc}$ and $\tilde{\sigma}^{ic}$ are the FCN and FICN complex eigenfrequencies, A and B constant terms having the dimension $(time)^{-1}$ and depending on several geodynamical parameters; the elastic coefficients (with bars) appearing in the numerator of (6) are combinations of different kinds of gravimetric factors previously introduced. For tidal waves of frequency σ close to the eigenfrequencies, there will be a resonant amplification in the tidal gravimetric factors.

The high quality of tidal measurements now available especially with SCG could be used to set constraints on the period and damping of the NDFW [Neuberg et al., 1987], as well as on the rotational pressure gravimetric factor ($\bar{\delta}_2^1$) [Hinderer et al., 1991c]. The major result was to show that there is a significant discrepancy between the observed eigenperiod and the theory.

This was independently confirmed by VLBI (very long baseline interferometry) measurements [Gwinn et al., 1986] and has been attributed to a nonhydrostatic contribution to the flattening of the CMB [Neuberg et al., 1990; Wahr and De Vries, 1989].

The detection of the new resonance effect in tidal gravity related to the FICN of nearly diurnal prograde eigenfrequency would be of great importance providing information on several poorly constrained parameters of the Earth's solid inner core like its flattening, the density jump with the liquid outer core as well as its elastic behaviour. However, its detection will be much more difficult essentially because of the weakness of the resonance strength (term in $\bar{\delta}_2^2 B$) as compared to the FCN one [Dehant et al., 1991]. The largest effect concerns the tidal wave which is the closest in frequency, namely the prograde annual solar wave S_1 and is found to be only a few parts in 10^4 with respect to the non-resonant static tide [De Vries and Wahr, 1991]. Moreover, the uncertainty related to the atmospheric forcing at this frequency [e.g., Warburton and Goodkind, 1977] will bring a further complication and make any observation difficult to interpret exclusively in terms of inner core rotational characteristics.

Conclusion

We have shown that high quality observations of surface gravity changes are able to provide useful information on core dynamics. A generalized Love number approach can be used in the computation of gravity changes related to boundary conditions (in pressure, shear stress, gravitational potential and radial derivative) located at the ICB and CMB originating from the core flow. It is thereby possible to treat the gravity effects of long-period buoyancy oscillations in the outer core by including the elastic deformation of the solid inner core and mantle. Especially, it can be demonstrated that two major approximations (Boussinesq and subseismic) of the core flow are appropriate (at least in the non-rotating case) when computing the resulting surface gravity perturbation, except for very-low degree modes. It appears that these oscillations are only weakly excited by earthquakes, but, once excited (whatever the cause), they will persist for a long time (several years) because of the small anelastic damping. Their detection in a gravity spectrum, which would reveal information on the stability of the outer core stratification, is clearly a difficult task, knowing that the residual gravity noise which can be extracted from a superconducting gravimeter record has presently a threshold of several nanogal in the frequency band of these eigenmodes. Another important use of precise gravity measurements relates to the differential rotations of the inner core and outer core (with respect to the mantle) which lead to a double resonance in the tidal volume gravimetric factor. The determination of the resonance parameters of the FCN (eigenperiod, damping and strength) hence provides constraints on models of the deep interior, especially on the flattening of the CMB. Similarly, the detection of the resonant contribution coming from the FICN on tidal gravity would be important for the knowledge of some physical or geometrical properties of the Earth's solid inner core, which are only poorly resolved from other geophysical observations. To improve this field of research, it is necessary to perform a stack of superconducting gravimeter records obtained over a common long time observational period (at least several years) and having a worldwide spatial distribution. In addition, there must be a concerted effort in the future to process the data in a reliable and standard way, in particular for the treatment of gaps, glitches, spikes, before applying a tidal analysis programme taking into account a very accurate theoretical tidal potential. Finally, improved models for oceanic loading are required, as well as atmospheric loading corrections including regional and global contributions.

Acknowledgments. The authors are grateful to D. Smylie for a critical review of this paper. We also thank M. Rochester for his assistance in clarifying the number of independent Love numbers. D. Crossley wishes to acknowledge funding through a Canadian NSERC Operating Grant A 4-240. J. Hinderer would like to thank NSERC for providing an International Research Fellowship. This study has been supported by INSU-CNRS (Dynamique et Bilan de la Terre).

References

Agnew, D. C. and J. Berger, Vertical seismic noise at very low frequencies, *J. Geophys. Res., 83*, 5420-5424, 1978.

Aldridge, K. and L. I. Lumb, Inertial waves identified in the Earth's fluid outer core, *Nature, 325*, 421-423, 1987.

Aldridge, K., L. I. Lumb and G. Anderson, Inertial modes in the Earth's fluid outer core, in *Structure and Dynamics of the Earth's Deep Interior, Geophys. Monogr. Ser., 46*, edited by Smylie, D. E. and Hide, R., 13-21, AGU, Washington, D. C., 1988.

Alterman, Z., H. Jarosch, and C. L. Pekeris, Oscillations of the Earth, *Proc. Roy. Soc. London, Ser. A, 252*, 80-95, 1959.

Crossley, D. J., Core undertones with rotation, *Geophys. J. Roy. astr. Soc., 42*, 477-488, 1975a.

Crossley, D. J., The free-Oscillation Equations at the Centre of the Earth, *Geophys. J. Roy. astr. Soc., 41*, 153-163, 1975b.

Crossley, D. J., The excitation of core modes by earthquakes, in *Structure and Dynamics of the Earth's Deep Interior, Geophys. Monogr. Ser., 46*, edited by Smylie, D. E. and Hide, R., 41-50, AGU, Washington, D. C., 1988.

Crossley, D. J., and D. E. Smylie, Electromagnetic and viscous damping of core oscillations, *Geophys. J. Roy. astr. Soc., 42*, 1011-1033, 1975.

Crossley, D. J., and M. G. Rochester, Simple core undertones, *Geophys. J. Roy. astr. Soc.*, *60*, 129-161, 1980.

Crossley, D. J., J. Hinderer, and H. Legros, On the excitation, detection and damping of core modes, *Phys. Earth Planet. Int.*, *68*, 97-116, 1991.

Cummins, P., J. Wahr, D. Agnew and Y. Tamura, Constraining core undertones using stacked IDA gravity records, *Geophys. J. Int.*, *106*, 189-198, 1991.

Dahlen, F. A., On the static deformation of an Earth's model with a fluid core, *Geophys. J. Roy. astr. Soc.*, *36*, 461-485, 1974.

Dehant, V., J. Hinderer, H. Legros and M. Lefftz, Analytical computation of the rotational eigenmodes of the inner core, Paper presented at Symposium U6 of the 20th IUGG Gen. Ass., Vienna, Austria, 1991.

De Vries, D., and J. Wahr, The effects of the solid inner core and nonhydrostatic structure on the Earth's forced nutations and Earth tides, *J. Geophys. Res.*, *96*, 8275-8293, 1991.

Dziewonski, A. M. and Anderson, D. L., Preliminary reference Earth model, *Phys. Earth Planet. Int.*, *25*, 297-356, 1981.

Florsch, N., J. Hinderer, D. Crossley, H. Legros and B. Valette, Preliminary spectral analysis of the residual signal of a superconducting gravimeter for periods shorter than one day, *Phys. Earth Planet. Int.*, *68*, 85-96, 1991.

Gilbert, F. and Dziewonski, A. M., An application of normal mode theory to the retrieval of structural parameters and source mechanisms from seismic spectra, *Phil. Trans. R. Soc. Lond. A*, *278*, 187-269, 1975.

Gwinn, C. R., T. A. Herring, and I. I. Shapiro, Geodesy by radiointerferometry: Studies of the forced nutations of the Earth, 2, Interpretation, *J. Geophys. Res.*, *91*, 4755-4765, 1986.

Hinderer, J., and H. Legros, Elasto-gravitational deformation, relative gravity changes and Earth dynamics, *Geophys. J.*, *97*, 481-495, 1989.

Hinderer, J., H. Legros, and M. Amalvict, A search for Chandler and nearly diurnal free wobbles using Liouville equations, *Geophys. J. Roy. astr. Soc.*, *71*, 121-132, 1982.

Hinderer, J., H. Legros and D. Crossley, Global Earth dynamics and induced gravity changes, *J. Geophys. Res.*, *96*, 20257-20265, 1991a.

Hinderer, J., D. Crossley and N. Florsch, Analysis of residual gravity signals using different tidal potentials, *Bull. Inf. Mar. Terr.*, *110*, 7986-8001, 1991b.

Hinderer, J., W. Zürn, and H. Legros, Interpretation of the strength of the 'Nearly Diurnal Free Wobble' resonance from stacked gravity tide observations, in *Proc. 11th Int. Symp. on Earth Tides*, edited by Kakkuri, J., 549-555, Schweitzerbart'sche Verlag., Stuttgart, 1991c.

Lumb, L., I. and K. Aldridge, On viscosity estimates for the Earth's fluid outer core and core-mantle coupling, *J. Geomag. Geoelectr.*, *43*, 93-110, 1991.

Mansinha, L., D. Smylie and B. Sutherland, Earthquakes and the spectrum of the Brussels superconducting gravimeter data for 1982-1986, *Phys. Earth Planet. Int.*, *61*, 141-148, 1990.

Mathews, P. M., B. A. Buffett, T. A. Herring, and I. I. Shapiro, Forced nutations of the Earth: Influence of inner core dynamics, I, Theory, *J. Geophys. Res.*, *96*, 8219-8242, 1991a.

Mathews, P. M., B. A. Buffett, T. A. Herring, and I. I. Shapiro, Forced nutations of the Earth: Influence of inner core dynamics, 2, Numerical results and comparisons, *J. Geophys. Res.*, *96*, 8243-8257, 1991b.

Melchior, P. and B. Ducarme, Detection of inertial gravity oscillations in the Earth's core with a superconducting gravimeter at Brussels, *Phys. Earth Planet. Int.*, *42*, 129-134, 1986.

Melchior, P., D. Crossley, V. Dehant and B. Ducarme, Have inertial waves been identified from the Earth's core, in *Structure and Dynamics of the Earth's Deep Interior, Geophys. Monogr. Ser., 46*, edited by Smylie, D. E. and Hide, R., 1-12, AGU, Washington, D. C., 1988.

Merriam, J., Toroidal Love numbers and transverse stress at the Earth's surface, *J. Geophys. Res.*, *90*, 7795-7802, 1985.

Neuberg, J., J. Hinderer, and W. Zürn, Stacking gravity tide observations in Central Europe for the retrieval of the complex eigenfrequency of the nearly diurnal free wobble, *Geophys. J. Roy. astr. Soc.*, *91*, 853-868, 1987.

Neuberg, J., J. Hinderer, and W. Zürn, On the complex eigenfrequency of the nearly diurnal free wobble and its geophysical interpretation, in *Variations in Earth Rotation, Geophys. Monogr. Ser., 59*, edited by McCarthy, D. D. and Carter, W. E., 11-16, AGU, Washington, D. C., 1990.

Poirier, J. P., Transport properties of liquid metals and viscosity of the Earth's core, *Geophys. J. Roy. astr. Soc.*, *92*, 99-105, 1988.

Sasao, T., S. Okubo, and M. Saito, A simple theory on dynamical motion of a stratified fluid core upon nutational motion of the Earth, in *Proceedings of the IAU Symposium 78: Nutations and the Earth's Rotation*, edited by Fedorov, E. P. , Smith, M. L. and Bender, P. L., 165-183, D. Reidel, Hingham, Mass., 1980.

Smith, M. L., Translational inner core oscillations of a rotating, slightly elliptical Earth, *J. Geophys. Res.*, *81*, 3055-3065, 1976.

Smylie, D.E., The inner core translational triplet and the density near Earth's center, *Science*, *255*, 1678-1682, 1992.

Smylie, D. E., and M. G. Rochester, Compressibility, core dynamics and the subseismic wave equation, *Phys. Earth Planet. Int.*, *24*, 308-319, 1981.

Smylie, D. E., A. M. K. Szeto, and K. Sato, Elastic boundary conditions in long period core oscillations, *Geophys. J. Int.*, *100*, 183-192, 1990.

Smylie, D., J. Hinderer, B. Richter, B. Ducarme and L. Mansinha, A comparative analysis of superconducting gravimeter data, this issue, 1992a.

Smylie, D. E., Xianhua Jiang, B. J. Brennan and K. Sato, Numerical calculation of modes of oscillation of the Earth's core, *Geophys. J. Int.*, *108*, 465-490, 1992b.

Wahr, J. M., and D. De Vries, The possibility of lateral structure inside the core and its implications for certain geodetic observations, *Geophys. J.*, *99*, 511-519, 1989.

Warburton, R. J., and J. M. Goodkind, The influence of barometric pressure variations on gravity, *Geophys. J. Roy. astr. Soc.*, *48*, 281-292, 1977.

Widmer, R., G. Masters and F. Gilbert, The spherical Earth revisited, Paper presented at 17th Int. Conf. Mathematical Geophysics, Blanes, Spain, 1988.

Zürn, W., B. Richter, P. A. Rydelek and J. Neuberg, Comments on Detection of inertial gravity oscillations in the Earth's core with a superconducting gravimeter at Brussels, *Phys. Earth Planet. Int.*, *49*, 176-178, 1987.

J. Hinderer and D. J. Crossley, Geophysics Laboratory, Department of Geological Sciences, McGill University, 3450 University Street, Montreal, H3A 2A7 Canada.

A Search for Evidence of Short Period Polar Motion in VLBI and Supergravimetry Observations

Keith D. Aldridge and W. H. Cannon

Centre for Research in Earth and Space Science, York University, North York, Ontario, Canada

In recent years the techniques of VLBI and the use of superconducting gravimeters have both been refined as high precision geodynamical research tools. The former instrument senses the relative motions of radio observatories with respect to inertial space while the latter instruments sense the variations in the magnitude of the local gravity vector. According to the principle of equivalence the inertial accelerations associated with the geodynamical motions measured by VLBI are indistinguishable from gravitational fields and, if sufficiently large, will be reflected in the observations of supergravimeters. The work presented reports on a search for evidence of short period polar motion using VLBI and supergravimetry data sets.

Two searches for evidence for short period polar motion were conducted. In the first effort the polar motion observations from VLBI Extended Research and Development Experiment (VLBI/ERDE) data sets were compared with simultaneous gravimetric signals recorded on a supergravimeter to seek evidence of the expected gravimetric signal. In the second effort the post-fit group delay residuals from the IRIS VLBI observations were analyzed for evidence of short period polar motion.

Although these searches were carried out near the limit of resolution of both instruments, encouraging results were obtained suggesting the presence, at the 1σ level, of short period polar motions of a few centimeters amplitude. Polar motions found at periods near 14 hours are consistent with observed gravimetric signals of 11 nanogals obtained over the same time interval if these are interpreted as mainly due to changes in latitude of the gravity meter. Large scale oscillations in the Earth's fluid core or global torques generated by atmospheric surface tractions or both may be major contributing factors to these apparent motions.

Introduction

In recent years the techniques of Very Long Baseline Interferometry (VLBI) and the use of superconducting gravity meters (supergravimeters) have both been refined as high precision geodynamical research tools. The former technique senses the changes in scale and orientation, relative to the directions of space-fixed extra-galactic radio sources, of a baseline vector between radio observatories in response to motions of the Earth relative to inertial space while the latter technique senses changes in the magnitude of the local gravity vector. According to the principle of equivalence the inertial accelerations associated with the geodynamical motions measured by VLBI are indistinguishable from gravitational fields and, if sufficiently large, will be reflected in the observations of supergravimeters. The work presented here explores the relationship between the polar motions inferred from VLBI observations and simultaneous data sets obtained from supergravimeters.

Dynamics of Earth's Deep Interior and Earth Rotation
Geophysical Monograph 72, IUGG Volume 12
Published in 1993 by the International Union of Geodesy and Geophysics and the American Geophysical Union.

VLBI Systems and Superconducting Gravity Meters

The technique of Very Long Baseline Interferometry has been used in recent years to investigate details of the Earth's rotational dynamics as well as the kinematics of the earth's crust including the phenomena of plate tectonics [Caprette, D.S., Ma, C., Ryan, J.W., 1990]. The basis of this measurement is the precise determination of the differential time of arrival or group delay between radio observatories, τ_g, of wide band cosmic noise emitted by quasars and other extragalactic objects at cosmological distances. The measurement of VLBI wide band group delay by bandwidth synthesis (BWS) was first proposed by Gold [1967] and the techniques described by Rogers [1970] form the basis of the geodynamical VLBI observations obtained by the MkIII system [Thompson et. al. 1986].

In simple terms BWS establishes the slope, $\tau_g = d\phi/d\omega$, where ϕ is the interferometer phase and ω is the interferometer observing frequency, across a wide band of RF frequencies. If one were to ignore details concerning such matters as the "effective bandwidth" of the interferometer which are related to the distribution of sampling of the RF frequency spectrum, the order of the precision with which τ_g can be determined by BWS is given by $\sigma_\tau = \delta\phi/(\omega_2 - \omega_1)$ where $\delta\phi$ is the rms error in interferometer phase determination and $\omega_2 - \omega_1$ is the width of the synthesized band. In most geodetic applications of the MkIII VLBI system the synthesized band $(\omega_2 - \omega_1)/2\pi$ is generally of the order of 360 MHz. The rms error in interferometer phase is dependent on a number of system parameters including: the strength of the observed radio source, the single channel band width of the BWS system, the collecting area and aperture efficiency of the interferometer antennas, the noise characteristics of the receivers of the interferometer, and the coherent integration time. For geodetic applications of the MkIII VLBI system in which strong quasar sources are observed $\delta\phi$ is of the order of a few degrees or 10^{-2} turns [Thompson et. al. 1986]. This implies that, under optimal conditions, individual group delay measurements can be made with the MkIII VLBI system with an rms precision approaching 2.5 ps. In the sub-optimal world of daily practice corrupted by a variety of instrumental instabilities and environmental sources of noise, the standard error on MkIII group delay measurements is of the order of 10 to 25 ps, an interval corresponding to light travel of the order of 3 to 8 mm [Thompson et. al. 1986].

A superconducting gravity meter senses the variations in position of a superconducting sphere suspended in a constant magnetic field arising as a result of variations in the weight of the sphere due to changes in the local gravitational field. The noise level on the observations of supergravimeters is typically of the order of 3-4 nanogals [Melchior and Ducarme, 1986].

A Joint Search For Short Period Polar Motion Using VLBI and Supergravimetry

It is well established that the Earth's rotation axis moves in a quasi-circular, prograde path of about ten meters in diameter relative to a fixed point on the Earth's surface. This large-scale, polar motion consists primarily of two components of approximately equal amplitude, an annual oscillation (the annual wobble) and a 14-month oscillation (the Chandler wobble).

Polar motion will, as a consequence of the resulting change in interferometer geometry, generate a contribution to the group delay, $\delta\tau_g$, the magnitude of which will depend on the relative direction of the body-fixed interferometer baseline vector $\vec{b}$ and the space fixed unit vector in the direction of the source, $\hat{s}$. The contribution to the group delay $\delta\tau_g$, between two stations at positions $\vec{x}_l$, $l = 1, 2$ resulting from polar motion m_1, m_2 measured in radians in the direction of Greenwich and $90°$ east longitude, respectively, is given by:

$$\delta\tau_g(\hat{s}_k, \vec{b}_j, \theta(t_i)) =$$
$$\frac{1}{c}\left[m_1[-b_j^3 \cos\delta_k \cos(\alpha_k - \theta(t_i)) + b_j^1 \sin\delta_k]\right.$$
$$\left. + m_2[-b_j^3 \cos\delta_k \sin(\alpha_k - \theta(t_i)) + b_j^2 \sin\delta_k]\right]$$

where $b_j^n = x_j^n - x_j^n$ is the nth body-fixed Cartesian component of the jth interferometer baseline $\vec{b}_j$, $\theta(t_i)$ is sidereal time at UT time t_i, α_k, δ_k are the right ascension and declination of the kth radio source lying in the direction $\hat{s}_k$ and c is the velocity of light. The effects of polar motion on interferometer group delay will lie in the range ± 30 ps/cm of polar motion.

Polar motion will alter the global distribution of the centrifugal force field generated by Earth rotation as the Earth's figure axis is displaced relative to the Earth's rotation axis giving rise to an observable gravimetric signal. The expected gravimetric signal for polar motion of m_1, m_2 radians is given by

$$\Delta g' = \delta_{eff}\Omega^2 R \sin 2\phi \left[m_1 \cos\lambda + m_2 \sin\lambda\right] \quad (1)$$

where $\delta_{eff} = 1.19$ is a gravimetric factor [Hinderer and Legros, 1989], $\Omega = 7.29 \times 10^{-5} s^{-1}$ is the Earth's rotation rate, $R = 6.37 \times 10^6$ meters is the Earth's radius, λ is the station longitude and ϕ is the station latitude. Polar motion with an amplitude of one centimeter will give rise to a change in acceleration near mid-latitudes of about 6 nanogals (1 gal $= 1$ cm/sec^2) on a gravity meter fixed on the Earth's surface.

The effects of large scale polar motion on the response of a long baseline interferometer as well as on the response of a gravity meter is readily observable [Richter, 1990]. Fig. 1 shows one component of large scale polar motion as measured by IRIS VLBI observations together with the resulting gravimetric signal measured by a supergravimeter which arises as a result of the altered centrifugal force field accompanying the changes in latitude of the gravimeter station.

A variety of geodynamical processes are in principle capable of generating a wide spectrum of short period variations in the position of the Earth's pole of rotation [Munk and MacDonald, 1960; Lambeck, 1988]. These include, among many others:

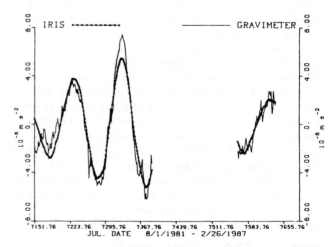

Fig. 1. Comparison of gravity observations using a Superconducting Gravimeter with gravity expected due to Annual and Chandler wobble measured using VLBI (IRIS data)(From Richter[1990]).

1. the dynamical response of the mantle and crust to the gravitational torques exerted by the Sun and Moon

2. changes in the inertia tensor of the mantle and crust arising from tidal deformations generated by the gravitational fields of the Sun and Moon

3. the dynamical response of the mantle and crust to torques generated by surface tractions on the exterior surface arising as a result of variations in the fluid motions of the atmosphere

4. the dynamical response of the mantle and crust to torques generated by surface tractions on the interior surface arising as a result of variations in the fluid motions of the outer core

The VLBI Extended Research and Development Experiments (VLBI/ERDE) are a program of VLBI observations in which high precision group delay measurements are made using bandwidth synthesis over 720 MHz instead of the usual 360 MHz. This fact combined with a significant number of other MkIII VLBI system improvements facilitates a factor two improvement in the precision of the VLBI group delay observables in the ERDE data. In the reduction and analysis of the VLBI ERDE data independent parameters for the pole position were introduced for every 1 hour of VLBI observational data analyzed.

In this paper we compare the recently obtained pole positions from the Extended Research and Development Experiment of the NASA Crustal Dynamics Project for the months of September, October and November of 1989. Shown in Fig. 2 are the pole positions found by J. Ryan (personal communication, 1991) at hourly intervals for the days 27 September and 12 October 1989. Units are in milliarcseconds (mas) and the graph plots the pole in terms of (x,y) where x is measured along the Greenwich Meridian (IRM, the Reference Meridian of the International Earth Rotation Service) and the y axis is in the direction of 90 degrees West Longitude. These two days were chosen to illustrate the large magnitude of polar motion that can be observed in a 24 hour period. The bars represent the formal error as determined in the least-squares recovery of the polar motion. The (x,y) values are those that remain after Bulletin B values have been subtracted and these in turn include tidal corrections from Yoder[1981].

In order to illustrate that the 24 hour polar motions may have more complex character, examples from October 18 and 30 are reproduced in Fig. 3. Here the bars again represent formal errors and at least for errors of this type the irregular excursions of the axis of Earth rotation appear to be significant for the earlier date.

We center our attention on two final examples of ERDE polar motion from November 10 and 14, 1989. These two days have been selected because corresponding gravimetry data is available from our superconducting gravimeter at Cantley, Quebec, Canada. Fig. 4 shows the results of pole position determinations, along with their 1σ formal error bars, determined every hour from the VLBI/ERDE obser-

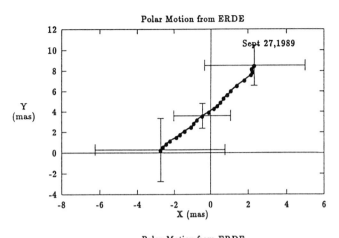

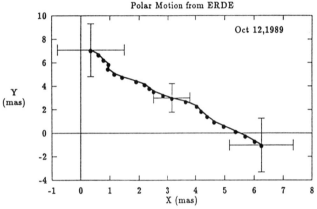

Fig. 2. Polar motion from ERDE of NASA Crustal Dynamics Project. Pole positions at one hour breaks for 24 hour period are shown by points with formal errors shown by bars. (J. Ryan, personal communication, 1991)

vations for November 10 and November 14 1989. It is apparent from these data that the position of the pole of rotation exhibits significant short term excursions on a time scale of several hours. Fig. 5 shows the theoretically predicted variations in gravity which would accompany the VLBI/ERDE observed variations in both the pole position and UT1 for November 10 1989 and November 14 1989 respectively. The 1σ error bars shown in Fig. 5 are determined analytically from the 1σ error bars on the pole positions shown in Fig. 4.

The expected values of gravity due to the observed polar motion can be compared with gravimetry observations after appropriate corrections and reductions of the raw gravity data. Removal of tides was done for more than 3000 terms using code written by Merriam (personal communication, 1991) based on the Xi Qinwen series. Pressure corrections were done using single point pressure readings and applying a Bouguer type of correction for the atmosphere. There remained a significant (5 microgal) periodic signal which is probably due to unmodelled ocean loading effects and this was removed by a notch filter whose center frequency was found from a MEM analysis of the residual gravity data.

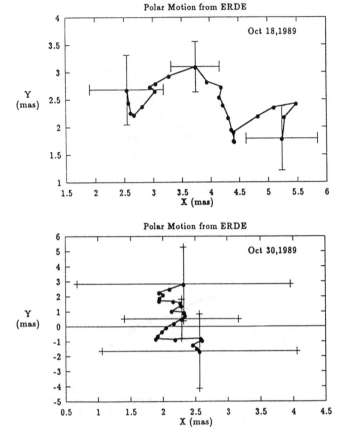

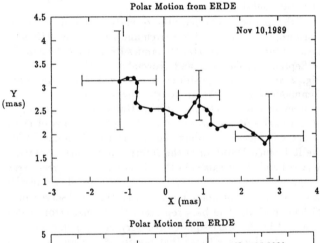

Fig. 3. ERDE data as in Fig. 2 but oscillatory properties are evident here for 18 and 30 October 1989. (J. Ryan, personal communication, 1991)

other sources. Accordingly it is important to note that the significant reduction in power seen for November 14 1989 occurs for a much larger predicted and observed gravity change than for the case of November 10 1989 where there is little reduction in power. Clearly more examples are needed to verify the gravity effects expected for observed polar motion.

A SEARCH FOR SHORT PERIOD POLAR MOTION USING VLBI GROUP DELAY RESIDUALS

The measured VLBI group delay is a consequence of the superposition of a large number of physical phenomena including: the changing relative geometry of the interferometer baseline $\vec{b}$ and the direction $\hat{s}$ to the source of radiation, the effects of atmospheric delays including both the troposphere and ionosphere, the effects of signal delays in the electrical paths of the VLBI system instrumentation, as well as mis-synchronism and drifts in the station clocks. In the reduction and analysis of VLBI data, the observed group delay measurements are subtracted from the predictions of a multi-parameter model and the parameters of the model are adjusted so as to minimize, in a least squares sense, the post-

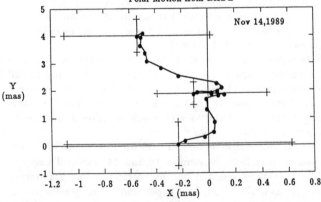

Fig. 4. Polar motion for two days observations 10 and 14 November 1989 from ERDE data (J. Ryan, personal communication, 1991).

What was left after the application of the notch filter is shown in Fig. 6 along with the expected gravity from polar motion given in Fig. 5 for the two days November 10 and 14, 1989.

We call the observed short period variations in gravity $\Delta g(t_i)$ and the theoretically predicted short period variations in gravity $\Delta g'(t_i)$ for November 10 1989 and November 14 1989, respectively, where the means of each time series have been removed. A useful measure of the degree to which these two time series resemble each other is provided by the normalized relative power R given by

$$R = \frac{\sum_{i=1}^{N} \Delta g^2(t_i) - (\Delta g'(t_i) - \Delta g(t_i))^2}{\sum_{i=1}^{N} \Delta g^2(t_i)} \qquad (2)$$

For November 10 1989 $R = 0.0022$ and for November 14 1989 $R = 0.1633$.

The gravity effect due to polar motion, given in equation (1), must be contained in the observed gravimetric signal. Our ability to detect the gravimetric effects of polar motion depends on the size of this signal compared to combined effects from instrument noise and gravimetric signals from

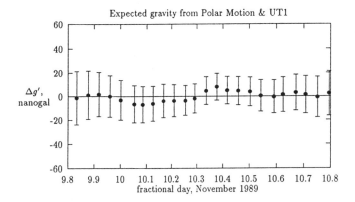

Expected gravity from Polar Motion & UT1

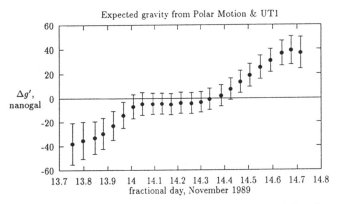

Expected gravity from Polar Motion & UT1

Fig. 5. Expected change in gravity derived from ERDE polar motion data for each of two days observations 10 and 14 November 1989.

fit group delay residuals. Any systematic variations in group delay which are present in the observations but which are not represented in the model will remain in the post-fit residuals. The reported precision of VLBI baseline determination by the MkIII system for an observing period extending over 24 hours for the data used in this study is of the order of 3 cm based on an rms post-fit group delay residuals for the entire period of the order of 100 ps. However since the standard error on the *individual* group delay measurements is in the range 10 to 25 ps, we were motivated to search the postfit residuals to seek evidence for short period (less than 24 hours) variations in the position of the Earth's pole of rotation which may remain in the postfit residuals following the standard reduction of a 24 hour period of MkIII VLBI data.

The motivation to conduct this search for short-period polar motion in VLBI observations is derived from the interpretation of observations made with a superconducting gravimeter at the time of the Mindanao earthquake (20 November 1984). Signals of approximately 11 nanogals amplitude at a period near 14 hours were attributed by some to internal motions in the Earth's outer core [Melchior and Ducarme, 1986; Zürn et al., 1987]. If a large scale fluid oscillation were excited in the core, the conservation of angular momentum

and coupling between the outer core and mantle would result in a periodic wobble of the mantle. A search for short period polar motion was conducted by examining the post-fit time delay residuals of the International Radio Interferometric Surveying (IRIS) data for the period 4 November 1984 to 29 December 1984, a time interval which brackets the Mindanao earthquake. A total of $l = 12$ data sets were used in the investigation. Each data set consisted of a number N_l of post-fit residual delays, $\Delta\tau_g(\hat{s}_k, \vec{b}_j, \theta(t_i))$, $i = 1, N_l$, and their one sigma standard deviations, $\sigma_{kji} = \sigma(\hat{s}_k, \vec{b}_j, \theta(t_i))$, $i = 1, N_l$, for each UT time of observation t_i, $i = 1, N_l$ within the lth 24 hour time period. The data set included observations obtained on interferometer baselines, $\vec{b}_j$, $j = 1, 6$ between the IRIS VLBI stations at Fort Davis, Texas; Westford, Massachusetts; Richmond, Florida; and Wettzel, West Germany and on 14 extra-galactic radio sources, $\hat{s}_k$, $k = 1, 14$. There were approximately 350 post-fit delay residuals in each of the 12 data sets; $N_l \approx 350$, $l = 1, 12$.

The short period polar motion being sought was represented by the following model:

$$m_1(t) = M_1 \cos(\omega t + \xi)$$
$$m_2(t) = M_2 \sin(\omega t + \xi)$$

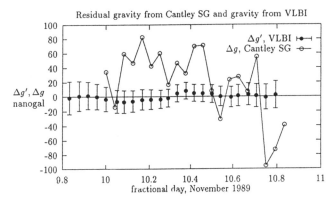

Residual gravity from Cantley SG and gravity from VLBI

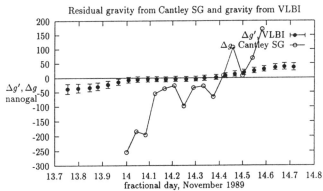

Residual gravity from Cantley SG and gravity from VLBI

Fig. 6. Comparison of gravity observations using a Superconducting Gravimeter at Cantley, Quebec, with gravity expected (from Fig. 5) due to polar motion measured using VLBI (ERDE data from J. Ryan, personal communication, 1991).

which allows for the existence of both prograde and retrograde, elliptically polarized polar motions with angular frequency ω and phase ξ. For a particular realization of the values of the parameters $\vec{a} = (M_1, M_2, \omega, \xi)$, the contribution to the group delay would be given by $\delta\tau_g(\hat{s}_k, \vec{b}_j, \theta(t_i) : \vec{a})$ where t_i is the UT time of the observation of the source $\hat{s}_k$ on baseline $\vec{b}_j$ and i ranges over $(1, N_l)$ where N_l is the number of points in the lth data set. The vector $\vec{a} = (M_1, M_2, \omega, \xi)$ is the group of parameters which will be fit to each data set. For each of the 12 data sets the sought after short-period polar motions are determined by minimization of

$$S_l^2(\vec{a}) = \sum_{i=1}^{N_l} \sum_{j=1}^{6} \sum_{k=1}^{14} \left[\frac{\Delta\tau_g(\hat{s}_k, \vec{b}_j, \theta(t_i)) - \delta\tau_g(\hat{s}_k, \vec{b}_j, \theta(t_i) : \vec{a})}{\sigma_{ijk}} \right]^2$$

with respect to the parameters defined by $\vec{a} = (M_1, M_2, \omega, \xi)$ following linearization of the equations about initial estimates for each of the parameters. The resulting set of linear equations is solved recursively for the values of the parameters $\vec{a}$. The procedure is considered to have converged to a solution when S_l^2 changes by less than a factor 0.1 between any two successive iterations. Formal error estimates are returned from this calculation under the assumption that the noise in the data is Gaussian. The search was conducted by successively incrementing the apriori values of the parameter ω in equations (3) such that the apriori period of the polar motion was incremented in 1/2 hour steps. Fig. 7 presents a histogram showing, for those recovered periods which were significant at the 1σ level or higher, the number of times the least squares procedure converged to a given solution as a function of period. Also presented in Fig. 7 is the same histogram obtained from an identical analysis of simulated group delay residuals consisting of a random sequence with the same variance as the observed group delay residuals. The short period polar motions whose recovered periods and amplitudes exceeded the 1σ standard error are shown in Table 1.

It appears that the two sets of results in Fig. 7 are drawn from different distributions thus giving support to the claim that the data used in our work was meaningful. A formal statistical test for this was done as follows. The total number of searches conducted in each of the two cases, one for the post-fit residual VLBI data and one for the pseudo-random data was $12 \times 38 = 456$. The aggregate number of recovered periods was 214 for the VLBI data and 134 for the pseudo-random data. Under the assumption that the recovered periods are independent in the 1/2 hourly intervals, the recovered aggregate frequencies (fractions) will be estimates of a binomial random variable. Let $\hat{p}_1 = 214/456$ and $\hat{p}_2 = 134/456$ and we test the null hypothesis $p_1 = p_2$ using the statistic $z = (\hat{p}_1 - \hat{p}_2)/\hat{S}$ where $\hat{S}^2 = 2\bar{p}(1-\bar{p})/456$ and $\bar{p} = (p_1 + p_2)/2$. For the data in Fig. 7 there is no evidence to support the hypothesis at the 1% level. Accordingly we reject the hypothesis and conclude there is no evidence at

TABLE 1. Short period polar motions recovered from post-fit residual VLBI data using Levenberg-Marquardt inversion in the range 6.0 -24.0 hours.

$period(hrs)$	$M1(meters)$	$M2(meters)$
6.61	0.0822	0.1014
±0.04	±0.0061	±0.0044
8.10	0.0891	0.0522
±0.08	±0.0073	±0.0057
8.32	0.0389	0.0918
±0.14	±0.0084	±0.0092
9.17	0.0529	0.0283
±0.19	±0.0061	±0.0061
9.60	0.0233	0.1123
±0.09	±0.0054	±0.0052
10.62	0.0584	0.0890
±0.14	±0.0055	±0.0051
13.93	0.0776	0.0282
±0.26	±0.0071	±0.0054
14.64	0.1125	0.0854
±0.18	±0.0058	±0.0044
16.09	0.0441	0.0438
±1.04	±0.0033	±0.0130
16.22	0.0333	0.0339
±0.093	±0.0056	±0.0085
21.29	0.1045	0.0876
±0.040	±0.0046	±0.0040
23.07	0.0819	0.0491
±0.70	±0.0083	±0.0069

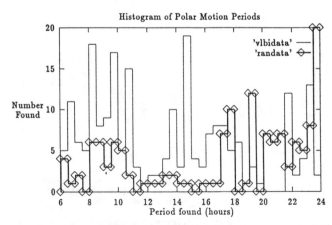

Fig. 7. Histogram showing number of times a given period is recovered from post fit residual delay times (IRIS data) and from pseudo-random delay times.

the 1% level that the VLBI data has the same distribution as that of the pseudo-random data.

Discussion and Conclusions

A joint analysis of simultaneous VLBI/ERDE and super-gravimetry observations suggest that short period polar motion components exist and may be reflected in the response of supergravimeters. These results strongly suggest that VLBI observations and supergravimetry observations complement each other and that the standard reduction procedures for supergravimetry data should include corrections for the effects of polar motion and UT1 variations as evidenced by VLBI.

A search of IRIS MkIII group delay, post-fit residuals for short period polar motion has revealed the possible existence of several periodic signals, each representing polar motion of a few centimeters amplitude. The apparent presence of these short period polar motion components in the VLBI data is consistent with signals appearing in simultaneous gravimetric observations and, if confirmed, would result in a new interpretation of that data. An alternative approach to the detection of short period polar motion, as suggested by a referee, would be to look for an associated nutation which would be very much larger than the polar motion. In the work presented here, however, we have studied records of only 24 hours duration and accordingly we would see only a very small fraction of one period in the associated nutation. Furthermore we are also using a gravimetric signal to attempt corroboration of the polar motion recovered from the geodetic observations.

Within the group of short period polar motions with recovered periods near 14 hours, the best resolved and largest is at 14.61 ($\pm$0.39) hours. For this short period polar motion gravimetric signal $\Delta g'$ is "postdicted" by equation (1) to have an amplitude of 11.9 nanogals with a formal error of 3.6 nanogals. Following the Mindanao earthquake, the superconducting gravimeter in Brussels ($\lambda = 4°21'29"E$, $\phi = 50°47'55"N$), showed a signal of about 11 nanogals with a noise level of 3-4 nanogals and a period near 13.9 hours after extensive corrections for tidal and atmospheric effects and large scale polar motion. These results are consistent with the hypothesis that the supergravimeter and the VLBI systems are sensing the same 14 hour short-period polar motion. An alternative explanation of the 14 hour signals observed here is the 0.61 days atmospheric normal mode described by Eubanks [1991]. This mode has not yet been detected in atmospheric equatorial angular momentum data but we may well be observing it here. Dehant et. al [1991] give further evidence of an atmospheric source of the observed signal and it also appears that the Mindanao earthquake which originally motivated our study has no significance in the signals we are observing.

It is important to note, however, that analysis of a gravimeter record from Bad Homberg for the same time interval as the Brussels record, did not reveal a significant signal near 14 hours [Zürn et al., 1987]. One explanation for this dis-crepancy is found in equation (1): the expected gravimetric signal from polar motion is longitude dependent so that a signal which is only just observable at one site could fall below a threshold of significance at a second site. Significance tests themselves depend on stationarity of the data and it is far from certain that the properties of the observed gravimetric signals at Bad Homberg and Brussels are independent of time. This apparent lack of stationarity in the gravimetric data has already been noted in the literature. [Aldridge et al.,1988; Dehant et al., 1991]

Several other short period polar motions are obtained from the analysis of VLBI group delay residuals, the largest of which appear in the neighborhood of a 10 hour period. This result may relate to a resonance between the Earth's core, mantle and atmosphere with a period predicted to be near 10 hours [Hinderer and Legros, 1989]; however further analysis will be required to substantiate this speculation.

The results of this work are considered by the authors, to be rather encouraging in spite of the fact that the data analysis was carried out near the limit of sensitivity of both instruments. Further work on this problem is certainly warranted.

Acknowledgements. We are grateful to D. Robertson and the National Geodetic Survey for supplying the IRIS data used in this work and to C. Kuehn, J. Ryan and the Goddard Space Flight Center for the results from ERDE of the NASA Crustal Dynamics Project. Financial support for this work came from the Natural Sciences and Engineering Research Council (NSERC) of Canada.

References

Aldridge, K. D. and L. I. Lumb, Identification of inertial Waves in the Earth's fluid outer core, *Nature, 325*, 421-423, 1987.

Aldridge, K. D., L. I. Lumb and G. A. Henderson, Inertial modes in the Earth's fluid outer core, in *Structure and Dynamics of the Earth's Deep Interior*, (eds.) by D. E. Smylie and R. Hide, *Geophysical Monographs / IUGG Series*, American Geophysical Union, 13-21, 1988.

Aldridge, K. D., L. I. Lumb, and G. A. Henderson, A Poincaré model for the Earth's fluid core, *Geophys. Astrophys. Fluid Dynamics, 48*, 5-23,1989.

Caprette, D.S., Ma, C., Ryan, J.W. *Crustal Dynamics Project Data Analysis - - 1990: VLBI Geodetic Results 1979-1989*, NASA Technical Memorandum 100765, December, 1990.

Dehant, V., B. Ducarme and P. Defraigne, New Analysis of the Superconducting Gravimeter Data of Brussels, 20th General Assembly, IUGG, Vienna, 1991.

Eubanks, M., T. M., Fluid normal modes and rapid variations in the rotation of the Earth, 20th General Assembly, IUGG, Vienna, 1991.

Gold, T. Radio Method for the Precise Measurement of the Rotation Period of the Earth, *Science, 157*, 302-304, 1967.

Hide, R., N. T. Birch, L. V. Morisson, D. J. Shea and A. A. White, Atmospheric angular momentum fluctuations and changes in the length of the day, *Nature, 286*,114-117, 1980.

Hinderer, J. and H. Legros, Elasto-gravitational deformation, relative gravity changes and Earth dynamics, *Geophys. J. R. astr. Soc., 97*, 481-495, 1989.

Lambeck, K., *Geophysical Geodesy* (Oxford University Press, Oxford, 1988), p396.

Melchior, P. and Ducarme, B., 'Detection of inertial gravity oscillations in the Earth's core with a superconducting gravimeter at Brussels,' *Phys. Earth planet. Int., 42*, 129-134, 1986.

Munk, W.H. and MacDonald, G.J.F. *The Rotation of the Earth: A Geophysical Discussion* (Cambridge University press, Cambridge, 1960).

Press, W. H., B. P. Flannery, S. A. Teukolsky and W. T. Vetterling, *Numerical Recipes* (Cambridge University Press, Cambridge, 1986), p521.

Richter, B., The long period elastic behaviour of the Earth, in *Variations in Earth Rotation,* (eds.) by D. D. McCarthy and W. E. Carter, Geophysical Monograph 59, IUGG Volume 9, 21-25, 1990.

Richter, B. and W. Zürn, Chandler effect and nearly diurnal free wobble as determined from observations with a superconducting gravimeter, *The Earth's Rotation and Reference Frames for Geodesy and Geodynamics,* (eds.) A. K. Babcock and G. A. Wilkins, Kluwer Academic Publishers, 309-315, 1988.

Robertson, D. S., W. E. Carter, J. A. Campbell and H. Schuh, Daily Earth rotation determinations from IRIS very long baseline interferometry, *Nature, 316,* 424-427, 1985.

Rochester, M. G., Causes of fluctuations in the rotation of the Earth, *Philos. Trans. R. Soc. London Ser. A., 313,* 95-105, 1984.

Rogers, A.E.E. Very Long Baseline Interferometry with Large Efective Bandwidth for Phase Delay Measurements,*Radio Sci., 5,* 1239-1247, 1970.

Thompson, A.R., Moran, J.W., Swenson, G.W. *Interferometry and Synthesis in Radio Astronomy* (J.Wiley and Sons, 1986)

Yoder, C. F., Williams, J. G. and Parke, M.E., Tidal Variations of Earth Rotation, *J. Geophys. Res. , 86,* 881-891, 1981.

Zürn, W., B. Richter, P. A. Rydelek and J. Neuberg, Comment on 'Detection of inertial gravity oscillations in the Earth's core with a superconducting gravimeter at Brussels', *Phys. Earth planet. Int., 49,* 176-178, 1987.

K. D. Aldridge and W. H. Cannon, Centre for Research in Earth and Space Science, York University, 4700 Keele Street, North York, Ontario, M3J 1P3, Canada.

IDA Tidal Data and the Earth's Nearly Diurnal Free Wobble

PHIL R. CUMMINS

Research School of Earth Sciences, Australian National University

JOHN M. WAHR

Dept. of Physics and CIRES, University of Colorado, Boulder

We consider the measurement of resonant enhancement of the diurnal body tides by the Nearly Diurnal Free Wobble (NDFW), using tidal gravity data from 11 IDA stations simultaneously. The use of such a global data set should serve to mitigate any regional bias which may have affected the results of previous studies of this type. We describe a procedure for extrapolating ocean loading corrections to frequencies outside the band including the main diurnal tides. After applying these corrections, a nonlinear least-squares solution is obtained for the period (T_{NDFW}) and Q of the wobble (Q_{NDFW}), as well as the strength of the NDFW resonance. This results in an estimated $T_{NDFW} = 428 \pm 12$ sidereal days, in agreement with previous studies, and an estimated Q_{NDFW} in the range 3340 - 36640. This latter estimate is in better agreement with VLBI nutation measurements than are the results of previous tidal studies, and suggests that previous estimates of Q_{NDFW} may have suffered from regional bias. Finally, we consider how incompatible some of the low Q_{NDFW} values obtained in tidal studies are with the recent VLBI nutation measurements. This comparison indicates that the 'low' estimates for Q_{NDFW} obtained in the tidal studies (including this one) lead to residuals for the nutation measurements which are an order of magnitude larger than the observational uncertainty.

INTRODUCTION

The presence of a liquid core in the Earth's interior leads directly to the existence of a rotational mode in the Earth's normal mode spectrum called the Nearly Diurnal Free Wobble (Toomre, 1974). This mode consists primarily of a rigid rotation of the mantle relative to the core about an equatorial axis. Because of the slight ellipticity of the core-mantle boundary, the misalignment of the instantaneous rotation axes of the mantle and core caused by this rotation enables the pressure of the fluid outer core to exert a restoring torque on the mantle. The resulting oscillatory motion has a period in inertial space of somewhat longer than 1 year, corresponding to a period of slightly less than one day as observed on the rotating Earth.

Because the displacement field for this free mode is predominantly that of a rigid rotation about an axis perpendicular to the rotation axis, it is often referred to as the free core nutation. This nutational component has been observed as a resonant enhancement of the lunisolar forced nutations by Her-

ring *et al.* (1986 and 1991) using Very Long Baseline Interferometry (VLBI). The change in gravity associated with this nutational motion, due to the incremental changes in the centrifugal and Coriolis forces, is very small. A larger change in gravitational acceleration (as observed on the Earth's surface) is generated by the deformational component of the eigenfunction, which has the character of a wobble. The NDFW causes an enhancement of the deformation associated with the diurnal body tide which is very similar to the enhancement of the forced nutations, and this has been observed in a number of studies of tidal gravity and strain (Richter and Zürn 1986, Neuberg *et al.* 1987, and Sato 1991).

Both types of studies, nutational and tidal, have suggested very similar values for the NDFW period. However, results for the imaginary part of the eigenfrequency, which can be expressed in terms of an apparent Q for the NDFW (Q_{NDFW}), are not always consistent. The VLBI results of Herring *et al.* (1986 and 1991) suggest a Q_{NDFW} of about 27000, while the tidal studies of Richter and Zürn (1986) and Neuberg *et al.* (1987) suggest a lower Q_{NDFW} of about 3000. Sato's (1991) more recent result for Q_{NDFW}, obtained from tidal strain data, lies somewhere between these at about 5100. While the consistent results for the period imply a dynamical ellipticity of the core in excess of that expected for a hydrostatic Earth, the

Dynamics of Earth's Deep Interior and Earth Rotation
Geophysical Monograph 72, IUGG Volume 12

inconsistent results for Q_{NDFW} make it difficult to answer questions about the extent to which mantle anelasticity, as well as electromagnetic and viscous coupling between mantle and core, affect the NDFW.

One of the main issues to be addressed by the present study is this discrepancy between the VLBI and tidal results. This study differs from the tidal studies already mentioned primarily in its use of data from the International Deployment of Accelerometers (IDA, see Agnew *et al.* 1986). This network consists of a large number of gravimeters distributed throughout the globe (see Figure 1). Simultaneous use of several gravimeters to study the NDFW has been considered before in the work of Neuberg *et al.* (1987), using methodology very similar to that used here. Their study , however, and that of Richter and Zürn (1986), used only gravimeters in northern Europe. While the results of both Neuberg *et al.* (1987) and Richter and Zürn (1986) for Q_{NDFW} are supported to some extent by the study of Sato (1991) using tidal strain data collected in Japan, the possibility exists that tidal studies in both of these locations are biased by errors in the ocean tide models used to calculate corrections to the tidal data. Though data from the globally distributed instruments used here are subject to this same source of bias, it seems unlikely that they will be biased in the same way for instruments located in widely separate regions of the globe. While we cannot claim that the use of such a global data set, some parts of which will inevitably be strongly influenced by the oceans, is always better than the use of data from a single region that may be weakly influenced by the ocean tides, the results from the global data set should provide an interesting comparison to the other studies because they are less sensitive to regional effects.

OCEAN CORRECTIONS

The tidal gravity signal observed at the Earth's surface is influenced by the effects of the oceans, whose response to the tidal potential contributes to the observed gravity through both loading and direct attraction. The combination of these effects can be quite large at coastal or ocean island stations, so they must be corrected for in order to compare the results of tidal analysis with the theory for the solid Earth tide. The ocean corrections used here were calculated by D. C. Agnew (personal communication, 1990), using the model for the ocean

tides described in Schwiderski (1980). This model represents the ocean response to the tidal potential at four diurnal frequencies (Q_1, O_1, P_1, and K_1) and four semidiurnal frequencies (N_2, M_2, S_2, and K_2).

Since the tidal analysis procedure described below requires corrections for each of 20 tidal groups, the Schwiderski corrections must be interpolated to obtain corrections for tidal groups within the frequency band Q_1–K_1, and they must be extrapolated for tidal groups exterior to this band. There are a number of ways in which one could perform such an inter/extrapolation, and it is difficult to judge which particular method works best. It does seem, however, that the procedure must take into account the systematic trends in the ocean response as a function of frequency. These generally include a monotonic increase or decrease in amplitude and phase with increasing frequency, and these trends appeared to be modeled well by a linear behavior of the in-phase and out-of-phase components of the ocean corrections. Therefore, for the tidal groups within the Q_1–K_1 frequency band, M_1 and π_1, we used a linear interpolation of the in-and out-of phase components of the ocean response inferred from the Schwiderski corrections for O_1 and P_1. For the extrapolation, however, the situation is more complicated. Similar linear extrapolation of the P_1 and K_1 corrections would not take into account the expected NDFW resonance in the ocean tide which affects these components (see Wahr and Sasao 1981). We correct for this effect by dividing the response at Q_1, O_1, P_1, and K_1 by the resonance factor $R(\omega)$ of Wahr and Sasao (1981, Table 2) before inter/extrapolation, and then multiplying the response by $R(\omega)$ for all tidal frequencies after inter/extrapolation.

This procedure is illustrated in Figure 2a-b. In Figure 2a, we show all of the diurnal corrections used, as well as the linear trend obtained by a least-squares fit to the in- and out-of-phase components of the 4 diurnal Schwiderski corrections, for stations GUA and BDF. From these figures it is readily seen that the estimated linear trend in the in- and out-of-phase components models the general trend in the corrections quite well, although it does not match the smaller features of the frequency dependence in the Schwiderski corrections. In particular it does not match the small but abrupt change in the values of the corrections for the P_1 and K_1 tides. This latter situation held for all of the stations, and we stress that this is the case even after applying the Wahr and Sasao (1981) corrections for the NDFW resonance. This suggests that the influence of the NDFW resonance in the Schwiderski corrections is larger than predicted, although we cannot be sure that it is due to some other effect. In any case, we consider it prudent to extrapolate Ψ_1 and Φ_1 corrections by using the linear fit to the general trend in the frequence dependence of the Schwiderski corrections, rather than extrapolating what appears to be the anomalous frequency dependence in the vicinity of the P_1 and K_1 tides. In Figure 2b are plotted the residuals of the Schwiderski corrections and the interpolated corrections, after subtracting the linear trend obtained by the least-squares fit. This figure shows that the deviations of the corrections from

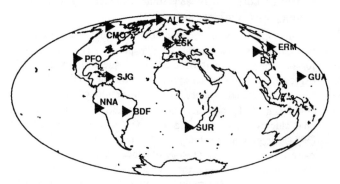

Fig 1. Map showing locations of the IDA stations used in this study

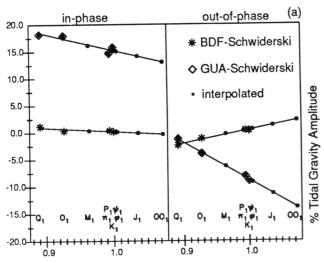 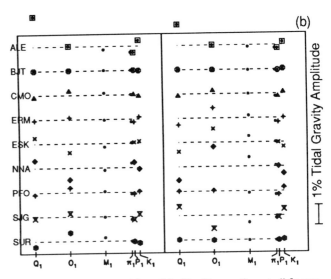

Fig 2. Illustration of the method for inter/extrapolation of the ocean corrections. (a) Ocean response inferred from Schwiderski corrections at all frequencies for stations GUA and BDF. Large symbols indicate the response calculated from the corrections, small filled circles indicate the inter/extrapolated responses, and the straight lines illustrate the fits to the in-phase and out-of-phase terms. (b) Ocean response inferred from corrections in the Schwiderski band (Q_1-K_1) for the remaining 9 stations, with the linear fit to the in-phase and out-of-phase terms subtracted.

the linear model are generally well below 1%. When it is considered that the ocean corrections are in most cases only a few percent of the gravity signal due to the solid Earth tide, the errors introduced by using the interpolated corrections should be no more than 0.1%. The extrapolated corrections may be more strongly biased, and we can only hope that the error introduced by extrapolation is much smaller than the percent error of measurement for these tides (which is usually large, because Ψ_1 and Φ_1 are so weak).

RESONANT TIDAL EXCITATION

Resonant tidal excitation can be observed in the tidal gravity data by comparing amplitude estimates of the diurnal tides. These estimates were obtained using the method of tidal analysis developed by Ishiguro *et al* (1983). A very good description of the method's application to tidal data is given in Tamura *et al* (1991). In order for the algorithm to achieve a numerically stable separation of the various tidal constituents when used with a finite amount of data, the 377 constituents in the tidal potential of Cartwright and Edden (1973) are grouped into 11 diurnal and 10 semidiurnal groups. Within each of these groups the amplitude and phase of the tidal admittance is assumed to be a constant function of frequency. Of the 11 diurnal constituents, S_1 was discarded because it is often strongly contaminated by the atmospheric S_1 tide. These leaves 10 diurnal admittances whose amplitudes and phases are expected to vary as a function of frequency in a manner consistent with the NDFW resonance.

The most complete quantitative description of this effect is that of Wahr (1981), whose results can be summarized in terms of a frequency-dependent diurnal gravimetric factor:

$$\delta_0(\omega) = \delta_0(\omega_{O_1}) + \delta_1 \frac{\omega - \omega_{O_1}}{\omega_{NDFW} - \omega} \qquad (1)$$

where ω_{NDFW} is the wobble frequency, and

$$\delta_0(\omega) = 1 - \frac{3}{2}k_0(\omega) + h_0(\omega) \quad , \quad \delta_1 = -\frac{3}{2}k_1 + h_1$$

and the Love numbers k_0, h_0, k_1, h_1 are described in Wahr (1981). The gravimetric factor is defined as the ratio of the theoretical tidal gravity signal on a deformable Earth to that on a rigid Earth. The term $\delta_0(\omega_{O_1})$ represents the enhancement of the tidal signal due to deformation associated with all free modes at the frequency of the O_1 tide. The values predicted according to Wahr (1981) varies very little from the value 1.155 for a number of realistic models of the internal structure of the Earth. Similarly, the value of δ_1 is predicted to be about -0.0006. While the theory used to predict these values is based on the assumption of an elastic Earth, the effect of weak anelasticity can be accommodated by allowing the three quantities ω_{NDFW}, $\delta_0(\omega_{O_1})$, and δ_1 to be complex.

The tidal admittances estimated for the IDA data are uncertain by an overall multiplicative factor. This calibration factor is estimated by IDA personnel by performing a tilt test of the gravimeters, but comparison with theoretical values for some of the major tidal constituents indicates that these are often inaccurate by a few percent. In order to reduce the influence of possible inaccuracies in the calibration factors in our search for values of the wobble parameters which best fit the data, we normalized the diurnal admittances by the admittance for the O_1 tide, so that the function we actually fit to the data is:

$$\frac{A(\omega_i)}{A(\omega_{O_1})} - 1 = \frac{\delta_0(\omega_i)}{\delta_0(\omega_{O_1})} - 1 = B \frac{\omega_i - \omega_{O_1}}{\omega_{NDFW} - \omega} \qquad (2)$$

where $A(\omega_i)$ is the (complex) admittance estimate for diurnal tidal constituent i, and

$$B = \frac{\delta_1}{\delta_0(\omega_{O_1})}$$

Furthermore, instead of fitting for the real and imaginary parts of ω_{NDFW}, we fit for the more physically meaningful quantities ω_0 and $1/Q_{NDFW}$. ω_0 can be expressed as $(1+1/T_{NDFW})$, where T_{NDFW} is the NDFW period as observed from inertial space, and Q_{NDFW} is the number of oscillations after which the wobble amplitude will decrease to $e^{-\pi}$ of it's initial value, if excited and left undisturbed. We performed a fit of the data from each station to obtain the results listed in Table 1. While the negative Q_{NDFW} estimates listed in Table 1 are probably biased, we have no guarantee that the positive estimates are not biased as well, and can only hope that by using all of the stations together the effects of such a regional bias will tend to cancel. On the other hand, the results for some of the stations do appear to be biased much more strongly than the others. For example, all of the stations have estimated periods which lie within the range 300-500 days, except for stations ERM, NNA, and PFO. Since the results for these three stations are so inconsistent with the results form the other stations, it seemed sensible to discard the results from these stations. We can then perform another least squares fit using the data from the remaining 8 stations simultaneously. The results for this 8-station fit constitute our preferred values for the wobble parameters, and they are listed in Table 1 along with a similar set of results obtained using all eleven stations. Our preferred values for T_{NDFW} and Q_{NDFW} are compared with the results from previous studies in Figure 3.

CONCLUSIONS

One result of this study, that the period of the NDFW is shorter than predicted by theory, is consistent with the results of previous studies (Herring *et al.* 1986 and 1991, Richter and Zürn 1986, Neuberg *et al.* 1987, and Sato 1991), as can be seen in Figure 3. This shortened period almost certainly indicates that the dynamical ellipticity of the fluid outer core is greater than would be predicted for an Earth in hydrostatic equilibrium. This fact has been used by Gwinn *et al.* (1986) to infer that the ellipticity of the core-mantle boundary (CMB) is greater than its hydrostatic value, with a maximum peak-to-peak deviation from the hydrostatic figure of about ½ km.

Our result of a value for Q_{NDFW} of about 6100, on the other hand, is most consistent with the result of Sato (1991). Our lower bound of 3300 falls just within the range of acceptable values suggested by the tidal studies in northern Europe (Richter and Zürn 1986, and Neuberg *et al.* 1987). Whether our preferred upper bound for Q_{NDFW} of 36638 agrees with the VLBI results is less clear. The study of Herring *et al.* (1986) does not estimate a value for Q_{NDFW}, but does estimate a value for the imaginary part of ω_{NDFW}, while the more recent study (Herring *et al.* 1991) does not estimate even the imaginary part of ω_{NDFW} but states that it can be done. We estimated a Q_{NDFW} value for the data of Herring *et al.* (1991) by performing a least-squares fit to the nutation observations, varying both T_{NDFW} and Q_{NDFW}. The result for T is 429.8 days, in agreement with the value obtained by Herring *et al.* (1991), and the optimum value for Q_{NDFW} is about 27000. This falls well within our preferred upper bound for Q_{NDFW} (see Figure 3).

It is interesting to speculate how well the nutation data are fit using the low Q_{NDFW} values which are estimated in the tidal

Table 1. Results for Wobble Parameters

Station	Period	Q	Br×10⁴	Bi×10⁴	Data span (months)
8 stations	428.4 (416.3 - 441.2)	6127 (3343 - 36638)	-5.2 (0.2)	-0.5 (0.2)	
All 11 stations	440.6 (426.1 - 456.1)	6414 (3223 - 651963)	-4.8 (0.2)	-0.5 (0.2)	
ALE	394.6 (350.1 - 451.9)	1033 (609 - 3116)	-11.6 (1.5)	-3.8 (1.5)	
BDF	404.4 (357.7 - 465.0)	-1553 (-1091270 - -777)	-5.2 (0.7)	1.7 (0.7)	51.1
BJT	401.1 (388.5 - 414.5)	6731 (3233 - -82211)	-4.8 (0.2)	-0.2 (0.2)	36.8
CMO	406.2 (367.4 - 454.1)	2241 (1039 - -14363)	-5.3 (0.6)	-0.3 (0.6)	72.3
ERM*	810.7 (516.0 - 1889.6)	-2563 (988 - -558)	-2.9 (1.6)	0.2 (1.6)	37.1
ESK	426.7 (404.3 - 451.8)	6704 (2447 - -9067)	-3.9 (0.2)	-0.8 (0.2)	44.9
GUA	375.4 (361.7 - 390.1)	-2388 (-3996 - -1703)	-12.3 (0.7)	-9.2 (0.8)	63.0
NNA*	-142.1 (-151.6 - -133.7)	3139 (820 - -1716)	4.2 (1.0)	2.3 (1.1)	47.4
PFO*	843.9 (560.8 - 1704.4)	-607 (-2150 - -353)	-2.2 (1.1)	1.4 (1.1)	49.3
SJG	404.7 (391.8 - 418.6)	-8991 (18754 - -3626)	-7.0 (0.3)	-0.2 (0.3)	84.5
SUR	433.5 (403.7 - 468.0)	2051 (1205 - 6873)	-5.8 (0.4)	-0.7 (0.4)	58.2
					41.6

* Station not used in 8-station solution.

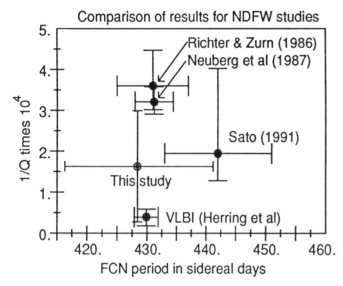

Fig 3. Comparison of the least-squares results obtained here for T_{NDFW} and Q_{NDFW} with results from other studies. Error bars indicate ± one standard deviation.

studies, especially when we also allow the strength-of-resonance (i.e., the quantity for the nutations which is analogous to B, in (2), for the body tides) to vary. We therefore performed least-squares fits to the nutation data always holding the period T_{NDFW} constant at 429.8 days, and holding Q_{NDFW} constant at 3500, 6100, and 26758, respectively, while allowing the strength-of-resonance to vary. The former values for Q_{NDFW} were chosen to represent an upper bound (3500) for the studies done in Europe (Richter and Zürn 1986 and Neuberg *et al.* 1987), as well as a value (6100) close to the best estimate

obtained in the present study, while the latter Q_{NDFW} value (26758) is the one which best fits the nutation measurements of Herring *et al.* (1991). The results for these fits are illustrated in Figure 4. There it can be seen that, if Q_{NDFW} is forced to have values as low as 3500 or even 6100, residuals of 0.5-2.0 mas would be evident in the retrograde annual and prograde semiannual terms. These residuals are very large, considering the 0.04 mas precision for the nutation measurements reported by Herring *et al.* (1991). For a Q_{NDFW} value as high as is suggested by their data, almost all of the nutation terms can be fit to within the observational uncertainty.

It seems very unlikely that systematic errors can bias the VLBI measurements to the extent indicated for the 'low' Q_{NDFW} values in Figure 4. The most useful nutation term for learning about the NDFW is the retrograde annual term (equivalent to the Ψ_1 body tide), and since there is likely to be considerable atmospheric contamination at annual periods it is natural to speculate that this contamination may affect the VLBI results significantly. But the prograde annual nutation is the annual term which actually follows the sun (equivalent to the S_1 body tide), and so if there were atmospheric contamination at the retrograde period then there should be even stronger contamination at the prograde period. Since this is not seen in the VLBI data, it seems that values for Q_{NDFW} which are as low as suggested in some of the tidal studies (including this one) are implausible, even when one allows for atmospheric contamination.

While the range of acceptable values for Q_{NDFW} found in this study agrees with the other tidal studies as well as the VLBI work, our preferred Q_{NDFW} value of 6100 is incompatible with the VLBI nutation measurements. There is still the possibility that our tidal study, even though it uses data from globally distributed stations, may still be biased due to errors in the ocean load corrections. Those corrections are often substantial. If the discrepancy is indeed due to inaccurate ocean corrections, then the bias introduced by this inaccuracy may be global in nature.

Acknowledgments We are grateful for the critical and constructive comments of Veronique Dehant and an anonymous reviewer. We also thank Duncan Agnew for his assistance in obtaining and analyzing the IDA data, for providing the ocean loading corrections, and for some very useful discussions, and also Yoshiaki Tamura for providing us with a version of the BAYTAP computer program.

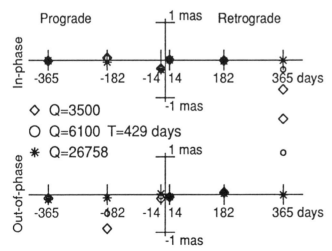

Fig 4. Residuals between the VLBI nutation measurements of Herring *et al.* (1991)

REFERENCES

Agnew, D., J. Berger, W. Farell, F. Gilbert, G. Masters, and D. Miller, Project IDA: a decade in review, *EOS*, 67, 203-212, 1986.

Cartwright, D.E., and A. C. Edden, Corrected tables of tidal harmonics, *Geophys. J. R. astr. Soc.*, 33, 253-264, 1973.

Gwinn, C. R., T. A. Herring, and I. I. Shapiro, Geodesy by radio interferometry: Studies of the forced nutations of the Earth, 2, Interpretation, *J. Geophys. Res.*, 91, 4755-4765, 1986,

Herring, T. A., J. L. Davis, and I. I. Shapiro, Geodesy by radio interferometry: Studies of the forced nutations of the Earth, 1, Data Analysis, *J. Geophys. Res.*, 91, 4755-4765, 1986.

Herring, T. A., B. A. Buffett, P. M. Mathews, and I. I. Shapiro, Forced nutations of the Earth: Influence of inner core dynamics 3. Very Long Baseline Interferometry, *J. Geophys. Res.*, 96, 8259-8273, 1991.

Ishiguro, M., H. Akaike, Ooe, M. and S. Nakai, A Bayesian approach to the analysis of Earth tides, *Proc. 9th Int. Sympos. Earth Tides*, New York, pp. 283-292, ed. Kuo, J. T., Schweizerbart'sche Verlangsbuchhandlung, Stuttgart, 1983.

Neuberg, J., J. Hinderer, and W. Zürn, Stacking gravity tide observations in central Europe for retrieval of the complex eigenfrequency of the nearly diurnal free wobble, *Geophys. J. R. astr. Soc.*, 91, 853-868, 1987.

Richter, B. and W. Zürn, Chandler effect and nearly diurnal free wobble as determined from observations with a superconducting gravimeter, *The Earth's Rotation and Reference Frames for Geodesy and Geodynamics*, eds. A. Babcock and G. Wilkins, Kluwer, Dordrecht, Holland, 1986.

Sato, T., Fluid core resonance measured by quartz tube extensometers at the Esashi Earth Tides Station, *Proc. 11th Int. Sympos. on Earth Tides*, Helsinki, pp. 573-582, ed. Kakkuri, J., Schweizerbart'sche Verlangsbuchhandlung, Stuttgart, 1991.

Schwiderski, E. W., On charting global ocean tides, *Rev. Geophys. Space Phys.*, 18, 243-268, 1980.

Tamura, Y., T. Sato, M. Ooe, and M. Ishiguro, A procedure for tidal analysis with Bayesian information criterion, *Geophys. J. Int.*, 104, 507-516, 1990.

Toomre, A., On the 'nearly diurnal wobble' of the Earth, *Geophys. J. R. astr. Soc.*, 38, 335-348, 1974.

Wahr, J. M., Body tides on an elliptical, rotating, elastic and oceanless earth, *Geophys. J. R. astr. Soc.*, 64, 677-703, 1981b.

Wahr, J. M., and T. Sasao, A diurnal resonance in the ocean tide and in the Earth's load response due to the resonant free 'core nutation', *Geophys. J. R. astr. Soc.*, 64, 747-765, 1981.

Phil R. Cummins, Research School of Earth Sciences, Australian National University, GPO Box 4, Canberra ACT 2601, Australia.

John M. Wahr, Department of Physics and CIRES, University of Colorado, Boulder, Colorado.

Inner Core Motions: Implications on Earth Rotation

Anthony M. K. Szeto[1]

School of Civil and Mining Engineering, University of Sydney, Sydney

Dynamical motions of the inner core are determined by a balance of several torques: electromagnetic, viscous, pressure and gravitational. The net torque can be analyzed into axial and equatorial components. Their action on the rest of the Earth gives rise respectively to changes in the length of day and polar motion. The symmetry axis of the inner core does not precisely coincide with that of the whole Earth. The obliquity between these two axes determine the pressure (or inertial) and gravity restoring (or figure-figure) torques coupling the inner core with the rest of the Earth. These torques have the same functional dependence on obliquity. We assert that the gravity torque dominates by several orders of magnitude. Recent efforts to include the effects of the inner core on the Earth's rotation are reviewed, with special attention on the role of the gravity torque. We point out the need for considering the dynamics of a realistic fluid core.

INTRODUCTION

Until recently the inner core has been neglected in analyses of Earth rotation, but several papers have recently addressed this problem [Mathews et al., 1991a, b; Buffett et al., 1991; de Vries and Wahr, 1991; Dehant et al., this issue]. We review some of this work, drawing attention to disagreements between these efforts and ours [Szeto and Smylie, 1989], and among themselves. For several years we have been considering a simplified form of inner core motion – regular precession, motivated by geomagnetism [Szeto and Smylie, 1984]. It is gratifying to see the recent attention paid to evaluating the role of the inner core in Earth rotation.

We begin with estimates of all torques on the inner core considered to date – pressure, viscous, electromagnetic and gravity. Analysis used in determining the regular precession of the inner core under a dominating gravity retoring torque is discussed, revealing up to four modes of motion. De Vries and Wahr [1991] and Mathews et al. [1991] have obtained periods of "free motions" within the Earth, evaluating the influence on these by the presence of the inner core. Their results for "forced motions" are also discussed. Finally we consider the need for taking into account realistic fluid core behaviour, since all existing work cited so far have all taken the outer core as an equivalent elastic solid, albeit deformable.

[1] On sabbatical from Department of Earth and Atmospheric Science, York University, Ontario, CANADA M3J 1P3.

Dynamics of Earth's Deep Interior and Earth Rotation
Geophysical Monograph 72, IUGG Volume 12
Published in 1993 by the International Union of Geodesy and Geophysics and the American Geophysical Union.

TORQUES ON THE INNER CORE

The torque acting on the inner core due to the rest of the Earth is the integral over the inner core volume of $\vec{r} \times \vec{F}$, where $\vec{F}$ is the force per unit volume at radius vector $\vec{r}$. (Surface traction can be included by transforming a relevant surface integral into a volume integral. Torques due to extra-terrestrial origin such as the lunar-solar attraction can also be easily incorporated into $\vec{F}$.) Four torques arise from this consideration: pressure, electromagnetic, viscous and gravity. The first two directly arise as a consequence of fluid flow in the outer core; the third is due to electromagnetic interaction between the inner core and the magnetic field, which is presumed to be generated by outer core motion. These three are "dynamic" torques, since their presence relies on outer dynamical motion. The gravity torque, however, is due to gravitational figure-figure interaction between the inner core and the rest of the Earth, and would not vanish even in the absence of an outer core flow, and may thus be classified "static".

Estimates of these torques are presented in Table 1, based on the work of numerous researchers [Kakuta et al., 1975; Rochester, 1968, 1970; Toomre, 1966; Szeto and Smylie, 1984; Guo, 1989]. The torques have been analyzed into axial and equatorial components, which respectively produce changes in rotation speed and orientation of the inner core (and of the rest of the Earth by equal and opposite reaction). We note the following salient features.

Being dependent on surfaces of equal pressure and density, inertial and gravitational torques act predominantly in an equatorial direction so as to restore any misalignment between the symmetry axis of the inner core and that of the mantle-outer core assembly. (i.e. $\hat{x}$ points along the

TABLE 1. Inner Core Torques

	Axial (Nt.m.)	Equatorial (Nt.m.)
Inertial	$O(e^2)$ equatorial	$\approx -10^{20} \cos\theta_i \sin\theta_i \hat{x}$
Viscous	$\approx 10^{17}$ (dissipative)	$\approx 10^{17} \cos\theta_i \sin\theta_i \hat{y}$
E.M.	$\approx 10^{19}$ (balanced by differential rotation)	$\approx 10^{19}(\hat{N} \cdot \hat{x})\hat{x}$, $\approx 10^{19}(\hat{N} \cdot \hat{y})\hat{y}$
Grav.	$O(e^2)$ equatorial	$\approx -10^{24} \cos\theta_i \sin\theta_i \hat{x}$

cross product of mantle and inner core symmetry axes in such a way that $-\hat{x}$ tends to reduce any misalignment between them.) The equatorial components of these torques are smaller than the axial ones by the square of the ellipticity of the internal surfaces. Thus these torques are more effective in inducing polar motion than in length-of-day changes.

The electromagnetic torque appears to have a significant axial component. This must be nearly cancelled, however, by differential rotation between the inner core and the outer core, as argued by Smylie et al. [1984] and others. $\hat{N}$ is the direction of the mean magnetic moment in the core, which has non-vanishing components in the equatorial plane. It is interesting to note that this torque can be significant along the $\hat{y}$ direction, which is normal to both the symmetry axis and $\hat{x}$, and thus can give rise to polar motion out of phase in relation to that due to inertial and gravitational torques. Viscous torques are likely to be less important in determining the dynamics of the inner core.

Finally we observe that inertial and gravity torques share the same angular dependence on θ_i, the obliquity between the inner core and mantle symmetry axes, but different by about 4 orders of magnitude. (The gravity torque is formally a second order quantity dependent on the product of the dynamical and geometric flattening. However, it is also proportional to the mass of the inner core and its moment of inertia, which are substantial quantities indeed.) The surprisingly large magnitude of the gravitational restoring torque was first mentioned by Szeto and Smylie [1984], but this has not been seriously recognized. We will return to this when we discuss the relatively low importance assigned to this torque in the recent papers of Mathews et al. [1991a, b] and de Vries and Wahr [1991]. Guo [1989] re-derived an estimate for this torque, and found it to be also very large, although his value turned out about four times smaller than ours. This difference has not yet been reconciled to date.

INNER CORE MOTIONS

We have previously investigated the dynamical motion of the iner core coupled to the combined mantle-outer assembly arising from the gravitational restoring torque. We skip details here, which can be found elsewhere [Szeto and

Smylie, 1984, 1989]. The Earth is treated as an isolated body, in which the inner core experiences a torque, and the rest of the Earth experiences an equal an opposite torque; essentially:

$$\frac{dL_{IC}}{dt} = \Gamma \tag{1}$$

$$\frac{dL_{M+OC}}{dt} = -\Gamma \tag{2}$$

In this simple two-layer system, L_{IC} and L_{M+OC} are the angular momenta of the inner core and the mantle-outer core assembly, Γ is the torque which couples the system. Any torque expression can be used in the above equations. (We prefer the gravitational restoring torque, since we assert that it is the dominant torque, at least in the equatorial component.) We found four inner core motions: one nearly diurnal prograde, one of arbitrarily long period, and two close to the Chandler period and straddling it on both sides. In our calculation the rotation rate of the inner core and its obliquity to the mantle symmetry axis are adjustable parameters. The long period mode can be either prograde or retrograde depending on the choice of these parameters. Numerical details can be found in our earlier papers cited above, but we consider our model so crude that a direct comparison with observed modes (inferred from VLBI, for instance) is not appropriate without a substantial improvement.

Recent Efforts

While a single-layer Earth has only a Chandler Wobble, a two-layer Earth has an additional mode, the Free Core Nutation (FCN). A three-layer model incorporating the inner core has yet another mode, dubbed Free Inner Core Nutation (FICN). Following Sasao et al. [1980] one can calculate FCN frequency, which appears to be relatively model insensitive. Its value in cycles/day relative to a rotating frame is $-1 - \frac{1}{460}$ (Model 1066A) or $-1 - \frac{1}{458}$ (PREM). (The notation here follows Mathews et al. [1991a], which differs from that used by de Vries and Wahr [1991].) As the inner core is introduced, the FCN frequency is somewhat altered. De Vries and Wahr found this value, known as $\omega_{FCN'}$, to be $-1 - \frac{1}{457}$. Proximity to ω_{FCN} indicates the relative insignificant influence the inner core has on this particular motion. FICN depends on a number of parameters relating to Earth structure. De Vries and Wahr gave a preferred value of $-1 + \frac{1}{471}$. (However, this may vary as greatly as $-1 + \frac{1}{499}$ if the mantle and inner core were made entirely rigid.) The basis of their calculation is a rotating, oceanless Earth with an elastic mantle and inner core, and a fluid core in hydrostatic equilibrium. (In other parts of their paper, they also considered departures of the core-mantle boundary and inner core-outer core boundary from hydrostatic equilibrium.)

Mathews et al. [1991a] identified a fourth motion associated with the wobble of the solid inner core, which may be named Inner Core Wobble (ICW). They found a frequency of

$\frac{1}{2400}$ cycles/day, based on model PREM. Their theory takes into account gravitational and pressure coupling among the inner core, outer core and the mantle. However, they estimated the contribution of gravitational coupling to be only 14%. Thus the pressure torque dominates their calculation. This is a major discrepancy between their theory and ours.

Their model and that of de Vries and Wahr are capable of predicting forced motions. While these generally agree, significant differences do exist. This points to the fact that not both of these calculations can be correct. Consider the prograde semi-annual nutation, for which Mathews et al. [1991a] derived a correction to Wahr's IAU series by 0.6 mas amplitude. De Vries and Wahr [1991] found, on the other hand, a correction of only 0.2 mas. Both of these corrections as well as the difference between them are greater than the claimed VLBI resolution limit of 0.04 mas. This is important to bear in mind, since it is through reconciling discrepancies between observed and predicted values that one is able to provide constraints on the Earth's internal structure (e.g. by modifying the flattening of the core-mantle or inner-outer core boundaries, or the deformability compliances of the mantle and the inner core). Dehant et al. [1992] have apparently succeeded in reproducing most of the forced nutation amplitudes. We assert, however, that the abovementioned procedures of inferring internal structure may be premature at this stage, since all of these efforts have adopted a relatively simple model of the fluid core with a Poincaré or equivalent flow. A more realistic fluid core must be incorporated before theoretical predictions can be meaningfully compared with observations.

SUMMARY

Recent research cited above suggests that the inner core does not significantly influence the relatively well established periods of the Chandler Wobble and Free Core Nutation. However, the presence of an inner core is found to produce additional modes (Inner Core Wobble and Free Inner Core Nutation) as well as measurable effects of a small number of forced nutation. Details of these calculations have been used to infer certain internal Earth parameters. We call into question the validity of these inferences on the basis that 1) the so-called gravitational restoring torque may not have been correctly treated so far, and 2) contrary to the findings of de Vries and Wahr [1991], a fluid outer core may alter the final outcome above present observational uncertainties.

REFERENCES

Buffett B. A., P. M. Mathews, and I. I. Shapiro, Forced Nutations of the Earth: Influence of Inner Core Dynamics 3. Very Long Interferometry Data Analysis, *J. Geophys. Res.* , *96*, 8259-8273, 1991.

Dehant V., J. Hinderer, H. Legros, and M. Lefftz, Analytical Computation of the Rotational Eigenmodes of the Inner Core, *this issue.*

De Vries D. and J. M. Wahr, The Effects of the Solid Inner Core and Nonhydrostatic Structure on the Earth's Forced Nutations and Earth Tides, *J. Geophys. Res.* , *96*, 8275-8293, 1991.

Guo J., On the Coupling Dynamics of the Inner Core – Outer Core – Mantle System, PhD Thesis, Université Catholique de Louvain, Louvain-La-Neuve, 1989.

Kakuta C., I. Okamoto, and T. Sasao, Is the Nutation of the Solid Inner Core Responsible for the 24-Year Libration of the Pole? *Publ. Astron. Soc. Jpn.*, *27*, 357-365, 1975.

Mathews P. M., B. A. Buffett, T. A. Herring, and I. I. Shapiro, Forced Nutations of the Earth: Influence of Inner Core Dynamics 1. Theory, *J. Geophys. Res.* , *96*, 8219-8242, 1991a.

Mathews P. M., B. A. Buffett, T. A. Herring, and I. I. Shapiro, Forced Nutations of the Earth: Influence of Inner Core Dynamics 2. Numerical Results and Comparisons, *J. Geophys. Res.* , *96*, 8243-8257, 1991b.

Rochester M. G., Perturbations in the Earth's Rotation and Geomagnetic Core-Mantle Coupling, *J. Geomag. Geoelectr.* , *20*, 387-402, 1968.

Rochester M. G., Core-Mantle Interactions: Geophysical and Astronomical Consequences, in *Earthquake Displacement Fields and the Rotation of the Earth,* edited by L. Mansinha et al., pp. 136-148, D. Reidel Publ. Co., 1970.

Sasao T., S. Okubo, and M. Saito, A Simple Theory on the Dynamical Effects of a Stratified Core upon Nutational Motion of the Earth, in *Proceedings of IAU Symposium 78,* edited by E.P. Federov et al., pp. 165-183, D. Reidel Publ. Co., 1980.

Smylie D. E., A. M. K. Szeto, and M. G. Rochester, The Dynamics of the Earth's Inner and Outer Cores, *Rep. Prog. Phys.*, *47*, 855-906, 1984.

Szeto A. M. K., and D. E. Smylie, Coupled Motions of the Inner Core and Possible Geomagnetic Implications, *Phys. Earth Planet. Int.* , *36*, 27-42, 1984.

Szeto A. M. K., and D. E. Smylie, Motions of the Inner Core and Mantle Coupled via Mutual Gravitation: Regular Precessional Modes, *Phys. Earth Planet. Int.* , *54*, 38-49, 1989.

Toomre A., On the Coupling of the Earth's Core and Mantle during the 26,000–Year Precession, in *Moon System,* edited by B. G. Marsden and A. G. W. Cameron, pp. 33-45, Plenum Press, 1966.

A. M. K. Szeto, Department of Earth and Atmospheric Science, York University, Ontario, CANADA M3J 1P3.

New Analysis of the Superconducting Gravimeter Data of Brussels

V. Dehant, B. Ducarme, and P. Defraigne

Observatoire Royal de Belgique,
3, avenue Circulaire,
B1180 Brussels, Belgium

In 1986, Melchior and Ducarme identified some spectral peaks, between the diurnal and semi-diurnal frequency bands in the power spectrum of the gravity variation measured by the superconducting gravimeter installed in 1982 at the Royal Observatory of Belgium. Using further spectral analysis techniques we show in this paper that the existence of these peaks, in particular the one detected at 13.9 hours, is confirmed. However its origin, given by Melchior and Ducarme as being core inertial waves excited by strong and deep earthquakes, is controversial and probably not correct.

We show indeed that the 13.9 hr peak which they associated with the Hindu Kush earthquake of 30 December 1983 appeared already before this earthquake. We also illustrate situations where spectral energy can be found between the diurnal and semi-diurnal frequency bands, without earthquake occurrence and other ones where the gravity data spectrum remains quiet although deep and strong earthquakes occurred. For these reasons, we looked for other origins of the excitation of the gravity spectrum between the diurnal and semi-diurnal frequencies. In this paper we compare a moving window spectral analysis applied to the superconducting gravimeter data, with the same analysis applied to atmospheric pressure and oceanic data. We obtain a good correlation between the time of occurrence of the episodic perturbations in all the spectra.

Introduction

From vertical gravity changes measured with a superconducting gravimeter in Brussels, Melchior and Ducarme [1986] found significant energy in the power spectrum of tidal data between the semi-diurnal and diurnal frequency bands. The record was analysed after two large and deep earthquakes. Melchior and Ducarme [1986] found that several peaks appeared after these earthquakes, in the period range from 13 to 17.5 hrs. The largest peak was identified at 13.9 hrs.

Mansinha et al. [1990] have reanalysed the data with the method of Welch [1967] in which the data are divided into overlapping, windowed segments. A 95% confidence interval has been indicated and used in order to conclude that no statistically significant non-tidal spectral peak is found in the band between 12 and 24 hrs. This work assumes stationarity in frequency and amplitude of the peaks. In the present work we do not assume stationarity because we want to see the

Dynamics of Earth's Deep Interior and Earth Rotation
Geophysical Monograph 72, IUGG Volume 12

time-evolution of the amplitudes and frequencies. Moreover, this reconciles the two works.

The peaks found by Melchior and Ducarme [1986] were interpreted, at that time, as normal modes of the Earth's core. Indeed, core undertones for a rotating, homogeneous, stratified fluid in a spherical container exhibit periods in that frequency band [see Aldridge and Lumb, 1987 or Melchior et al., 1989]. The most general theoretical solutions for a spheroid of arbitrary ellipticity filled completely with fluid, was obtained by Kudlick [1966, see also Guo, 1990]. As indicated in Aldridge and Lumb [1987] or Melchior et al. [1989], the solutions for the frequencies can be obtained by finding the roots of a simple combination of Associated Legendre Functions of degree n, azimuthal order k, and radial number m. When indefinitely increasing n, k, and m, the inertial wave spectrum in fact becomes infinitely dense. Thus at any observed frequency in the spectrum of the superconducting gravimeter data, one can find a corresponding theoretical period in the core mode eigenspectrum.

Melchior et al. [1989] made the assumption that the largest amplitude gravity peaks correspond to the most simple geometry of the mode, i.e. to the lowest radial number.

However, with this assumption, the matching of the observed eigenspectrum and the computed one is not very good.

Melchior et al. [1989] did also consider the effect of a possible influence of the inner core on the core undertones. The presence of an inner core would, following them, produce a shift in the frequencies. They therefore proposed to displace the theoretical frequency spectrum with respect to the observed one to simulate the inner core correction; but this process did not significantly improve the matching of the frequencies.

Another possible explanation for the poor match with the inertial waves would be some interaction between the core modes themselves or of the core modes with the inner core modes.

Inner core translational modes have also been reviewed by Melchior et al. [1989]. Their computations showed that without rotation, these modes can have almost any period between about 5 hrs and 60 hrs depending on the Earth model which is mainly affected by stratification. With the addition of rotation, the periods become severely constrained to lie below an upper limit of 13.3 hrs, thus ruling out a correspondence with the observed peak at 13.9 hrs.

Of course if one relaxes the assumption on the spatial distribution, there are certainly sufficient inertial modes, so that the question of identification of the observed peaks is still open.

Using a Love number formulation, Crossley [1989] computed the effects at the Earth's surface from the core undertones as excited by an earthquake, and found that the effects for one single spectral line were too small by one order of magnitude to be seen. This still does not exclude the explanation foreseen above, because one can imagine that for some reason, the energy developed by the core modes distributed all over the spectrum is concentrated at one frequency (i.e., via non-linear interaction). Thus, again, the question of identification of the observed spectral peaks with core undertones remains open.

Another important fact is that the 13.9 hr peak observed at Brussels after the Hindu Kush earthquake was not detected by the superconducting gravimeter of Bad Homburg in Germany at 320 kilometres from Brussels [Zürn et al., 1987]. This could be explained by trapping of core normal mode waves in some regions of the core, and thus, the energy measured at the surface would be higher or lower over regions where they are respectively trapped or not. But Bad Homburg is so close to Brussels that this explanation would be not plausible.

In this paper, we reanalyse the Brussels data in order to verify the existence of those particular peaks and search for other possible explanations of their presence.

We will show that their existence is confirmed and that other explanations for their origin are possible.

Data

We use gravity data measurements obtained with the superconducting gravimeter installed at the Royal Observatory of Belgium. The instrument has been continuously operating since 1982. It is a GWR instrument containing a 2.5 cm-diameter sphere of superconducting material levitated by an electromagnetic force. By immersing the whole system in a helium bath at $3.2°K$, the intensity and the geometry of the magnetic field is kept essentially constant. With an additional thermostatisation at the 10^{-5} °K level, the short term stability is $10^{-11}g$ and the long term one around $10^{-9}g$. The gravity variation is recorded by measuring the variation of the electric current needed to maintain the sphere at a constant level.

The data are carefully controlled and missing data due to big earthquakes or helium refilling are interpolated; the hourly data are corrected for the local atmospheric pressure by the impulse response relation [De Meyer and Ducarme, 1987, and Ducarme et al., 1987]

$$a(t) = -0.2628p(t) - 0.0779p(t-1) - 0.0167p(t-2) \quad (1)$$

where $a(t)$ is the correction at the instant t for local pressure p measured at times t, $t-1$ and $t-2$. The units are choosen such that if the pressure is in $mbar$ ($1mbar = 100\ Pa$), and the correction is expressed in μgal ($1\ \mu gal = 10\ nms^{-2}$).

The data corrected for the local atmospheric pressure in the time domain are then corrected for the tides using an extended tidal potential [Tamura, 1987]. The resulting residuals are expressed in $0.01\ \mu gal$.

New Spectral Analysis of the Superconducting Gravimeter Data after the 1983 Hindu Kush Earthquake

We first try to verify whether the 13.9 hr oscillation detected after the 1983 Hindu Kush earthquake could not be explained by some artifacts of the spectral analysis method itself. We limit our analysis to the data collected around the Hindu Kush earthquake of December 30, 1983 (depth 209 km, magnitude 6.2).

Five different methods of spectral analysis were used; they were applied to the data by moving a 6-month window in steps of 15 days :

- the classical FFT (Fast Fourier Transform) with a Parzen window;

- the Thomson method; this method is very similar to the FFT but uses a set of spheroidal wave functions called the "eigentapers" in order to offer a better resistance to leakage;

- the classical harmonic analysis; this method fits each amplitude of a sine and a cosine at a frequency which is a harmonic of the fundamental frequency; the advantage of this method is that a statistical test of significance can be performed;

- the periodic regression method which fits a set of frequencies globally and iteratively;

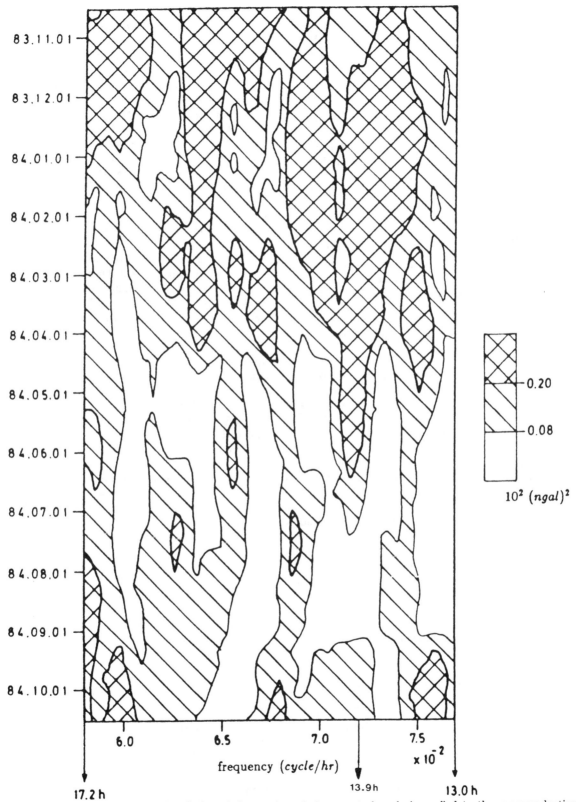

Fig. 1. Projection of the 3-dimensional plot of the moving window spectral analysis applied to the superconducting gravimeter data around the Hindu Kush earthquakes (date = center of window).

- the maximum entropy method, which computes analytically the spectrum associated with a realization of a moving average stochastic process (the variable at a time t depends on the previous variables plus white noise at the instant t); the computation is performed in such a way that no information is added (maximum entropy).

All the results of these spectral analyses are detailed by Defraigne [1991] who concludes that the 13.9 hr peak is not an artefact of the method used by Melchior and Ducarme [1986]. Figure 1 shows the results of the FFT method for gravity data around the time of the earthquake of Hindu Kush. The peak at 13.9 hrs is a real peak. Moreover it is not the only one in the frequency band between 13 and 17.5 hrs. Most of the methods show that harmonic energy is existing in this frequency band.

We also performed a spectral analysis on simulated data obtained from a theoretical tide with white or red noise; the details for which can be found in Defraigne [1991]. One concluded that no artificial peak was generated by the method of tidal analysis itself.

ANALYSIS OF THE DATA AROUND STRONG AND DEEP EARTHQUAKES

We then perform moving window spectral analysis on the data selected after some strong and deep (large seismic moment) earthquakes. In Table 1, we reproduce the list of strong and deep, or not deep but very strong, earthquakes that occurred since the installation of the superconducting gravimeter at Brussels.

Figure 1 shows the results of the moving window spectral analysis described above, for a 6-month window shifted fifteen days at each step, and applied on data around the 1983 Hindu Kush earthquake.

Figures 2 and 3 show the results of the moving window spectral analysis for two sets of data: (1) from June 2, 1982 to October 15, 1986, and (2) from July 22, 1987 to December 30, 1989. The windows in these analyses are 6 months long and shifted by a step of one month for each successive analysis.

In Figure 1 we observe that the peak at 13.9 hrs really appears in the data around the 1983 Hindu Kush earthquake. In Figure 2 we observe that the 13.9 hr peak already appears before this same earthquake.

In order to have a more detailed picture, Figure 4 presents a projection of a three-dimensional plot of the power spectrum, as a function of frequency and time, for the period around the 1983 Hindu Kush earthquake. This graph clearly indicates that the peak appears before the earthquake. We conclude that the Hindu Kush earthquake is not the cause of the 13.9 hr peak detected in the spectrum of the superconducting gravimeter data.

Let us note that as the Parzen window on a six-month data set reduces the influences of the first and last quarters of the data, the effect of the earthquake of December 30, 1983 could be seen (if the Earth's response is instantaneous) only from the spectrum of data beginning 4 1/2 months before the earthquake, i.e., beginning on August 15, 1983 or centered on October 30, 1983.

In Figure 3, we observe that the spectrum has an increase of energy at the 13.9 hr period, although there is neither a strong and deep earthquake nor a very strong earthquake around January 1989. On the other hand in the spectrum of data collected around the strong and deep earthquake which happened in Brazil in May 1989, no increase of energy appears in the frequency band between 13 and 17.5 hrs.

Background noise evaluation has been obtained from the spectral analysis results between the diurnal and the semi-diurnal tidal band [P. Melchior and B. Ducarme, 1986, Table 1]. It was evaluated at 3 nanogal level for 6-month windowing. This background noise level puts the 13.9 hr peak above the 95% confidence interval when we speak about its detection or its occurrence.

We have thus a series of arguments against the fact that, in the gravimetric data, the peaks detected in the frequency band in between the diurnal and semi-diurnal frequencies are due to excitation of core normal modes induced by a strong and deep, or by a very strong, earthquake. In consequence either the core modes are excited by something else or the peaks are not related to core modes.

OTHER ORIGIN OF THE PEAKS BETWEEN 13 AND 17.5 HRS

The gravimetric data have been corrected only for local atmospheric pressure effects. But if there are other uncorrected perturbations or if local pressure does not sufficiently

Table 1: List of strong and deep earthquakes since end of 1983

location	date Year Month Day	latitude	longitude	depth (km)	magnit.	Seismic Moment (Nm)
Hindu Kush	1983 12 30	36.42 N	70.75 E	209	6.2	$1.5 \ 10^{20}$
South of Honshu	1984 3 6	9.50 N	138.92 E	454	6.1	$1.4 \ 10^{20}$
Mindanao	1984 11 20	5.15 N	125.12 E	202	6.4	$2.1 \ 10^{20}$
Hindu Kush	1985 7 29	36.19 N	70.89 E	101	6.7	$1.5 \ 10^{20}$
Mexico	1985 9 19	18.18 N	102.57 W	33	7.0	$1.1 \ 10^{21}$
South Fiji	1986 5 26	20.07 S	178.72 E	553	6.8	$6.5 \ 10^{19}$
Brazil	1989 5 5	8.40 S	71.20 W	600	6.5	$5.0 \ 10^{19}$

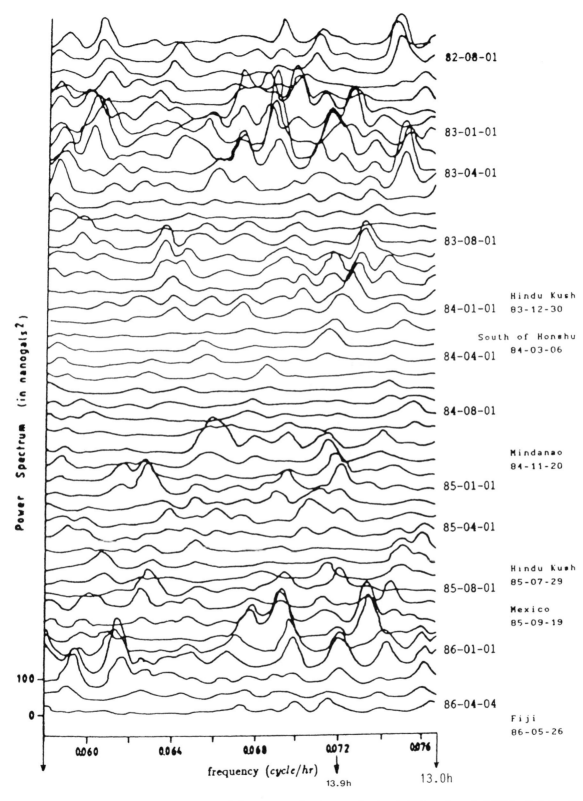

Fig. 2. Moving window spectral analysis of the superconducting gravimeter data from June 2, 1982 to October 15, 1986 (date = beginning of window).

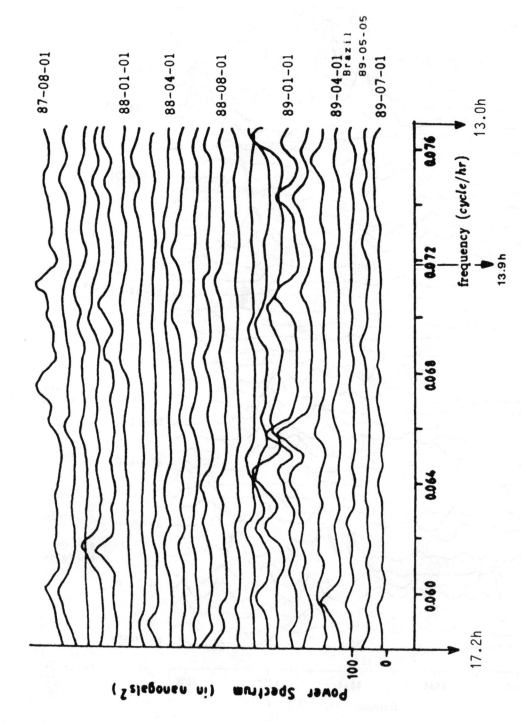

Fig. 3. Moving window spectral analysis of the superconducting gravimeter data from July 22, 1987 to December 30, 1989 (date = beginning of window).

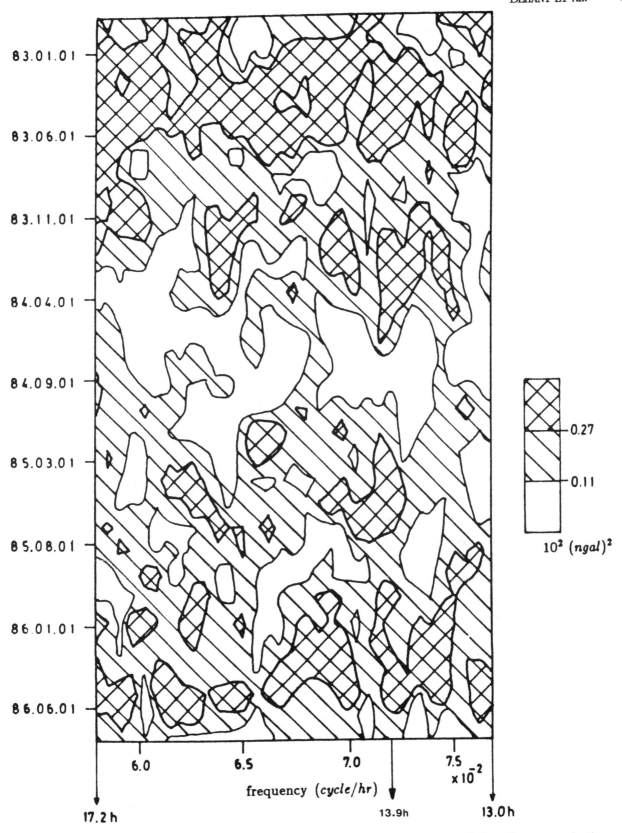

Fig. 4. Projection of the 3-dimensional plot of the moving window spectral analysis applied to the superconducting gravimeter data for the first set of data from June 2, 1982 to October 15, 1986 (date = center of window).

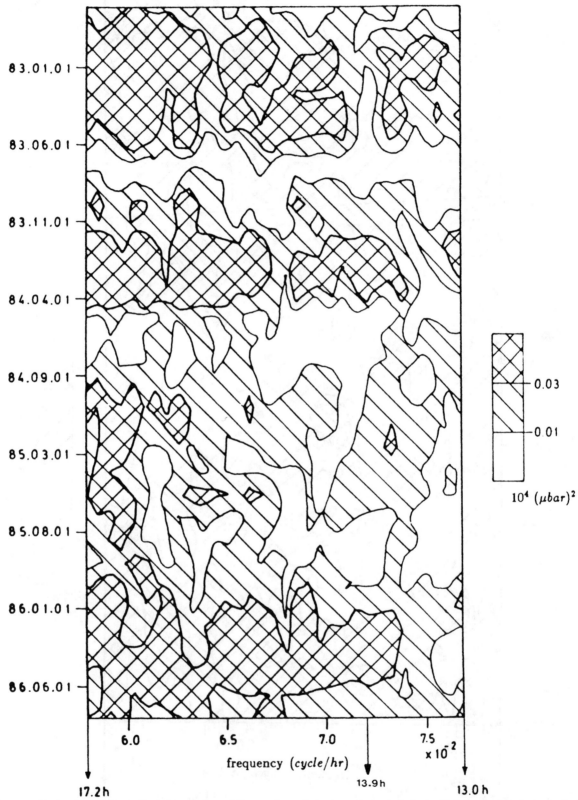

Fig. 5. Projection of the 3-dimensional plot of the moving window spectral analysis applied to the local atmospheric pressure for the same period as in Figure 4 (date = center of window).

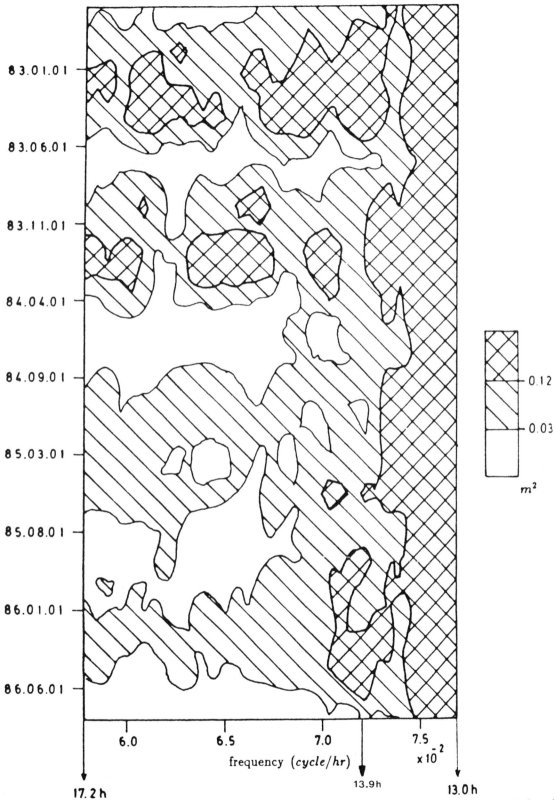

Fig. 6. Projection of the 3-dimensional plot of the moving window spectral analysis applied to the oceanic data (tidal gauge at Ostend) for the same period as in Figure 4 (date = center of window).

represent regional atmospheric effects, the residues should contain the associated effects.

As the Brussels station is close to the North Sea, oceanic loading effects are important (1.8 μgal on M_2 wave). Gravimetric records are corrected only for the tidal constituents, leaving effects outside the tidal bands. These effects are sea level variations due to atmospheric pressure [Van Dam and Wahr, 1987] or possibly resonance modes of the North Sea [Denis et al., 1988].

As we compute the impulse response of pressure on gravity from the observed data, the effects of sea level variations produced by the atmospheric pressure are largely taken into account but not completely. This suggested to us that we analyse, by the same methods as above, the atmospheric pressure and sea level data. Figure 5 and 6 are moving window spectral results of the atmospheric pressure at Brussels and oceanic tidal gauge data at Ostend (100 km away from Brussels). Comparison of these two graphs with Figure 4, which gives the results for the same period of the superconducting gravimeter data, shows immediately that the three spectra are excited during the same epochs. The frequencies do not exactly coincide but when peaks are excited in one spectrum, the energy increases in the other. In all spectra for example, a seasonal effect appears around November.

Discussion and Conclusions

We have shown in this paper that whatever spectral analysis method is used, the existence of the 13.9 hr peak in the Brussels superconducting gravimeter data is confirmed and is not an artefact.

Concerning its origin, we have shown that the peak can appear before an earthquake or in a period during which no strong and deep or very strong earthquake occurs; in other cases no peak at all appears although a strong and deep earthquake occurred. Hence it appears that there is no direct relation between earthquake occurrences and energy in the spectrum between the semi-diurnal and diurnal frequencies.

Rather we think that the origin of such harmonic energy in the superconducting gravimeter data could be related to a regional effect, not yet taken into account in the computation of the tidal residues; the moving window spectral analysis also indicates that in Brussels the spectra of the gravity, atmospheric pressure and Ostend sea level data are correlated and for example, simultaneously perturbed in November each year. This result stresses the necessity, on the one hand, of improving regional pressure correction procedures, and on the other hand, of stacking the data of a worldwide superconducting gravimeter network for the study of global phenomena such as normal modes of both the inner and outer core. As an example, a first attempt of stacking of superconducting gravimeter data from Belgium, Canada, France, and Germany is presented in Smylie et al. [1992].

Acknowledgment. Two of us (V. Dehant and B. Ducarme) are supported by the Belgian National Funds for Scientific Research.

References

Aldridge, K., and Lumb, I., Inertial waves identified in the Earth's fluid outer core, *Nature, 325,* 421-423, 1987.

Cartwright, D.E., and Tayler, R.S., New computations of the tide-generating potential, *Geophys. J. R. astron. Soc., 23,* 45-74, 1971.

Cartwright, D.E., and Edden, A.C., Corrected tables of tidal harmonics, *Geophys. J. R. astron. Soc., 33,* 253-264, 1973.

Crossley, D.J., The excitation of core modes by earthquakes, *Geophys. Monograph, 46* (1), IUGG/AGU Publ., eds. D.E. Smylie and R. Hide, pp. 41-50, 1989.

Defraigne, P., Analyses spectrales des données du gravimètre à supraconductivité, Mémoire de Licence, Université Catholique de Louvain, Louvain-la-Neuve, Belgium, in French, 117 pp., 1991.

Denis, C., Denis-Karafistan, A.I., Hinderer, J., and Ronday, F., Have core oscillations been observed?, in *Comptes Rendus des Journées Luxembourgeoises de Géodynamique,* 68ième session, 5 and 6 Dec. 1988, ed. P. Poitevin, pp. 11-12, 1988.

De Meyer, F., and Ducarme, B., Input-output of the observations of the superconducting gravimeter, in *Proc. 10th Int. Symp. on 'Earth Tides',* Madrid, Spain, 1985, eds. R. Vieira and Consejo Superior de Investigationes Cientificas, pp. 531-554, 1987.

Ducarme, B., van Ruymbeke, M., and Poitevin, C., Three years of registration with a superconducting gravimeter at the Royal Observatory of Belgium, in *Proc. 10th Int. Symp. on 'Earth Tides',* Madrid, Spain, 1985, eds. R. Vieira and Consejo Superior de Investigationes Cientificas, pp. 113-130, 1987.

Guo, J., Les développements analytiques de Kudlik sont applicables au noyau liquide de la Terre, Annex Ph. D. thesis, Université Catholique de Louvain, Louvain-la-Neuve, Belgium, in French, 47 pp., 1991.

Kudlik, M.D., On transient motions in a contained rotating fluid, Ph. D. thesis, M.I.T., 1966.

Melchior, P., and Ducarme, B., Detection of inertial gravity oscillations in the Earth's core with a superconducting gravimeter at Brussels, *Phys. Earth planet. Inter., 42,* 129-134, 1986.

Melchior, P., Crossley, D.J., Dehant, V., and Ducarme, B., Have inertial waves been identified from the Earth's core?, *Geophys. Monograph, 46* (1), IUGG/AGU Publ., eds. D.E. Smylie and R. Hide, pp. 1-12, 1989.

Qin-Wen, X., A new complete development of the tide generating potential for the epoch J 2000.0, *Bull. Inf. Marées Terrestres, 99,* 6786-6812, 1987.

Qin-Wen, X., The precision of the development of the tidal generating potential and some explanatory notes, *Bull. Inf. Marées Terrestres, 105,* Special issue Meeting of WG on 'High Precision Data Processing', October 1988, 7396-7404, 1989.

Smylie, D.E., Hinderer, J., Richter, B., Ducarme, B., and Mansinha, L., A comparative analysis of superconducting gravimeter data, this issue, 1992.

Tamura, Y., A harmonic development of the tide generating potential, *Bull. Inf. Marées Terrestres, 99,* 6813-6855, 1987.

Van Dam, T.M., and Wahr, J.M., Displacements of the Earth's surface due to Atmospheric Loading: effects on Gravity and Baseline Measurements, *J. Geophys. Res., 92* (B2), 1281-1286, 1987.

Zürn, W., Richter, B., Rydelek, P.A., and Neuberg, J., Comments on 'Detection of inertial gravity oscillations in the Earth's core with a superconducting gravimeter at Brussels' by P. Melchior and B. Ducarme, *Phys. Earth planet. Inter., 49* (1-2), 176-178, 1987.

V. Dehant, Observatoire Royal de Belgique, 3, Avenue Circulaire, B-1180 Brussels, Belgium.

VARIATION OF J_2 AND INTERNAL LOADS

MARIANNE LEFFTZ AND HILAIRE LEGROS

*Institut de Physique du Globe de Strasbourg 5 rue René Descartes
67084 Strasbourg Cedex - FRANCE*

The temporal variation of the terrestrial gravity potential coefficients, and particularly of the J_2 coefficient of the degree 2 zonal term, has often been attributed to the viscous response of the Earth to superficial loads, especially to the one due to the last deglaciation. We show here that for a Maxwell model of rheology, such temporal variation can also result from internal loads, located at various discontinuities of the Earth model, and in particular at the 670 km depth discontinuity and at the core-mantle boundary.

INTRODUCTION

While the influence of mantle viscosity has already been investigated, especially to explain some observations related to the last deglaciation, the effect of the core or of dynamical phenomena acting at the core-mantle boundary (CMB), such as pressure or loading potential, has never been taken into account simultaneously with the viscoelasticity of the mantle.

We intend to show, in this paper, that internal loads at the CMB or at the 670 km depth discontinuity between the upper and the lower mantle, can produce secular variation of the geoid coefficient because of the viscoelasticity of the mantle, reaching to an order of magnitude of the one produced by a surface load.

The study of the relaxation of a viscous Earth according to a Maxwell model of rheology supposes an evolution from a given state to a final state of hydrostatic equilibrium: the initial state of deformation is considered as being close to hydrostatic equilibrium (that is being a perturbation with respect to the hydrostatic equilibrium), and in the final state, all the stresses are relaxed. If we suppose that, at the initial state, a thermal stress is added to the hydrostatic pre-stress, the relaxation of a Maxwell viscous planet would taken into account an evolution to this more complete state of stress. In this study, no initial thermal stress is considered.

If, on the one hand, the physics of the surface loads is well known (ice loading, sea level variation at long time scale, atmospheric mass distribution ...), it will be more difficult on the other hand to find physical processes to account for the density variations appearing and vanishing progressively within the mantle or at the CMB. The most likely candidate to modify the densities within the Earth seems to be the material transport by convection; because of the convective motion, there has been a deformation at the various interfaces within the Earth, which may be considered as internal loads evolving towards a hydrostatic state.

In section 2, using the equations describing the elastogravitational deformation and the correspondence principle, we calculate the relaxation modes associated with the viscoelastic Love numbers, for a four-layer Earth model (elastic lithosphere, viscoelastic upper and lower mantle, inviscid fluid core). Then, we compute (section 3), the viscoelastic deformations due to a surface load and to internal loads, and we finally show (section 4) that the resulting perturbations in the J_2 geoid coefficient may be slowly relaxed and in good agreement with the observations.

THEORY

Because of self-attraction, there is an important stress at the initial state, where we suppose the Earth as being in hydrostatic equilibrium.

Dynamics of Earth's Deep Interior and Earth Rotation
Geophysical Monograph 72, IUGG Volume 12
Published in 1993 by the International Union of Geodesy and Geophysics and the American Geophysical Union.

To calculate the viscoelastic deformations of a planet subjected to a exciting source, we use the viscoelastogravitational set of equations (e.g. Alterman, Jarosch and Pekeris, 1959) which include:

- conservation of momentum;
- conservation of mass;
- Poisson's equation;
- a rheologic law (i.e. a stress-strain relation).

For this last equation, we have choosen a linear viscoelastic Maxwell model of rheology. This is because at short time-scales, the Earth has a quasi-elastic behaviour, whereas at long time-scales (geological), its behaviour is well characterized by that of a fluid (the flattening of the Earth is quasi-hydrostatic and explained by the fluid axial rotation). Denoting by σ_{ij} and ϵ_{ij} the stress and strain tensors respectively, we have:

$$\dot{\sigma}_{ij} + \frac{\mu}{\nu}\left(\sigma_{ij} - \sigma_{kk}\frac{\delta_{ij}}{3}\right) = 2\mu\dot{\epsilon}_{ij} \qquad (2.1)$$

when the planet is supposed to be incompressible.

If we apply the Fourier transformation to this relation, we will obtain :

$$\bar{\sigma}_{ij}(\lambda) = 2\mu(\lambda)\bar{\epsilon}_{ij}(\lambda) \quad with \quad \mu(\lambda) = \mu_E \frac{i\lambda}{i\lambda + \frac{\mu}{\nu}} \qquad (2.2)$$

where μ_E is the elastic rigidity of the planet and ν its Newtonian viscosity. In the Fourier domain, we have the Hookean law, but the Lamé parameter $\mu(\lambda)$ is function of the frequency.

In the Fourier domain, the viscoelastic equations and the boundary conditions are the same as those for an elastic body with the same geometry. Consequently, we may use the correspondence principle: we solve, for different frequencies, the elastic problem in order to build the viscoelastic solutions (Lee, Radoc and Woodwards, 1959; Peltier, 1974).

If the Earth is multilayered, we may have, in each homogeneous incompressible layer (with a density ρ_i, a rigidity μ_i and a viscosity ν_i), an analytical solution (Kelvin, 1863) in terms of the displacements $-u_r(r)$ and $u_t(r)$- and the mass redistribution potential $\chi(r)$. To completely determine the solutions in the whole Earth, we need to write, at each interface between two layers, precise boundary conditions. They are given, in absence of any exciting source, at the interface $r = a_i$, by:

- In stress :

$$\sigma_{rt}^i(a_i) = \sigma_{rt}^{i+1}(a_i)$$

$$\sigma_{rr}^i(a_i) + (\rho_{i+1} - \rho_i)\{-g(a_i)u_r(a_i) + \chi(a_i)\} = \sigma_{rr}^{i+1}(a_i)$$

- In displacements :

$$u_r^i(a_i) = u_r^{i+1}(a_i) \qquad (2.3)$$

$$u_t^i(a_i) = u_t^{i+1}(a_i)$$

with the mass redistribution potential $\chi(a_i)$:

$$= \frac{3ga}{5}\frac{\rho_1}{\rho}\left(\frac{a_i}{a}\right)^2\left\{\sum_{k=0}^{i}\frac{u_r(a_k)}{a_k}\beta_k + \sum_{k=i+1}^{N-1}\frac{u_r(a_k)}{a_k}\beta_k\left(\frac{a_k}{a_i}\right)^5\right\} \qquad (2.4)$$

where $g(a_i)$ is the gravity at $r = a_i$, β_k the density contrast $\left(\beta_k = \frac{\rho_{k+1} - \rho_k}{\rho_1}\right)$, and a mean radius of the Earth.

The potential $\chi(a_i)$ is due to the mass density appearing at each interface resulting from the radial displacement.

The second term in this equation (2.3) is a transport pressure term which arises as a consequence of the pre-stress in the initial state of the planet.

The frequency dependence comes from the viscoelastic stress $\sigma_{rr}^i(a_i)$ and $\sigma_{rt}^i(a_i)$, which are proportional to the rigidity $\mu_i(\lambda)$.

Therefore, for a N-layer planet, we have a time-dependent set of (4N-6) equations to which we need to add the action of the exciting sources. Before doing that, we compute the determinant of the system and we find a frequency-dependent polynomial whose zeros are called relaxation modes (Peltier, 1974). These relaxation modes are generated at each interface by the discontinuity of geometric and physical parameters. We describe them for a four layered Earth model with an elastic lithosphere, a viscoelastic upper and lower mantle, and an inviscid fluid core. The parameters of this model are given in the Table 1. We have various relaxation modes:

- due to a density jump:

M_o, due to the density contrast at r=a

M_1, due to the density jump between the upper and the lower mantle

C, because of the density discontinuity between the mantle and the fluid core;

- due to a discontinuity in rigidity or in viscosity:

L, due to the viscosity contrast between the elastic lithosphere and the viscoelastic mantle

T_1, and T_2, due to a Maxwellian time jump between the upper and the lower mantle, and called tran-

TABLE 1. Geometrical and physical parameters of the four layered Earth model, with an elastic lithosphere, a viscoelastic upper and lower mantle and an inviscid fluid core.

Radius in km	Density in $kg.m^{-3}$	Rigidity in Pa	Viscosity in Pa.s
$a_1 = 6221 < r < a$	$\rho_1 = 3232$	$\mu_1 = 6.11 \times 10^{10}$	$\nu_1 = inf$
$a_2 = 5701 < r < a_1$	$\rho_2 = 3666$	$\mu_2 = 9.17 \times 10^{10}$	$\nu_2 = 10^{21}$
$a_3 = 3480 < r < a_2$	$\rho_3 = 4904$	$\mu_3 = 2.22 \times 10^{11}$	$\nu_3 = 5 \times 10^{21}$
$0 < r < a_3$	$\rho_4 = 10987$	$\mu_4 = 0.$	$\nu_4 = 0.$

sition modes (Peltier, 1974), because they relaxe rapidly and are weakly excited.

The values of these relaxation modes are given in Table 2.

VISCOELASTIC DEFORMATIONS DUE TO INTERNAL LOADS

Solving the viscoelasto-gravitational set of equations, we may demonstrate that, in the frequency domain, the radial displacement and the mass redistribution potential are proportional to the excitation sources (Love, 1909; Peltier, 1974).

$$u_r(r,\lambda) = h(r,\lambda) \; \frac{V(\lambda)}{g}$$
$$\chi(r,\lambda) = k(r,\lambda) \; V(\lambda) \qquad (3.1)$$

TABLE 2. Relaxation modes associated with the viscoelastic Love numbers for the Earth's model defined in Table 1.

	Relaxation modes in s^{-1}	Relaxation times in year
T_1	0.90×10^{-10}	350
T_2	0.85×10^{-10}	370
M_o	1.84×10^{-11}	1720
C	4.85×10^{-12}	6530
L	1.68×10^{-12}	18800
M_1	2.58×10^{-14}	1.23×10^6

where $k(r,\lambda)$ and $h(r,\lambda)$ are viscoelastic Love numbers. Peltier (1974) or Wu and Peltier (1982), have shown that they can be expressed as following (for example, $k(r,\lambda)$):

$$k(r,\lambda) = k^E(r) + \sum_{i=1}^{M} \frac{r_i}{i\lambda + 1/\tau_i} \qquad (3.2)$$

where $1/\tau_i$ are the M relaxation modes, relative to the model of the planet, and $k^E(r)$ the elastic Love numbers. We emphasize that the $k^E(r)$ are independent on the frequency λ, and are consequently a Dirac function in the temporal domain: the elastic response in deformation is instantaneous.

The coefficients r_i are dependent on the excitation sources.

We compute now these Love numbers when the Earth is submitted to internal loads, located at the core-mantle interface, as well as at the 670 km depth discontinuity between the upper and the lower mantle, and at the Earth's surface.

We suppose a surficial mass density σ at the r_o depth. It creates two effects:

— a pressure

$$P(t) = -\sigma g(r_o)$$

— an attraction potential

$$V(t) = \frac{4\pi G}{5} \sigma r_o \begin{pmatrix} (\frac{r_o}{r})^3 & if \; r \geq r_o \\ (\frac{r}{r_o})^2 & if \; r \leq r_o \end{pmatrix} \qquad (3.3)$$

The mass redistribution potential $\chi(a)$ and the radial displacement $u_r(a)$ at the Earth's surface can be written as a function of these combined sources, using the loading Love numbers :

$$\chi(a,t) = k(a,t) * V(t) + \bar{k}(a,t) * \frac{P(t)}{\rho} = k'(a,t) * V(t)$$

$$u_r(a,t) = h(a,t) * \frac{V(t)}{g} + \bar{h}(a,t) * \frac{P(t)}{\rho g} = h'(a,t) * \frac{V(t)}{g} \tag{3.4}$$

$*$ denotes the temporal convolution, and ρ is the mean density of the Earth and g the gravity at the surface. We have noted $k'(a,\lambda)$ and $h'(a,\lambda)$ the loading Love numbers.

For a constant internal load, which provides $\sigma(t) = \sigma_o H(t)$ or $V(t) = V_o H(t)$, we have, in the temporal domain:

$$\chi(a,t) = \left(k'^f - r_i \tau_i e^{-\frac{t}{\tau_i}} \right) \ V_o H(t)$$

$$u_r(a,t) = \left(h'^f - s_i \tau_i e^{-\frac{t}{\tau_i}} \right) \ \frac{V_o H(t)}{g} \tag{3.5}$$

where

$$k'^f = k'^E + \sum_{i=1}^{M} r_i \tau_i \quad and \quad h'^f = h'^E + \sum_{i=1}^{M} s_i \tau_i \tag{3.6}$$

We have represented in the figures 1-a, 1-b and 1-c respectively, the percentage of excitation of each relaxation modes (i.e. the ratio $\dfrac{r_i \tau_i}{\sum_{j=1}^{M} r_j \tau_j}$) when the internal load is located at the Earth's surface, at 670 km depth and at the core-mantle boundary.

From the elasto-gravitational set of equations, for an internal load located at an interface with a density jump, the following relations hold, in the fluid limit :

$$k'^f(a) = -\left(\frac{r_o}{a}\right)^3 \ \delta(t) \tag{3.7}$$

$$h'^f(a) = \begin{pmatrix} 0, & if & r_o \neq a \\ -\frac{5}{3}\frac{\rho_1}{\rho}, & if & r_o = a \end{pmatrix} \tag{3.8}$$

If the load is located within the Earth, at r_o depth (where there is a density discontinuity), there is a radial displacement (that is a deflection of the interface where the load is located) which exactly compensates the mass density, so that the displacement at the Earth surface is zero.

RESULTING PERTURBATION IN THE J_2 GEOID COEFFICIENT

From the mass redistribution potential, we may easily compute the perturbation in the inertia moment C_{33} due to zonal internal loads, given by $V_{20}(t)$.

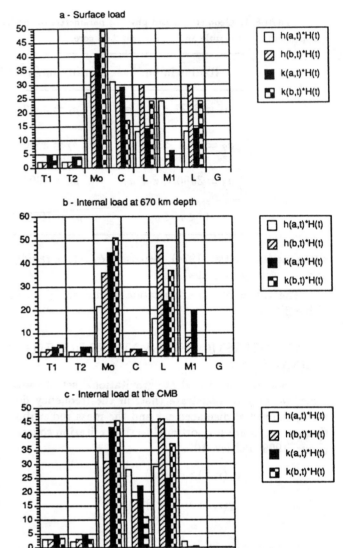

Fig. 1. The percentages of excitation of the relaxation modes $\dfrac{r_i \tau_i}{\sum_{j=1}^{M} r_j \tau_j}$ are presented in these charts for various excitation sources: (a) a surface load, (b) an internal load at 670 km depth, (c) an internal load at the CMB. The vertical axe indicates the respective percentages of each relaxation mode written in the horizontal axe, for various Love numbers: h(a,t) and h(b,t) proportional to the radial displacement at the surface and at the CMB (r=b), and k(a,t) and k(b,t) proportional to the mass redistribution potential at r=a and r=b.

$$C_{33} = -\left\{ \left(\frac{r_o}{a}\right)^3 \ \delta(t) + k'(a,t) \right\} * V_{20}(t) \tag{4.1}$$

where the first term within the braces is the direct effect, and the second, the one of deformation. $k'(a,t)$ is negative, whatever r_o, and consequently the deformational effect is opposite to the direct mass contribution

. From C_{33}, we can express the geoid coefficient J_2, and its time derivative $\dot{J}_2$. We find, for a constant zonal load $\sigma_{20}(t) = \sigma_o\, H(t)$:

$$\dot{J}_2 = J_{2o}\left\{ \left(\frac{r_o}{a}\right)^3 \delta(t) + k'(a,t) \right\} * \delta(t) \qquad (4.2)$$

$$\dot{J}_2 = J_{2o}\left[\left\{ \left(\frac{r_o}{a}\right)^3 + k'^E \right\} \delta(t) + \sum_{i=1}^{M} r_i e^{-\frac{t}{\tau_i}}\, H(t) \right] \qquad (4.3)$$

with $J_{20} = -\frac{3}{2}\frac{V_{20}}{ma^2}$ where m is the Earth's mass.

In the study of the secular variation of $\dot{J}_2$, we may neglect the instantaneous terms. The temporal variation of $\dot{J}_2$ is consequently proportional to the coefficients r_i and the relaxation modes $\frac{1}{\tau_i}$. Depending on the depth where the load σ is located, the modes $\frac{1}{\tau_i}$ are more or less excited and $\dot{J}_2$ is more or less rapidly equal to zero.

We have represented in the figure 2, the $\dot{J}_2$ coefficient for a zonal internal load located respectively at

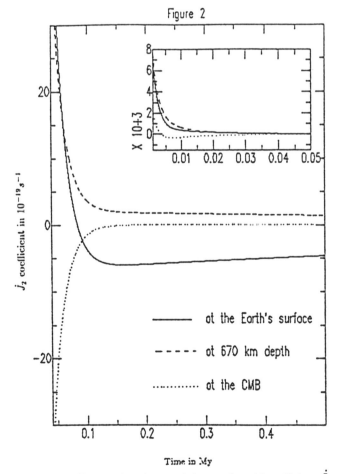

Figure 2

Fig. 2. Temporal variation of the zonal geoid coefficient $\dot{J}_2$ for a load located at the Earth surface (black line), at 670 km depth (dashed line) and at the CMB (dotted line).

the Earth's surface, at 670 km depth, and at the CMB. The observed $\dot{J}_2$ coefficient is $-7.9 \times 10^{-19} s^{-1}$ (Cheng et al., 1989).

The load is supposed to be the same for all depths (with a numerical value $\sigma_{20} = 10^7 kg/m^3$), in order to investigate the relative influences on the $\dot{J}_2$ geoid coefficient. Comparing the curves of the figure 2, we note that there is a secular variation in $\dot{J}_2$, for a load at 670 km depth, at long-time scale; this happens because the mode M_1, which is slowly relaxed (with a relaxation time of about one million years), is excited for a load located at such depth.

For a load at the Earth's surface, 5 % of the deformation is due to the M_1 mode excitation; it can consequently produce a secular variation of the $\dot{J}_2$ geoid coefficient.

For a load located at the core-mantle interface, the value of the $\dot{J}_2$ coefficient can be of the same order of magnitude than the observations, if the load has taken place about 5 or 6 thousands years ago. The mode C (with a relaxation time of about 6530 years, for our Earth's model) being predominantly excited in this case, the secular variation in the $\dot{J}_2$ coefficient will vanish in a time of some thousands years.

CONCLUSION

Not only a surface load (due to the last deglaciation) but also internal loads (at the CMB or at 670 km depth, for example, due to perturbations in the interfaces because of convective motions) can contribute to the temporal variation of J_2.

REFERENCES

Alterman, Z., Jarosch, H. & Pekeris, C.H., 1959. Oscillation of the Earth. Proc. R. Soc. London, A-252, 80-95.

Cheng, M.K., Eanes, R.J., Shum, C.K., Schutz, B.E. & Tapley, B.D., 1989. Temporal variations in low degree zonal harmonics from Starlette orbit analysis. Geophys. Res. Lett., 16, 393-396.

Lee, E.H., Radok, J.R.M., & Woodwards, W.B., 1959. Stress analysis for linear viscoelastic materials. Transactions of the Society of Rheology, III, 41-59

Love, A.E.H., 1909. The yielding of the Earth to disturbing forces. Proc. R. Soc. London, A-82, 73-88.

Peltier, W.R., 1974. Impulse response of a Maxwell Earth. Geophys. Space Physics, 12, 649-669.

Wu, P. & Peltier, W.R., 1982. Viscous gravitational relaxation. Geophys. J. R. astr. Soc., 70, 435-485.

Marianne Lefftz and Hilaire Legros, Institut de Physique du Globe de Strasbourg, 5 rue Rene Descartes, 67084 Strasbourg Cedex, France.

A Generalized 'Core Resonance' Phenomenon: Inferences from a Poincaré Core Model

L. IAN LUMB, KEITH D. ALDRIDGE AND GARY A. HENDERSON

Department of Earth and Atmospheric Science, York University,
North York, Ontario, Canada.

It has been demonstrated that tidal forcing of the Earth's fluid outer core, at nearly diurnal periods, leads to a resonantly enhanced gravimetric factor due to the existence of the nearly diurnal free wobble. However, the core mode which participates in this is but one of an infinite set of oscillations permissible in a rotating fluid like the Earth's core. Furthermore, the similarity in azimuthal spatial structure of wavenumber-two ($k = 2$) Earth tides and core modes, at semi-diurnal periods, suggests the possibility of a '$k = 2$ core resonance'. Results from a Poincaré core model indicate that the relevant inertial waves are more spatially complex and more densely clustered in period than those for the $k = 1$ case. Because the M_2 tide lies in the semi-diurnal band, and has an amplitude about 1.7 times that of the largest tide in the diurnal band, possibilities for the excitation of core modes through core-mantle boundary deformation appear to be good. We thus expect that gravimetric factors at semi-diurnal periods will be resonantly enhanced due to this $k = 2$ resonance effect. Since the tidal deformation is a continual forcing, it is also expected that wave interactions will exist and lead to elliptic instability. These complex fluid processes can be studied through laboratory experiments on $k = 2$ inertial waves in a rotating, deformable, spheroidal shell of fluid.

INTRODUCTION

Coupling between the Earth's fluid outer core and mantle gives rise to a nearly diurnal free wobble (NDFW), which has been shown to resonantly enhance tidal gravimetric factors at nearly diurnal periods, and is therefore included in tidal modeling. The tidal deformation of the solid Earth deforms the core-mantle boundary (CMB), and therefore excites the participating core mode at a period of about 24 hours [e.g., Hinderer and Legros, 1989; Hinderer et al., 1987; Sasao et al., 1980; Wahr, 1981a]. Since this core mode reciprocally enhances the tidal deformation at that period, this phenomenon has become known as a 'core resonance', with the core possessing a single, simple fluid motion. Although more recent work by Hinderer et al. [1991] has attempted consideration of a 'dynamical' core, the primary effect of rotation is neglected due to complications arising from Coriolis coupling between harmonics of different degrees [Smylie, 1974; Johnson and Smylie, 1977; but see Rieutord, 1991, for a more recent discussion]. The extension of the 'core resonance phenomenon' to include modes in addition to that which participates in the NDFW was not considered by Hinderer et al. [1991].

To a space-based observer, this type of core-mantle coupling also leads to the free core nutation (FCN), a rotational mode which is accounted for in high-precision lunar-laser ranging (LLR), satellite-laser ranging (SLR) and very-long-baseline interferometry (VLBI). Since the FCN appears to have the form of a sinusoidal perturbation, it is usually incorporated as a 'correction' to the IAU 1980 nutation series, although Scherneck (1991) has recently proposed an approach similar to that used in tidal studies. It is important to note that the NDFW and the FCN are essentially the same rotational mode as viewed, respectively, from an Earth-based or space-based reference frame. In the current context, we will thus refer to the NDFW when discussing gravimetric observations, while the FCN will be referenced in connection with VLBI.

In contrast to such geodynamical efforts, researchers studying core hydrodynamics are concerned with the infinite but denumerable set of modes permissible in a contained, rotating fluid like the Earth's core. This immediately suggests the possibility that the fluid core can contribute to the Earth's rotational spectrum through modes in addition to the one which participates in the NDFW or FCN [e.g., Aldridge et al., 1989; Lumb & Aldridge, 1991; Smylie, 1990]. We therefore forward the following twofold generalization of the 'core

Dynamics of Earth's Deep Interior and Earth Rotation
Geophysical Monograph 72, IUGG Volume 12

resonance phenomenon': When a core model which considers the primary effects of rotational and pressure-gradient forces is used, it is clear that there are a number of $k = 1$ core modes whose periods lie in close proximity to those of Earth tides in the diurnal tidal band [Lumb and Aldridge, 1991]. It is expected that more realistic core models, which include compressibility and density stratification of the core plus elasticity of the mantle [e.g., Smylie et al., 1992], entail only perturbations to the periods defined by the simple Poincaré core model [e.g., Aldridge and Lumb, 1987; Aldridge et al., 1989]. Since this same core model also provides $k = 2$ oscillations whose periods are close to those of semi-diurnal Earth tides, there exists the possibility of a core resonance at semi-diurnal periods as well. This suggests that the tidal models used in long-period gravimetry (LPG), and the nutation series used by space-based measurement techniques, may need to be modified to account for this new conceptual framework for core resonances.

On an elementary level, there are two criteria which must be met for a resonance to occur. Namely, the spatial form of the tide must match that of the core mode, and their periods must lie sufficiently close. Since the tides and core modes are both decomposed into spherical harmonics, we consider the matching of spatial form in terms of the degree n and order k of the associated Legendre polynomials. Thus, the previous work for the tesseral core resonance is reviewed in this context, and appended for other core modes lying in the diurnal tidal band. This core resonance, whose effects have been measured by several different tidal gravimeters [Levine et al., 1986; Neuberg et al., 1987; Richter and Zürn, 1989] and by VLBI [Gwinn et al., 1986], has been shown to alter gravimetric factors and the Earth's nutation series. Presently, we make similar claims for what we have termed the '$k = 2$ core resonance', and suggest that its effects may be measured by LPG and VLBI. The spatial and temporal criteria suggest that there are indeed core modes that could lead to resonantly enhanced deformation of the solid Earth, but now at semi-diurnal periods. Since fluid oscillations in containers which are deformed in a $k = 2$ fashion can be subject to elliptic instability, our current efforts are concerned with a laboratory study of this geophysically motivated problem.

In the next section, a brief description of the Poincaré core model is given. The application of this model for internal core modes in the case of the $k = 1$, $k = 2$ and other core resonances is then possible. Using the well-known Poincaré model as a reference, we then review the mathematical framework for elliptically unstable oscillations of tidally deformed, rotating fluids like the Earth's core. The discussion closes with suggestions for future theoretical and observational initiatives which, together with our own experimental work, will aid in characterizing the core resonance phenomenon.

THE POINCARÉ CORE MODEL

The motions of a homogeneous, incompressible, inviscid and contained fluid, rotating at a constant angular velocity, can be expressed by the near balance between pressure-gradient and Coriolis forces; the resulting inertial oscillations do not possess a net circulation [Greenspan, 1968, § 2.8]. Whereas, in the case of an exact balance, the resulting geostrophic flow does possess a net circulation [see Greenspan, 1968, § 2.6]. The simplified form of the Navier-Stokes equations of motion, subject to the boundary condition of vanishing normal velocity, is known as the Poincaré problem [e.g., Greenspan, 1968]. Separable solutions may be found by use of an 'oblate spheroidal' coordinate system [Bryan, 1889], in which the eigenfunctions are represented by associated Legendre polynomials of degree n and order k.

Eigenfrequencies for spherical [e.g., Greenspan, 1968] and spheroidal [Kudlick, 1966] cavities may be calculated through the use of recurrence relations for the associated Legendre polynomials. For any n and k, the mth root of this period equation corresponds to the mth largest eigenvalue [after Aldridge and Lumb, 1987]. Physically, k represents the azimuthal wavenumber, while n and m roughly indicate spatial complexity parallel and perpendicular to the rotational axis, respectively. Subsequent applications of these findings from rotating fluids in a geophysical context assume a hydrostatically flattened CMB; i.e., $f = 1/392.7$ [see § 3 of Aldridge et al., 1989, for more details].

The Bryan transformation, [see Greenspan, 1968, equation (2.12.6)], can be inverted to obtain the image variables, (η, θ, μ), in terms of the original variables, (r, θ, z). Thus, the reduced pressure eigenfunction, Ψ, can be evaluated at a grid of points in the first quadrant, again using recurrence relations for the associated Legendre polynomials. For a certain phase and scale, the radial, u, and vertical, w, velocities are given by

$$
\begin{aligned}
u &= \frac{1}{4 - \lambda^2} \left(\lambda \frac{\partial \Psi}{\partial r} + 2k \frac{\Psi}{r} \right) \\
w &= -\frac{1}{\lambda} \frac{\partial \Psi}{\partial z},
\end{aligned}
\tag{1}
$$

and centered finite difference versions of these equations can be used to evaluate the velocity at a grid of points. In (1), the ratio of the perturbation frequency to the rotation rate is given by the eigenvalue λ.

When the frequency and spatial form of the forcing closely matches that of a resonance, a maximal response will be obtained. However, this response will also vary with the placement of the detecting system. These subtleties led to an ad hoc amplitude scaling scheme, first used by Aldridge et al. [1988], for the amplitudes

$$
A_{nmk} = \frac{A_{fund}}{n + m + k}
\tag{2}
$$

of retrograde modes. For a given value of k, (2) suggests that the response varies inversely with the spatial complexity; i.e., with increasing values of n and m. In this paper, for a given k the normalization factor $A_{fund} = A_{M_2}(n + m + k)$ is the sum of the indices of the lowest-order mode multiplied by the amplitude of the largest tide – namely, M_2. For example, the fundamental $k = 1$ mode, which has $n = 2$ and $m = 1$, thus has $n + m + k = 4$ and $A_{M_2} = 908175$ arcsec, and therefore $A_{fund} = 3.63 \times 10^6$.

'CORE RESONANCES'

Like the inertial waves mentioned in the previous section, Earth tides are conveniently represented in terms of spherical harmonics, and therefore the following three types of tides are recognized: zonal tides, which may be equatorially odd or even, have $k = 0$ and are therefore axisymmetric; sectoral tides, which are always symmetric about the equator, have $n = k$; while the general case of the tesseral tides, which may be equatorially even or odd, have $0 < k < n$ [e.g., Stacey, 1977]. Therefore, as resonance possibilities for $k = 1$ core modes, only tesseral Earth tides need to be considered; whereas for the $k = 2$ and higher modes, both sectoral and tesseral tides are involved. Zonal tides are related to $k = 0$ core modes, and this case is mentioned briefly with the $k = 3$ and $k = 4$ possibilities in the last part of this section.

$k = 1$ Core Resonance

The spatial distribution of $k = 1$ Earth tides is such that an observer on the Earth experiences a high tide and a low tide as the Earth revolves through the tidal-deformation field in one sidereal day (see Fig. 1). Since one wavelength is contained in one circumference of the planet, this is a $k = 1$ disturbance. Thus the azimuthal spatial form of these Earth tides matches that of $k = 1$ internal core modes. Coupling between these core modes and Earth body tides would result in a wobble for an Earth-based observer, while the associated nutation would be observed from space.

Shown in Fig. 2 are the reduced pressure eigenfunctions for several $k = 1$ core modes, including the $(2,1,1)$ mode, which may participate in a $k = 1$ core resonance. A meridional quadrant is shown for each of the four modes, with the rotation axis appearing as the vertical bounding line on the left of each panel, while the horizontal bounding line represents the equatorial plane. The relevance of these particular modes is based on their azimuthal spatial form and closeness in period to Earth tides in the diurnal band. It is readily observable from Fig. 2(a) that the equatorially odd structure of the $(2,1,1)$ mode very nearly duplicates the spatial form of the tesseral tidal forcing. Indeed, this is the ideal situation for a resonant enhancement of the Earth's response by a core mode at nearly diurnal periods.

Of the remaining three modes depicted in Fig. 2, the equatorially odd $(8,3,1)$ and $(10,4,1)$ modes are better coupled for tesseral tidal excitation than is the equatorially even $(5,2,1)$ mode. Since degree 2, 3 and 4 tesseral Earth tides are the only significant ones, one may argue that excitation of the higher-degree core modes is unlikely. However, it is important to note that there will be no coupling between Earth tides and core modes only in the case where their respective eigenfunctions are precisely orthogonal. Furthermore, experience with $k = 1$ inertial waves in the laboratory [e.g., Lumb

and Aldridge, 1988; Stergiopoulos and Aldridge, 1984], suggests that spatially complex modes are also readily excited by careful frequency matching and the same form of $k = 1$ perturbation. Therefore, resonance possibilities appear to exist for the $(8,3,1)$ and $(10,4,1)$ modes.

Shown in Fig. 3 are instantaneous velocity fields corresponding to the pressure eigenfunctions of Fig. 2. Note that the fluid oscillates largely in a direction parallel to the rotation axis, but also at right angles to it. Given that inertial waves are oscillatory disturbances, every half cycle the directions of these arrows changes by 180°. Clearly the 'spin over' or $(2,1,1)$ mode allows for coupling over the largest global extent. The displacement vector field for this mode in calculations of more-realistic Earth models [see Fig. 5 of Smylie et al., 1992] indicates an analogous result, even though the periods are somewhat higher in the latter case. It may also be noted that the best-coupled modes have velocity fields which are equatorially asymmetric, as is intuitively expected for wobble-type motions. Differences in spatial patterns will also affect surface expressions of these modes, and this has been noted by other workers who have been concerned with the direct detectability of core modes with gravimeters [Cummins et al., 1991; Zürn et al., 1987].

Neuberg et al. [1987] were able to retrieve the complex eigenfrequency of the NDFW by stacking the records from several tidal gravimeters in central Europe, as were Richter and Zürn [1988] who used a single GWR superconducting gravimeter in Bad Homburg, Germany. Neuberg et al. [1987] obtained an estimate of $[1 + 1/(434.2 \pm 7)]$ for the eigenfrequency of the NDFW, while the comparable result of $[1 + 1/(431.2 \pm 3.2)]$ was reported by Richter and Zürn [1988]. The associated nutations, as seen from space, would have periods of 434.2 and 431.2 days, respectively. In both of these cases, the existence of the NDFW was inferred from its effect on the gravimetric factors of several Earth tides at nearly diurnal periods. For this situation, the gravimetric factor comprises static and dynamical components, the latter of which owes its existence to resonant enhancement by the NDFW. Thus observations of the gravimetric factor, δ, as a function of frequency, σ, for several $k = 1$ tesseral Earth tides [see, e.g., Fig. 4 of Hinderer and Legros, 1989], provide discrete values for the Earth's response at nearly diurnal periods. In this model of the core resonance, the response at the frequency of the NDFW is infinite since there is no form of dissipation.

According to Herring et al. [1991], the IAU 1980 nutation series is based on the rigid-Earth series by Kinoshita [1977] and Wahr's [1981a] normalized response function. Gwinn et al. [1986] found that their error residuals, from a least-squares fit of VLBI data to the adopted IAU 1980 nutation series, were reduced by the inclusion of a rotational mode whose period converged to 434.6 days. More recently, Herring et al. [1991] found a period of 429.8 days based on their analysis of 798 VLBI experiments spanning almost nine years.

Interestingly, the consistency between the results from tidal gravimetry and VLBI is striking, while comparisons

Fig. 1. Spatial form of the $k = 1$ tesseral core resonance for $k = 1$ Earth tides and core modes.

[e.g., Dehant, 1990] with the theoretical estimates, 459.9 days from Earth model 1066A [Gilbert and Dziewonski, 1975] or 457.8 days from PREM [Dziewonski and Anderson, 1981], reveals a discrepancy of some 26 or 24 days, respectively. Much effort has been directed at attempts to account for this difference, with the favored explanation calling for a significant extra flattening of the CMB [Gwinn et al., 1986;

Neuberg et al., 1990], although there has been some discussion about the viability of this as an explanation [e.g., de Vries and Wahr, 1991; Lumb et al., in preparation, 1992; Souriau and Poupinet, 1991].

We stated at the outset that, in models of the core resonance used for identification of the NDFW, the core possesses but a single mode of oscillation [Hinderer et al., 1987;

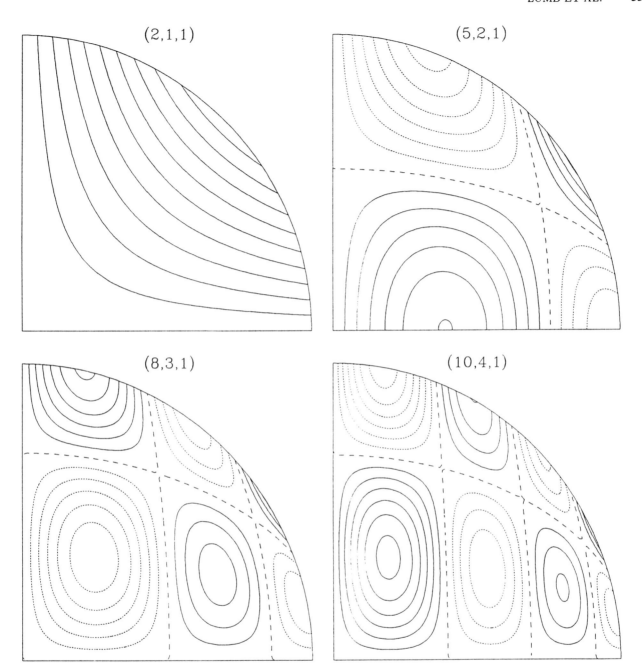

Fig. 2. Reduced pressure eigenfunctions for several of the $k = 1$ inertial modes which may participate in the $k = 1$ tesseral core resonance.

Hinderer and Legros, 1989]. The same has been true for work on the FCN with VLBI [e.g., Gwinn et al., 1986; Scherneck, 1991]. However, experimental studies [e.g., Lumb and Aldridge, 1988; Stergiopoulos and Aldridge, 1984] have clearly demonstrated the existence of an infinite, but discrete, spectrum of inertial modes in a spheroidal shell geometry akin to that of the Earth's core. It therefore seems reasonable to conjecture that the existence of a set of core

modes affords the possibility of resonantly enhanced gravimetric factors at periods in addition to that for the core mode which participates in the NDFW [Lumb and Aldridge, 1991].

The above statements can be made more meaningful by the graphical comparisons of Fig. 4, and quantitative comparisons in Table 1. In Fig. 4 the relative amplitudes of the $k = 1$ tesseral Earth tides from Q. Xi's compilation

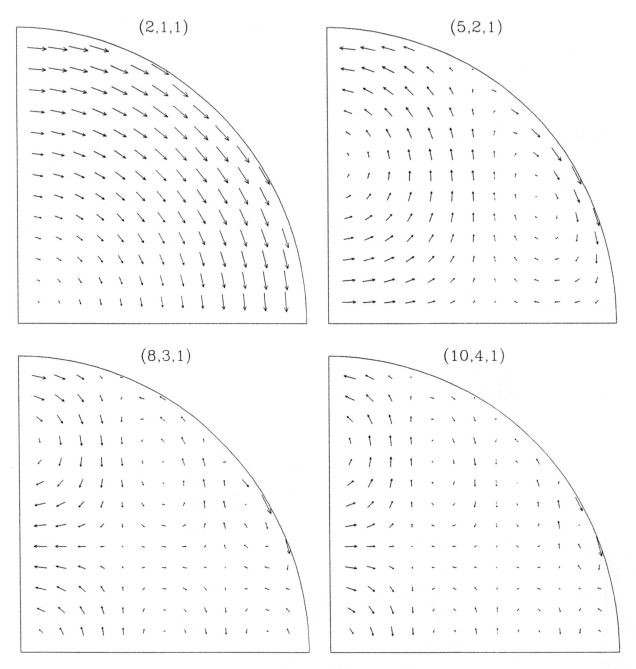

Fig. 3. Instantaneous velocity fields corresponding to the modes in Fig. 2.

have been plotted as a function of period with solid lines, and prominent tidal constituents have been indicated. The dashed lines represent $k = 1$ inertial waves for the Poincaré core model discussed in the previous section, and the lowest-order modes have been indicated. The first column of Table 1 lists the most-prominent $k = 1$ tesseral Earth tides [e.g., Melchior, 1978; Lambeck, 1988], as well as their amplitudes and periods in the second and third columns, respectively. Here the periods and relative tidal amplitudes

are taken from Q. Xi's catalog, while the inertial wave amplitudes are calculated as in (2).

Table 1 also indicates, with superscripted daggers, the tidal modes used in the study by Neuberg et al. [1987], plus the theoretical, T_{NDFW}^{thy}, and LPG or VLBI observed periods, T_{NDFW}^{obs}, of the NDFW. It is clear from both of these comparisons that the (2,1,1) inertial mode is nested in the K_1 tidal band. This proximity in period, as well as its simple, $k = 1$, spatial structure renders it the best-suited fluid mode to participate in the NDFW. Calculations by Smylie et al. [1992], using a finite-element method for the subseismic wave equation, suggest a somewhat higher period (see $(2,1,1)^{SS}$ in the Table) of 23.939 hours when a more realistic core model is used. It is important to note that the simple Poincaré core model for a hydrostatically flattened CMB accounts for over 99% of Smylie et al.'s [1992] realistic Earth period for the (2,1,1) mode; this demonstrates that the effects of rotation and boundary shape are primary ones for this fundamental mode, and that other effects may be treated as perturbations.

Note that the $(2,1,1)$ inertial mode has been implied as the sole core mode in previous discussions of the core resonance phenomenon [e.g., Hinderer et al., 1987; Hinderer and Legros, 1989]. It is also clear from Fig. 4 and Table 1 that there are other modes – e.g., the (5,2,1) and (8,3,1) modes in the J_1 band and the (10,4,1) mode in the σ_1 band – which appear to satisfy the two criteria for resonance. As the modal nomenclature indicates, these modes are more spatially complex than the fundamental (2,1,1) mode. Thus Fig. 4 and Table 1 clearly illustrate that there are several other core modes whose periods lie very close to those of tesseral Earth tides, suggesting the first generalization of the core resonance concept for nearly diurnal periods [Lumb and Aldridge, 1991].

When one plots the dimensionless eigenvalue, λ, [see Fig. 5] as a function of the degree of the associated Legendre polynomial from the period equation [see equation (4) of Aldridge et al., 1989], it is clear that all of the modes which may be relevant in the $k = 1$ tesseral core resonance are clustered about $\lambda \approx 1$. A dimensionless eigenfrequency of about one corresponds to modes which have a period of about one sidereal day for the Earth. For a given value of the degree, n, the modes are relatively spread out in frequency near this value of λ, which contrasts with the relatively dense clustering around $|\lambda| \leq 2$. Thus it may be reasonable to conjecture that even in the case where these resonances are broadened by a molecular viscosity, these additional modes may have little effect on the period of the FCN or NDFW. Since these rotational modes require coupling between the core and mantle, the use of an effective viscosity [Lumb and Aldridge, 1991] leads to increased viscous broadening, and this may allow for viscous coupling between the (2,1,1) and other modes.

$k = 2$ Core Resonance

The $k = 2$ Earth tides cause an observer on the Earth to pass through two high tides and two low tides in one sidereal day. Since two wavelengths are contained in a single circumference of the planet, these are $k = 2$ tidal modes. If the degree, n, coincides with the order, k, then the resulting sectoral Earth tides have the spatial form depicted in Fig. 6. Although there are also tesseral Earth tides in this band, in terms of amplitude the sectoral tides are substantially dominant. The $k = 2$ deformation can also excite internal waves in the liquid core of similar spatial character.

Reduced pressure eigenfunctions for some $k = 2$ inertial modes are shown in Fig. 7. Although its period of 35.8 hours lies well outside the semi-diurnal tidal band, the eigenfunction for the fundamental (3,1,2) inertial mode is illustrated in Fig. 7(a) to show its spatial similarity with the (2,1,1) mode. The remaining panels of this figure depict other $k = 2$ inertial modes which may affect Earth tides at semi-diurnal periods. Those inertial modes with even degree are symmetric about the equator, and are better suited to couple with the equatorially symmetric $k = 2$ sectoral Earth tides. In contrast, odd degree core modes are anti-symmetric about the equator, and therefore are likely to couple with $k = 2$ tesseral Earth tides. Since the semi-diurnal tides are predominantly sectoral in terms of amplitude, the best possibilities for a core resonance will likely involve even-degree inertial modes. Quite dissimilar to the case of the $k = 1$ core resonance, all of the modes relevant here appear to be related by the simple stacking of cells parallel to the rotation axis, and this in turn corresponds simply to increasing the harmonic degree since $m = 1$ in each case. The reason for this spatial relationship will become clear if the eigenfrequencies of inertial modes are considered below.

Instantaneous velocity fields for the same inertial modes of Fig. 7 are presented in Fig. 8. The spatial complexity of these modes is clearly evident in this figure. The stacking of cells parallel to the rotation axis seems to favor velocity fields whose primary oscillations are in a direction perpendicular to the rotation axis. Note that this contrasts with the $k = 1$ case, where the fluid oscillations are largely parallel but also perpendicular to the rotation axis. In all of these figures, the strong cylindrical symmetry of the rotation is evident, even though the container is decidedly spherical.

The period comparison is illustrated graphically in Fig. 9, and quantitatively in Table 2. Relative amplitudes of $k = 2$ Earth tides appear in Fig. 9 as solid lines, and major components have been indicated. The dashed lines represent Poincaré core modes plotted as a function of period, and amplitudes have been scaled in the manner of (2). It is important to note that there are several $k = 2$ inertial modes whose periods lie close to those of Earth tides in this semi-diurnal band. In particular, the spatially complex (14,1,2) and (13,1,2) inertial modes lie on either side of the M_2 band, the (12,1,2) mode is between this and the N_2 band, the (11,1,2) mode is in the N_2 band, the (10,1,2) mode is at the lower end of the $\mu_2 - 2N_2$ band, while lower-order modes can be found close to other bands. As was noted above, since this tidal band is predominantly sectoral, the (14,1,2), (12,1,2) and (10,1,2) inertial modes are the best candidates for a $k = 2$ core resonance. A more quantitative comparison is given in Table 2, which lists the most prominent $k = 2$ Earth tides [e.g., Melchior, 1978; Lambeck, 1988] by name in the first column, while the second and third columns give their amplitudes and periods, respectively. Again the tidal periods and relative amplitudes are from Q. Xi's compilation, whereas the inertial-mode amplitudes are scaled according to (2).

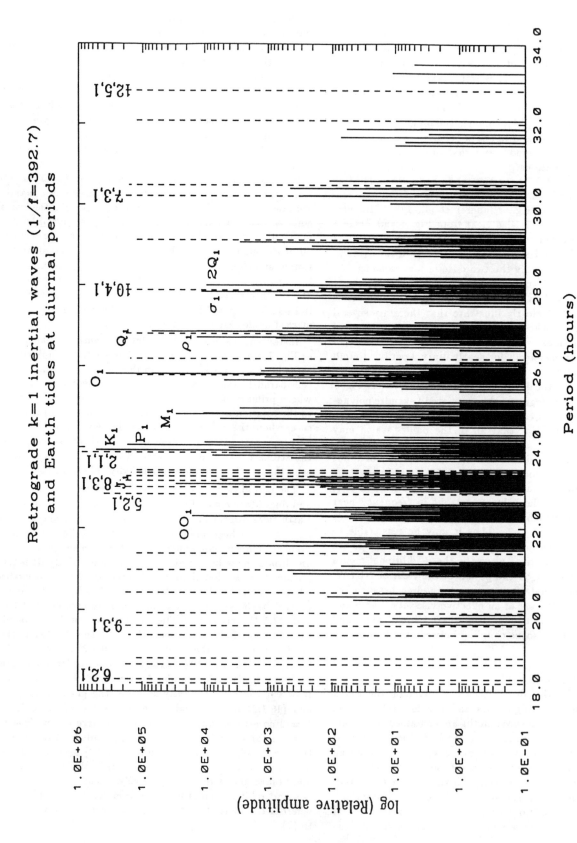

Fig. 4. Semilogarithmic plot of the relative amplitudes of $k = 1$ tesseral Earth disturbances as a function period.

TABLE 1. $k = 1$ tesseral Earth tidal modes and $k = 1$ inertial modes with nearly diurnal periods.

Mode	Amplitude (arcsec)	Period (h)
–	−10386	22.303002
OO_1	−16216	22.306074
$(5, 2, 1)$	454088	22.846
$(8, 3, 1)$	302725	23.019
J_1	−29363	23.098477
τ_1	−5666	23.206930
ϕ_1	−7545	23.804476[†]
ψ_1	−4145	23.869300[†]
$(2, 1, 1)$	908175	23.874
T_{NDFW}^{obs}		23.879
T_{NDFW}^{thy}		23.883
–	1541	23.927434
–	−71843	23.930927
$m_{K_1} - s_{K_1}$	−529929	23.934469[†]
–	10494	23.937963
$(2, 1, 1)^{SS}$		23.939
S_1	−4145	23.999997
P_1	175316	24.065891[†]
π_1	10249	24.132140
M_1	−29635	24.833248
O_1	376801	25.819341[†]
–	71058	25.823400
ρ_1	13706	26.727405
Q_1	72144	26.868356
–	13603	26.872736
σ_1	11526	27.848365
$(10, 4, 1)$	245207	27.887

It is noteworthy that, these inertial modes are spatially complex, and this is in direct contrast to the case of the $k = 1$ core resonance. It is also significant that the M_2 tide has the largest amplitude of all tidal modes, and is about 1.7 times the amplitude of the combined $m_{K_1} - s_{K_1}$ tide in the diurnal band. This almost twofold amplitude factor may aid to offset the spatial complexity, and thus allow excitation of a semi-diurnal resonance. The other significant difference from the $k = 1$ case is also evident from Fig. 9; namely, there are a large number of densely spaced modes in the semi-diurnal band. This finding can be understood by consideration of Fig. 10, which plots the dimensionless eigenvalue, λ, against the degree, n, of the associated Legendre polynomial from the Kudlick period equation [see equation (4) of Aldridge et al., 1989]. Periods of about 12 hours correspond to $\lambda \le 2$; it is clear from this figure that for a given degree, n, the spacing of modes in frequency, λ, decreases as $\lambda \to 2$ [Greenspan, 1968].

Although the details of the $k = 2$ case are somewhat different from those for the $k = 1$ case, it appears that a second generalization of the core resonance phenomenon is a tidally enhanced response in the gravimetric factor at semi-diurnal periods, which could in principle be measured by tidal gravimeters. Since VLBI can also resolve solid-Earth deformation [e.g., Herring et al., 1983], measurement and interpretation with reference to Scherneck's (1991) work appears promising.

Other Core Resonance Possibilities

Zonal Earth tides have $k = 0$ and thus the same azimuthal spatial form as so-called axisymmetric inertial modes, which have been studied previously [e.g., Aldridge and Toomre, 1969; Aldridge, 1972]. The only $k = 0$ inertial modes which have periods close to those of zonal Earth tides are of high spatial degree; i.e., $n \ge 11$. Aldridge and Toomre [1969, Fig. 1] analytically calculated the streamfunctions for several low-order modes in a sphere, while Aldridge [1975, Fig. 1] numerically deduced eigenfunctions for a spherical shell geometry. More recent work by Smylie et al. [1992] has considered these same modes in descriptions of a weakly stable, compressible core bounded by an elastic mantle. The only location in period where zonal tides may resonate with $k = 0$ inertial modes appears to be in the 85 to 170 hour range, where there are several high-degree inertial modes which have periods close to those of zonal tides. While the $(15, 7, 0)$, $(17, 8, 0)$ and $(21, 10, 0)$ lie within zonal bands at 115, 135, and 175 hours, respectively, the $(11, 5, 0)$, $(13, 6, 0)$ and $(19, 9, 0)$ are in between these bands. Since the amplitudes of these zonal tides are more than two orders of magnitude smaller than that of the tides in the diurnal and semi-diurnal bands, resonance possibilities here are of secondary concern.

It is reasonable to conjecture that there might exist tidal resonance possibilities at terdiurnal and quarterdiurnal periods as well. These latter tidal modes have azimuthal numbers of $k = 3$ and $k = 4$, respectively, and can therefore interact with similar core modes. Thus Earth-based observers experience three or four tidal cycles in one sidereal day. Because the Poincaré core model does not allow for inertial modes with periods of less than 12 hours, core resonances in this model at 8 and 6 hours are not possible. Although there are some $k = 3$ tidal modes with periods between 23 and 29 hours, these are so small, $\approx 1/A_{M_2}$, that resonant amplification possibilities appear to be insignificant. When the core is allowed to deviate significantly from an adiabatic state, periods below 12 hours are possible [e.g., Smylie and Rochester, 1981].

ELLIPTIC INSTABILITY

In the case of the $k = 1$ tesseral core resonance, perturbations to the Poincaré core model to include more realistic Earth properties likely form an adequate description of the dynamics of a rotating fluid like the Earth's core. However, in the case of the $k = 2$ sectoral core resonance the situation is more complicated, since a $k = 2$ boundary deformation can render the fluid elliptically unstable. This is a far more complex problem, as the homogeneous partial differential equation (PDE) describing the free modes of the core now includes a term for the elliptical boundary deformation. To initiate applications of this type to the spheroidal shell geometry of the Earth's fluid outer core, we review some recent developments in rotating fluids.

Recall that the Poincaré core model describes the free motions of a homogeneous, inviscid, incompressible fluid in rotation. If the situation is now altered so that such a fluid

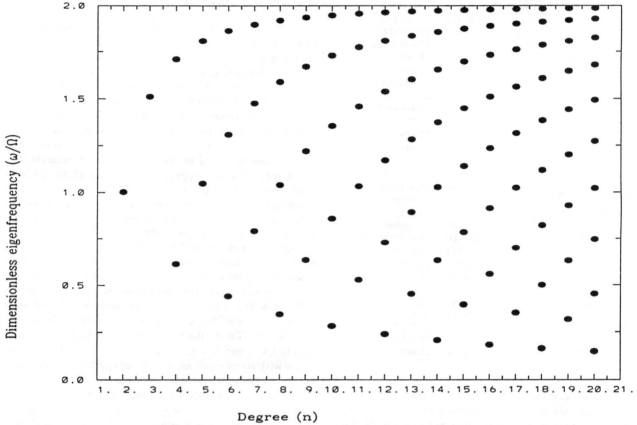

Fig. 5. Inertial-wave eigenfrequency as a function of Legendre-polynomial degree for the $k = 1$ modes [after Smith, 1977].

undergoes a deformation which forces an elliptical equatorial cross-section, then a perturbed Poincaré equation results. While there exist several studies of this type of problem [e.g., Bayly et al., 1988; Malkus, 1989; Vladimirov and Vostretsov, 1986; Vladimirov et al., 1983; Waleffe, 1990], the only work specifically tailored for spherical containers is due to Vladimirov and Vostretsov [1986], and we therefore review it here.

Recall that the Poincaré problem consists of Poincaré's equation, subject to the condition of a vanishing normal velocity on any boundary. Note that this is a homogeneous PDE, and therefore solutions of this problem yield eigenfrequencies and eigenfunctions for the free modes of a contained, incompressible, homogeneous, rotating fluid. If the container is an elastic one, then a $k = 2$ boundary perturbation will give rise to a modified version of the Poincaré equation whose functional form can be written as

$$\left[\mathcal{D}^2 (\nabla)^2 + 4(\widehat{\mathbf{k}} \cdot \nabla)^2 \right] p \propto \varepsilon e^{2i\theta} p, \qquad (3)$$

where $\mathcal{D}$ is the differential operator $\mathcal{D} = -i\omega + \partial/\partial\theta$, with ω a dimensionless eigenfrequency akin to λ in the case of the Poincaré equation. Here the ellipticity, ε, is defined as

$$\varepsilon = \frac{a^2 - b^2}{a^2 + b^2}$$

in terms of the semi-major, a, and semi-minor, b, axes of the elliptical cross-section. The ellipticity, ε, represents the magnitude of the perturbation and scales as the dimensionless Rossby number, which is a gross measure of the relative importance of nonlinear and Coriolis effects [Greenspan, 1968]. The transcendental term illustrates the $k = 2$ azimuthal dependence in terms of azimuthal angle, θ.

Thus, the sphere considered in the classical Poincaré problem will have an equatorial cross-section that is slightly elliptical in the present problem. Using the method of successive approximations, Vladimirov and Vostretsov [1986] first retrieved the solution for zero ellipticity, $\varepsilon = 0$; i.e., the inertial-wave solution discussed above. The next successive approximation illustrates that there will be modes which are

Sectoral Tide

Fig. 6. Spatial form of the $k = 2$ sectoral core resonance for $k = 2$ Earth tides and core modes.

unstable; i.e., the imaginary parts of their complex eigenfrequencies are negative. This form of instability is known as elliptic instability in the rotating-fluids literature, and has been interpreted as resulting from a triad of interacting inertial waves. It has been suggested [Malkus, 1989] that this instability may play a role in the dynamo process responsible for the Earth's magnetic field.

DISCUSSION

The present effort has been partly concerned with a preliminary assessment of the contribution of the Earth's fluid outer core to the Earth's rotational spectrum at periods of about a day. Such a characterization will eventually form the basis of detailed tidal models for use in LPG, as well as nutation series for use by space-based methods like VLBI. The

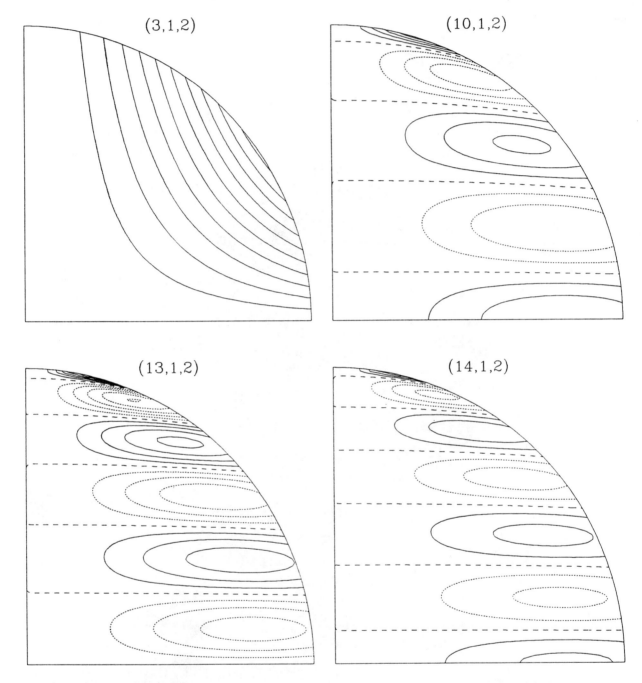

Fig. 7. Reduced pressure eigenfunctions for several of the $k = 2$ inertial modes which may participate in the $k = 2$ sectoral core resonance.

most important conclusions of our work relate to a twofold generalization of the core resonance phenomenon, which results from our consideration of the core as a contained, rotating fluid. In addition to the core mode which participates in the NDFW, we claim that there are several other modes which can produce resonance effects at nearly diurnal periods. Furthermore, the similarity in azimuthal spatial structure and proximity in period, of a number of $k = 2$ sectoral Earth tides and core modes, suggests the possibility of a $k = 2$ sectoral core resonance at semi-diurnal periods. A further extension of this conceptual framework for tides and core modes at zonal, terdiurnal and quarterdiurnal periods has also been presented. In addition to the already documented $k = 1$ tesseral core resonance, it seems that good

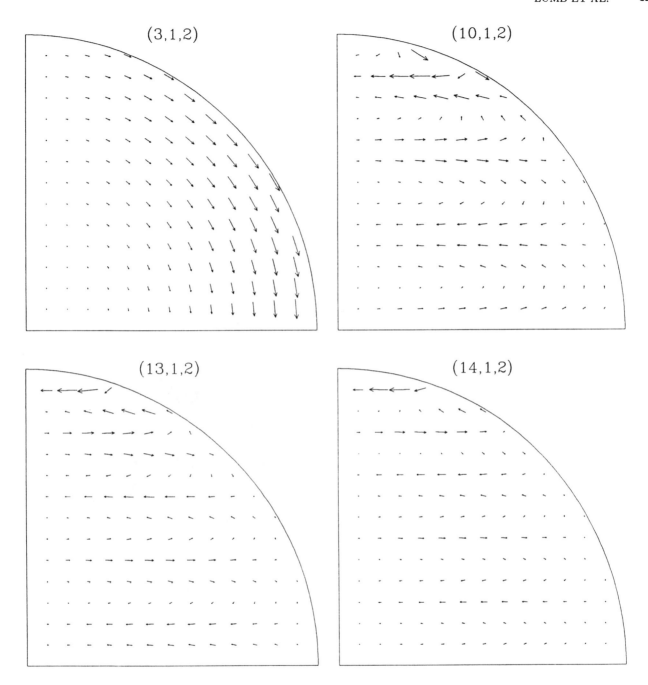

Fig. 8. Instantaneous velocity fields for several of the $k = 2$ inertial modes which may participate in the sectoral core resonance.

possibilities exist for a $k = 2$ sectoral core resonance, but more remote are chances for $k = 0, 3, 4$ resonances. Thus it seems possible that the fluid core not only contributes to the often cited decade-scale exchanges of axial angular momentum with the mantle [Jault and Le Mouël, 1991], but also to Earth's rotational spectrum at periods of about 12 hours and longer.

From the perspective of rotating fluids, elliptical deformation of a container significantly modifies the classical Poincaré problem. In the case of the Poincaré problem for a spheroidal shell of fluid, experimental studies [e.g., Stergiopoulos and Aldridge, 1984; Lumb and Aldridge, 1988] have shown that modes of oscillation do exist even though there is an inner boundary and the associated mathematical

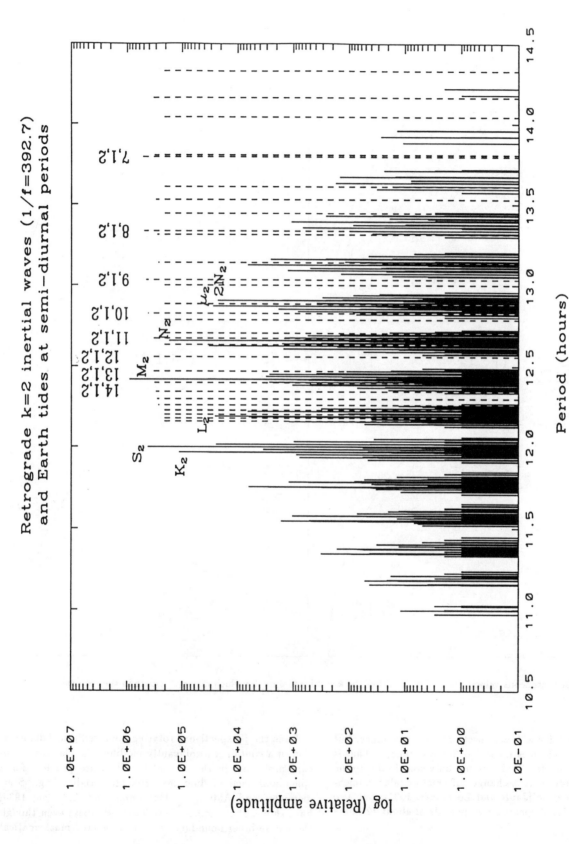

Fig. 9. Semilogarithmic plot of the relative amplitudes of $k = 2$ Earth disturbances as a function period.

TABLE 2. $k = 2$ sectoral Earth tidal modes and $k = 2$ inertial modes with semi-diurnal periods.

Modes	Amplitude (arcsec)	Period (h)
$s_{K_2} - m_{K_2}$	114874	11.967235
R_2	−3536	11.983596
S_2	422555	12.000000
T_2	24702	12.016449
L_2	−25667	12.191620
λ_2	−6697	12.221774
(14,1,2)	317861	12.400
M_2	908175	12.420601
−	−33853	12.421553
(13,1,2)	345107	12.470
(12,1,2)	363270	12.558
ν_2	33028	12.626004
N_2	173883	12.658348
(11,1,2)	390515	12.673
(10,1,2)	417761	12.825
μ_2	27774	12.871758
$2N_2$	23015	12.905374
(9,1,2)	454088	13.035
(8,1,2)	499496	13.336
(7,1,2)	544905	13.794

problem is said to be ill-posed [e.g., Stewartson, 1978; Stewartson and Rickard, 1969; Stewartson and Walton, 1976]. In the case of elliptically deformed spheres, solutions are found to be unstable [e.g., Vladimirov and Vostretsov, 1986]. Thus the inclusion of an elliptic boundary, as is relevant in the case for the Earth's deep interior, will likely involve complexities due to posedness and instability. It is therefore appropriate to develop an experimental analog for this important geophysical problem to guide the theoretical developments, and indeed this experimental work is currently under way in our laboratory.

In addition to its effect on the posedness of the mathematical problem, the presence of an inner boundary also affects the periods, eigenfunctions and amplitudes of the contained modes of oscillation. Aldridge [1967] conducted a number of experiments on the axisymmetric inertial waves of a rotating spherical shell of fluid. In his study, the radius, r_I, of the inner boundary was varied in relation to that, r_O, of the outer boundary, and thus the thickness of the fluid layer was systematically varied. The inclusion of successively larger inner spheres introduced a shift in period to lower values,

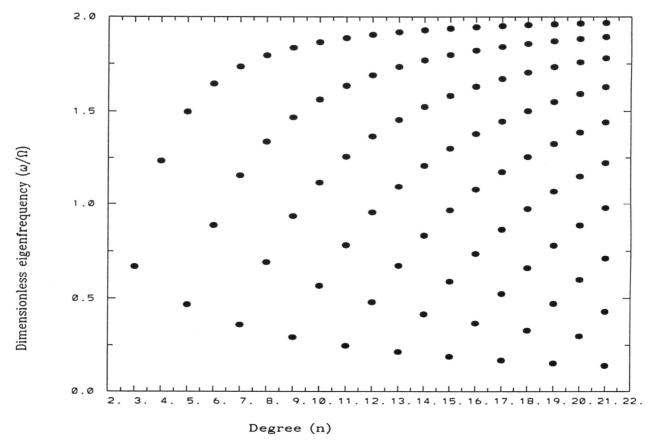

Fig. 10. Inertial-wave eigenfrequency as a function of Legendre-polynomial degree for the $k = 2$ modes [after Smith, 1977].

as is clearly demonstrated in Table 3, for several low-order axisymmetric modes. This has also been confirmed by calculations using a spherical-harmonics method [see Figs. 11-13 of Rieutord, 1991]. It may also be noted that this period shift increases with the spatial complexity of the mode. The streamfunction cells of low-order modes for the full sphere seem to contort and align with characteristic surfaces in order to accommodate the somewhat less ideal geometry of the spherical shell [see, e.g., Fig. 4 of Aldridge, 1975, and Figs. 4 and 5 of Rieutord, 1991]. Similarly, the numerical method of Henderson and Aldridge [1992] has revealed that eigenfunction features of low-order modes for a frustum tend to align with characteristic surfaces, although this is not so for a cylinder. In the inviscid limit the fluid velocity may even become discontinuous across a characteristic surface.

Aldridge [1967] also found that the amplitude of the fundamental axisymmetric inertial mode for this geometry, the $(4, 1, 0)$ mode, diminished as the thickness of the fluid layer decreased. He studied the cases presented in Table 4, which only vary in the ratio of the inner to outer radii; i.e., r_I/r_O. In this Table, the amplitude is defined in terms of the pressure coefficient

$$C_P = \frac{\Delta P}{\rho \epsilon \omega^2 r_O{}^2}, \qquad (4)$$

where ΔP is the differential pressure measured by a pressure transducer, ρ is the fluid density, and ϵ is the amplitude of the axisymmetric perturbation at frequency ω [Aldridge, 1967]. From Table 4 it is clear that the amplitude response for the fundamental mode decreases as the radius ratio increases, and in the last case the amplitude of the mode has been decreased to about 75% of its value for the full sphere. Although the amplitude response for other modes also varies somewhat with fluid-layer thickness, the effect is not nearly as striking as that just mentioned. Aldridge's [1967] experimental work was supported by his own theoretical calculations using a variational principle [see also Aldridge, 1972], and more recently by Rieutord's [1991] spherical-harmonics method. While it is true that these findings pertain directly to the $k = 0$ modes of oscillation, it is our expectation that analogous results would be arrived at for other azimuthal wavenumbers.

In addition to these purely geometrical effects, the presence of the inner core also has dynamical consequences. In particular, coupling between the outer core and inner core, gives rise to another rotational mode [de Vries and Wahr, 1991; Mathews et al., 1991a,b]. This free inner core nuta-

TABLE 4. Loss of a principal axisymmetric resonance in a rotating spherical shell due to the presence of an inner boundary [after Aldridge, 1967].

r_I/r_O	C_P
0.000	0.158
0.254	0.147
0.349	0.082
0.510	0.038

tion (FICN) has a prograde frequency of $(1 - 1/471)$ cpsd [de Vries and Wahr, 1991], or a period of 23.985 hrs for its associated nearly diurnal free inner core wobble (NDFIW). Since the inner core's moments of inertia are about 1/1400 that of the whole Earth, this mode will likely not have a direct effect on the Earth's rotation or on mantle deformation; the possibility of NDFIW having some effect on tidally induced nutations also appears remote [see Table 4 of de Vries and Wahr, 1991].

Although the Poincaré core model works well as a first-order approximation for the core resonance phenomenon, it is clear that more realistic Earth models are required. Indeed, this presents a theoretical challenge, as any viable model should combine the effects of a realistic fluid core [e.g., Smylie et al., 1992], with a proper description of tidal deformation analogous to that developed for the $k = 1$ tesseral core resonance [e.g., Hinderer and Legros, 1989; Hinderer et al., 1987]. Thus a counterpart of the Neuberg et al. [1987] experiment could be performed for a fully dynamical core model in the case of the $k = 1$ tesseral core resonance.

Resonantly enhanced gravimetric factors also seem a viable way of characterizing the $k = 2$ core resonance using LPGs. Since the $k = 2$ Earth tides are predominantly sectoral, and therefore equatorially symmetric, this resonance will not directly affect the Earth's nutation series. VLBI can also measure deformation of the solid Earth [e.g., Herring et al., 1983], and therefore an independent characterization of resonantly enhanced deformation would be useful in establishing the effects of the $k = 2$ sectoral core resonance. Scherneck's (1991) approach for studying tidally induced deformation via VLBI, which parallels that currently used in tidal gravimetry [e.g., Hinderer et al., 1987; Hinderer and Legros, 1989], may help to establish the existence of a $k = 2$ core resonance.

Theoretical predictions by Wahr [1981b] for the body tides of an elliptical, rotating, elastic and oceanless Earth, suggested that the gravimetric factor should vary with latitude. Merriam (1983) attempted to confirm this experimentally, concluding that he could not substantiate the latitude dependence to any significant level with his results. More recent studies by Dehant and Ducarme [1987] and by Dehant [1990] have shown that the gravimetric factor does indeed have a latitude dependence in both the diurnal and semidiurnal bands. The gravimetric factor for the M_2 sectoral Earth tide can be expressed as

$$\delta_{M_2} = 1.1613 - 0.0009 \frac{\sqrt{3}}{2}(7\cos^2\theta - 1) \qquad (5)$$

TABLE 3. The effect of increasing the size of the inner boundary on several low-order, axisymmetric inertial modes [after Aldridge, 1967].

r_I/r_O	(6,1)	(4,1)	(6,2)	(8,3)	(10,4)	(12,5)
0.00	14.53	18.28	25.46	32.92	40.36	–
0.25	14.48	18.28	25.35	32.52	39.11	46.20
0.35	14.76	18.19	25.01	31.12	37.81	44.08
0.51	15.18	18.05	22.16	28.49	33.57	40.70

[Dehant and Ducarme, 1987, Table III], from which it is clear that the maximum value is reached at the equator, and thus the core will similarly be forced most strongly in these regions. Interestingly, the $k = 2$ core modes of even degree have the largest components of their motion (see Fig. 8) at high latitudes, and this motion is predominantly perpendicular to the rotation axis. Thus it is reasonable to conjecture that although these modes probably derive most of their excitation in equatorial regions, instruments located in high latitudes will likely experience a larger resonant enhancement, and observational possibilities may be best in these regions.

Attention to a fully dynamical description of the fluid core in the case of the $k = 1$ tesseral core resonance might help to resolve the differences in period of the FCN [Lumb and Aldridge, 1991]. Since the $k = 2$ sectoral core resonance is also sensitive to the shape of the CMB, analyses of global Earth data from LPG or VLBI should lead to independent estimates of CMB flattening, and thus aid in the resolution of the significant discrepancy between observation and theory.

Acknowledgements. We thank J. B. Merriam for supplying us with Q. Xi's catalog of tidal constituents and for several helpful discussions. We also thank V. A. Vladimirov for stimulating discussions regarding elliptic instabilities. We are grateful to both anonymous referees for their comments that helped us improve the clarity of this paper. This work was supported by generous operating and equipment grants from the Natural Sciences and Engineering Research Council of Canada (NSERC) to KDA, and by an Ontario Graduate Scholarship to LIL.

REFERENCES

Aldridge, K. D., *An experimental study of axisymmetric inertial oscillations of a rotating liquid sphere*, Ph.D. Thesis, M.I.T., 1967.

Aldridge, K. D., Axisymmetric inertial oscillations of a fluid in a rotating spherical shell, *Mathematika, 19*, 163-168, 1972.

Aldridge, K. D., Inertial waves and the Earth's outer core, *Geophys. J. R. astr. Soc., 42*, 337-345, 1975.

Aldridge, K. D. and L. I. Lumb, Inertial waves identified in the Earth's fluid outer core, *Nature, 325*, 421-423, 1987.

Aldridge, K. D., L. I. Lumb and G. A. Henderson, Inertial modes in the Earth's fluid outer core, in *Structure and Dynamics of the Earth's Deep Interior*, (eds.) by D. E. Smylie and R. Hide, Geophysical Monographs / IUGG Series, American Geophysical Union, 13-21, 1988.

Aldridge, K. D., L. I. Lumb, and G. A. Henderson, A Poincaré model for the Earth's fluid core, *Geophys. Astrophys. Fluid Dyn., 48*, 5-23, 1989.

Aldridge, K. D. and A. Toomre, Axisymmetric inertial oscillations of a fluid in a rotating spherical container, *J. Fluid Mech., 37*, 307-323, 1969.

Bayly, B. A., S. A. Orszag and T. Herbert, Instability mechanisms in shear-flow transition, *Ann. Rev. Fluid Mech., 20*, 359-391, 1988.

Bryan, G. H., The waves on a rotating liquid spheroid of finite ellipticity, *Phil. Trans. R. Soc. Lond., A180*, 187-219, 1889.

Cummins, P., J. M. Wahr, D. C. Agnew and Y. Tamura, Constraining core undertones using stacked IDA gravity records, *Geophys. J. Int., 106*, 189-198, 1991.

Dehant, V., Tidal parameters and nutations: Influence from the Earth's interior, in *Variations in Earth Rotation*, (eds.) by D. D. McCarthy and W. E. Carter, Geophysical Monographs / IUGG Series, American Geophysical Union, 69-77, 1990.

Dehant, V. and B. Ducarme, Comparison between the theoretical and observed tidal gravimetric factors, *Phys. Earth planet. Int., 49*, 192-212, 1987.

de Vries, D. and Wahr, J. M., The effects of the solid inner core and nonhydrostatic structure on the Earth's forced nutations and Earth tides, *J. Geophys. Res., 96*, 8275-8293, 1991.

Dziewonski, A. M. and D. L. Anderson, Preliminary reference Earth model, *Phys. Earth planet. Int., 25*, 297-356, 1981.

Gilbert, F. and A. M. Dziewonski, An application of normal mode theory to the retrieval of structural parameters and source mechanisms from seismic spectra, *Phil. Trans. R. Soc. Lond., A278*, 187-269, 1975.

Greenspan, H. P., *The Theory of Rotating Fluids*, Cambridge University Press, 1968.

Gwinn, C. R., T. A. Herring and I. I. Shapiro, Geodesy by radio interferometry: Studies of the forced nutations of the Earth, 2. Interpretation, *J. geophys. Res., 91*, 4755-4765, 1986.

Henderson, G. A. and K. D. Aldridge, A finite-element method for inertial waves in a frustum, *J. Fluid Mech., 234*, 317-327, 1992.

Herring, T. A., B. A. Buffet, P. M. Mathews and I. I. Shapiro, Forced nutations of the Earth: Influence of inner core dynamics, 3. Very long interferometry data analysis, *J. geophys. Res., 96*, 8259-8273, 1991.

Herring, T. A., B. E. Corey, C. C. Counselman, I. I. Shapiro, A. E. E. Rogers, A. R. Whitney, T. A. Clark, C. A. Knight, C. Ma, J. W. Ryan, B. R. Schupler, N. R. Vandenburg, G. Elgered, G. Lundqvist, B. O. Rönnäng, J. Campbell, and P. Richards, Determination of tidal parameters from VLBI observations, *Proceedings of the Ninth Symposium on Earth Tides*, (ed.) by J. T. Kuo, E. Schweizerbart'sche Verlagsbuchhandlung, 205-214, 1983.

Hinderer, J. and H. Legros, Elasto-gravitational deformation, relative gravity changes and Earth dynamics, *Geophys. J. R. astr. Soc., 97*, 481-495, 1989.

Hinderer, J., H. Legros and M. Almavict, Tidal motions within the Earth's fluid core: resonance process and possible variations, *Phys. Earth planet. Int., 49*, 213-221, 1987.

Hinderer, J. H. Legros and D. Crossley, Global Earth dynamics and induced gravity changes, *J. geophys. Res., 96*, 20257-20265, 1991.

Jault, D. and J.-L. Le Mouël, Flow at the top of the core and exchange of angular momentum between core and mantle, *J. Geomag. Geoelectr., 43*, 111-129, 1991.

Johnson, I. M. and D. E. Smylie, A variational approach to whole Earth dynamics, *Geophys. J. R. astr. Soc., 50*, 35-54, 1977.

Kinoshita, H., Theory of the rotation of the rigid Earth, *Celestial Mech., 15*, 277-326, 1977.

Kudlick, M. D., On the transient motions in a contained, rotating fluid, *Ph.D. Thesis*, M.I.T., 1966.

Lambeck, K., *Geophysical Geodesy: The Slow Deformations of the Earth*, Oxford University Press, 1988.

Levine, J., J. C. Harrison and W. Dewhurst, Gravity tide measurements with a feedback gravimeter, *J. geophys. Res., 91*, 12835-12841, 1986.

Lumb, L. I. and K. D. Aldridge, An experimental study of inertial waves in a spheroidal shell of rotating fluid, in *Structure and Dynamics of the Earth's Deep Interior*, (eds.) by D. E. Smylie and R. Hide, Geophysical Monographs / IUGG Series, American Geophysical Union, 35-39, 1988.

Lumb, L. I. and K. D. Aldridge, On viscosity estimates for the Earth's fluid core and core-mantle coupling, *J. Geomag. Geoelectr., 43*, 93-110, 1991.

Malkus, W. V. R., An experimental study of global instabilities due to the tidal (elliptical) distortion of a rotating elastic cylinder, *Geophys. Astrophys. Fluid Dyn., 48*, 123-134, 1989.

Mathews, P. M., B. A. Buffett, T. A. Herring and I. I. Shapiro, Forced nutations of the Earth: Influence of inner core dynamics, 1. Theory, *J. geophys. Res., 96*, 8219-8242, 1991a.

Mathews, P. M., B. A. Buffett, T. A. Herring and I. I. Shapiro, Forced nutations of the Earth: Influence of inner core dynamics, 2. Numerical results and comparisons, *J. geophys. Res., 96*, 8243-8257, 1991b.

Melchior, P. J., *The Tides of the Planet Earth*, Pergamon Press, 1978.

Merriam, J. B., Inertial and ellipsoidal Earth effects on gravity tide observations, *Proceedings of the Ninth Symposium on Earth Tides*, (ed.) by J. T. Kuo, E. Schweizerbart'sche Verlagsbuchhandlung, 137-144, 1983.

Neuberg, J., J. Hinderer, and W. Zürn, Stacking gravity tide observations in central Europe for the retrieval of the complex eigenfrequency of the nearly diurnal free wobble, *Geophys. J. R. astr. Soc., 91*, 853-868, 1987.

Neuberg, J., J. Hinderer and W. Zürn, On the complex eigenfrequency of the 'nearly diurnal free wobble' and its geophysical interpretation, *Variations in Earth Rotation*, (eds.) D. D. McCarthy and W. E. Carter, Geophysical Monographs / IUGG Series, American Geophysical Union, 11-16, 1990.

Richter, B. and W. Zürn, Chandler effect and nearly diurnal free wobble as determined from observations with a superconducting gravimeter, *The Earth's Rotation and Reference Frames for Geodesy and Geodynamics*, (eds.) A. K. Babcock and G. A. Wilkins, Kluwer Academic Publishers, 309-315, 1988.

Rieutord, M., Linear theory of rotating fluids using spherical harmonics, Part II: Time-periodic flows, *Geophys. Astrophys. Fluid Dyn., 59*, 185-208, 1991.

Sasao, T., S. Okubo and M. Saito, A simple theory on the dynamical effects of a stratified fluid core upon the nutational motion of the Earth, in *Nutation and the Earth's Rotation*, by E. P. Federov, M. L. Smith and P. L. Bender (eds.), International Astronomical Union, 165-183, 1980.

Scherneck, H.-G., A parameterized solid Earth tide model and ocean tide loading effects for global geodetic baseline measurements, *Geophys. J. Int., 106*, 677-694, 1991.

Smith, M. L., Wobble and nutation of the Earth, *Geophys. J. R. astr. Soc., 50*, 103-140, 1977.

Smylie, D. E., Dynamics of the outer core, *Veröff. ZentInst. Phys. Erde Akad. Wiss. D.D.R., 30*, 91-104, 1974.

Smylie, D. E., Variational calculation of wobble modes of the Earth. *Variations in Earth Rotation*, Geophysical Monographs / IUGG Series, (eds.) D. D. McCarthy and W. E. Carter, American Geophysical Union, Washington, D. C., 5-10, 1990.

Smylie, D. E., X. Jiang, B. J. Brennan, and K. Sato, Numerical calculation of modes of oscillation of the Earth's core, *Geophys. J. Int., 108*, 465-490, 1992.

Smylie, D.E. and M. G. Rochester, Compressibility, core dynamics and the subseismic wave equation, *Phys. Earth planet. Int. 24*, 308-319, 1981.

Souriau, A. and Poupinet, G., A study of the outermost liquid core using differential travel times of the *SKS, SKKS* and *S3KS* phases, *Phys. Earth planet. Int., 68*, 183-199, 1991.

Stacey, F. D., *Physics of the Earth*, Second Edition, John Wiley & Sons, 1977.

Stergiopoulos, S. and K. D. Aldridge, Ringdown of inertial waves in a spheroidal shell of rotating fluid, *Phys. Earth planet. Int., 36*, 17-26, 1984.

Stewartson, K., Homogeneous fluids in rotation, *Rotating Fluids in Geophysics*, (eds.) P. H. Roberts and A. M. Soward, Academic Press, 67-103, 1978.

Stewartson, K. and J. A. Rickard, Pathological oscillations of a rotating fluid, *J. Fluid Mech., 35*, 759-773, 1969.

Stewartson, K. and I. C. Walton, On waves in a thin shell of stratified rotating fluid, *Proc. R. Soc. Lond., A349*, 141-146, 1976.

Vladimirov, V. A., V. F. Tarasov and L. Ya. Rybak, Stability of elliptically deformed rotation of an ideal incompressible fluid in a Coriolis force field, *Izv. Atmos. Ocean. Phys., 19*, 437-442, 1983.

Vladimirov, V. A. and D. G. Vostretsov, Instability of steady flows with constant vorticity in vessels of elliptic cross-section, *Prikl. Matem. Mekan. U.S.S.R., 50*, 279-285, 1986.

Wahr, J. M., The forced nutations of an elliptical, rotating, elastic and oceanless Earth, *Geophys. J. R. astr. Soc., 64*, 705-727, 1981a.

Wahr, J. M., Body tides on an elliptical rotating elastic and oceanless Earth, *Geophys. J. R. astr. Soc., 64*, 677-704, 1981b.

Waleffe, F., On the three-dimensional instability of strained vortices, *Phys. Fluids A, 2*, 76-80, 1990.

Zürn, W., B. Richter, P. A. Rydelek and J. Neuberg, Comment on 'Detection of inertial gravity oscillations in the Earth's core with a superconducting gravimeter at Brussels', *Phys. Earth planet. Int., 49*, 176-178, 1987.

K. D. Aldridge, G. A. Henderson and L. I. Lumb, Department of Earth and Atmospheric Science, York University, 4700 Keele Street, North York, Ontario M3J 1P3, Canada.

Analytical Computation of Modes Using Boundary-Layer Theory

K. Degryse and V. Dehant

Koninklijke Sterrenwacht van België, Brussels, Belgium

An examination of oscillations of a layered sphere which is composed of a homogeneous, incompressible mantle, a homogeneous, incompressible inner core and a homogeneous, incompressible "soft" outer core [Mochizuki, 1990] is carried out. The rigidity of the outer core, which is much smaller than that of the mantle and the inner core, gives rise to boundary layers near the boundaries of the outer core. Boundary-layer theory then provides analytical solutions.

Introduction

In this paper, we describe a formalism for determining analytically modes and frequencies of oscillations for a spherically symmetric Earth that consists of three homogeneous and incompressible layers. We assume that the rigidity of the outer core is much smaller than the rigidity of mantle and inner core, so that the introduction of two boundary layers with a small shear modulus in the outer core near the inner core-outer core boundary and near the core-mantle boundary is appropriate.

In every region, the solution vector is expanded as a sum of vector spherical harmonics with coefficients that depend on the radius only. The partial differential equations (the momentum equation, the continuity equation, Poisson's equation and the stress-strain relation) are then used to derive a set of ordinary differential equations for these coefficients. These equations, together with the appropriate boundary conditions are then solved. Continuity conditions on the unique solutions lead to the determination of the eigenfrequencies.

The plan of the paper is as follows. In section 2, we recall the equations governing a solid Hooke body with a bulk modulus and a shear modulus profile. In section 3, we start the treatment of the eigenvalue problem in the case of a rigidity which is negligible in the outer core. Section 4 is devoted to the treatment of the eigenvalue problem for the case in which the rigidity in the outer core is small but not negligible.

Dynamics of Earth's Deep Interior and Earth Rotation
Geophysical Monograph 72, IUGG Volume 12

Formulation of the Problem and Governing Equations

We restrict ourselves to models of the Earth which are non-rotating and spherically symmetric. We assume the evolution to be so slowly that it can be described by successive quasi-static states of hydrostatic equilibrium. For the quasi-static state mentioned, the governing equations are invariant with respect to a translation in time so that one may seek normal modes depending on time by $\exp(i\omega t)$ as basic solutions [Chandrasekhar, 1961]. Assuming that there is no externally applied body force (free oscillations), the complete system of linearized equations takes the form

$$\rho_0 D_t^2 \overline{s}_L = -\rho_0 \overline{\nabla}\phi_1^B - \rho_1^B \overline{\nabla}\phi_0 - \overline{\nabla}(\overline{s}_L . \rho_0 \overline{\nabla}\phi_0) + \overline{\nabla}.\overline{\underline{T}}_L \quad (1)$$

$$\rho_1^B = -\overline{\nabla}.(\rho_0 \overline{s}_L) \quad (2)$$

$$\nabla^2 \phi_1^B = 4\pi G \rho_1^B \quad (3)$$

$$\overline{\underline{T}}_L = k(\overline{\nabla}.\overline{s}_L)\underline{\overline{I}} + \mu[\overline{\nabla}\underline{s}_L + (\overline{\nabla}\underline{s}_L)^T] \quad (4)$$

[Dahlen, 1974].

Here, $\rho_0(r)$ and $\phi_0(r)$ denote, respectively, the density and the gravitational potential in the spherically symmetric equilibrium configuration. The quantities ρ_1^B and ϕ_1^B represent, respectively, the incremental part of the Eulerian mass density and of the Eulerian gravitational potential. D_t denotes a simple partial differentiation with respect to time of the Lagrangian quantity $\overline{s}_L(\overline{x}, t)$, which is the Lagrangian particle displacement of particle $\overline{x}$ at time t. $\overline{\underline{T}}_L$ is the symmetric Lagrangian stress tensor, $\underline{\overline{I}}$ is the identity tensor, and the superscript T means transpose. Furthermore, G is the Newtonian gravitational constant, k is the bulk modulus and μ the shear modulus of the material.

For a spherical Earth, the partial differential equations given by Eqs. (1)-(4) can be transformed into a set of ordinary scalar differential equations of the first order in d/dr by using spherical harmonic expansions. For example, ϕ_1^E takes the following form

$$\phi_1^E = \sum_{l=0}^{\infty} \sum_{m=0}^{l} \phi_{1l}^{Em}(r) Y_l^m(\theta, \varphi), \qquad (5)$$

and one defines

$$\bar{s}_L = \sum_{n=-}^{+} \sum_{l=0}^{\infty} \sum_{m=-l}^{+l} S_l^{mn}(r) D_{mn}^l(\theta, \varphi) \bar{e}_n, \qquad (6)$$

$$U_l^m = S_l^{m0}, \qquad (7)$$

$$V_l^m = S_l^{m+} + S_l^{m-}, \qquad (8)$$

where $D_{mn}^l(\theta, \varphi)$ are the Generalized Spherical Harmonics of Phinney and Burridge [1973], and the $\bar{e}_n$ are the basis vectors as defined in Smith [1974]. Since we consider an incompressible ($\bar{\nabla}.\bar{s}_L = 0, k \to \infty$) Earth consisting of three homogeneous layers, Eqs. (1)-(4) can be transformed into

$$\frac{dU_l}{dr} = -\frac{2}{r} U_l - \frac{L}{r} V_l, \qquad (9)$$

$$\frac{dV_l}{dr} = \frac{2L}{r} U_l + \frac{1}{r} V_l + \frac{1}{\mu} Q_l, \qquad (10)$$

$$\frac{dP_l}{dr} = -\frac{L}{r} Q_l + \frac{12\mu}{r^2} U_l + \frac{6L\mu}{r^2} V_l + \rho_0 g_l - \frac{4}{r} \rho_0 g_0 U_l$$
$$- \frac{L}{r} \rho_0 g_0 V_l + 4\pi G \rho_0^2 U_l - \omega^2 \rho_0 U_l, \qquad (11)$$

$$\frac{dQ_l}{dr} = -\frac{3}{r} Q_l + \frac{2L}{r} P_l + \frac{12L\mu}{r^2} U_l + \frac{6L^2\mu}{r^2} V_l + \frac{2L'^2\mu}{r^2} V_l$$
$$- \frac{2L\rho_0}{r} \phi_l - \frac{2L}{r} \rho_0 g_0 U_l - \omega^2 \rho_0 V_l, \qquad (12)$$

$$\frac{d\phi_l}{dr} = g_l, \qquad (13)$$

$$\frac{dg_l}{dr} = -\frac{2}{r} g_l + \frac{2L^2}{r^2} \phi_l, \qquad (14)$$

with L defined by

$$L = \sqrt{\frac{l(l+1)}{2}}, \qquad (15)$$

L' by

$$L' = \sqrt{\frac{(l-1)(l+2)}{2}}, \qquad (16)$$

and where U_l and V_l are the spherical harmonic coefficients of the Lagrangian displacement field. P_l and Q_l are, respectively, the radial and tangential components of the stresstensor. ϕ_l is the Eulerian potential, to which g_l is related by means of Eq. (13).

The separation of the angular variables thus yields a system of equations belonging to a single spherical harmonic. The azimutal number m does not appear in Eqs. (9)-(14) so that the eigenfrequencies are m-fold degenerate [see Dahlen, 1968; Backus and Gilbert, 1961; Smith, 1974]. For the sake

of simplification of the notations we drop the subscript l in Eqs. (9)-(14).

The associated linearly independent boundary conditions are
in $r = 0$:

$$\lim_{r \to 0} P(r) = 0 \qquad (17)$$

$$\lim_{r \to 0} Q(r) = 0 \qquad (18)$$

$$\lim_{r \to 0} \phi(r) = 0 \qquad (19)$$

in $r = r_s$:

$$P(r_s) = 0 \qquad (20)$$

$$Q(r_s) = 0 \qquad (21)$$

$$g(r_s) + \frac{(l+1)}{r_s} \phi(r_s) = 0. \qquad (22)$$

The boundary conditions for $r \to 0$ must be satisfied for $l \geq 3$ (see section (4.1)).

Since the rigidity of the outer core is much smaller than that of the mantle and the inner core, a first approximation of the eigenfunctions can be obtained by setting $\mu = 0$ in the outer core. By this approximation, the system of differential equations reduces to a fourth-order system. The reduction of the order of the system of differential equations implies that particular solutions are removed from any general solution. However, it cannot be ruled out a priori that the particular solutions removed by the approximation are dominant in the case $\mu \neq 0$. Therefore, the full system of Eqs. (9)-(14) must be used for the construction of approximations of solutions in case $\mu \neq 0$ in the outer core. Our aim is to do this by applying boundary-layer theory [Kevorkian and Cole, 1981].

Rigidity in the Outer Core = 0

A first approximation of the solutions of the outer core is sought by dropping the rigidity terms in Eqs. (9)-(14) and thus by considering the equations

$$\frac{dU}{dr} = -\frac{2}{r} U - \frac{L}{r} V \qquad (23)$$

$$Q = 0 \qquad (24)$$

$$\frac{dP}{dr} = -\frac{4}{r} \rho_0 g_0 U - \frac{L}{r} \rho_0 g_0 V + \rho_0 g + 4\pi G \rho_0^2 U$$
$$- \omega^2 \rho_0 U \qquad (25)$$

$$0 = \frac{2L}{r} P - \frac{2L\rho_0}{r} \phi - \frac{2L}{r} \rho_0 g_0 U - \omega^2 \rho_0 V \qquad (26)$$

$$\frac{d\phi}{dr} = g \qquad (27)$$

$$\frac{dg}{dr} = -\frac{2}{r} g + \frac{2L^2}{r^2} \phi. \qquad (28)$$

The solutions of Eqs (23)-(28) take the following form

$$U = A_3'' r^{l-1} + A_4'' r^{-l-2} \qquad (29)$$

$$V = -A_3'' \frac{(l+1)}{L} r^{l-1} + A_4'' \frac{l}{L} r^{-l-2} \qquad (30)$$

$$P = A_1'' \rho_0 r^l + A_2'' \rho_0 r^{-l-1} - A_3'' \frac{\omega^2 \rho_0}{l} r^l$$

$$+ \ A_4'' \frac{\omega^2 \rho_0}{(l+1)} r^{-l-1} + A_3'' \rho_0 g_0 r^{l-1}$$

$$+ \ A_4'' \rho_0 g_0 r^{-l-2} \qquad (31)$$

$$Q = 0 \qquad (32)$$

$$\phi = A_1'' r^l + A_2'' r^{-l-1} \qquad (33)$$

$$g = A_1'' l r^{l-1} - A_2'' (l+1) r^{-l-2}, \qquad (34)$$

where the A'' are integration constants.

In the mantle and the inner core ($\mu \neq 0$), Eqs. (9)-(14) are transformed into one differential equation of the fourth order with a singularity of the first kind in $r = 0$:

$$-\frac{r^2 \mu}{2L^2} \frac{d^4 U}{dr^4} - \frac{4r\mu}{L^2} \frac{d^3 U}{dr^3} + (\mu - \frac{5\mu}{L^2} + \frac{L'^2 \mu}{L^2}) \frac{d^2 U}{dr^2} +$$

$$(\frac{2\mu}{rL^2} + \frac{2\mu}{r} + \frac{2L'^2 \mu}{rL^2}) \frac{dU}{dr} +$$

$$(\frac{4\mu}{r^2} - \frac{2\mu}{r^2 L^2} - \frac{2L'^2 \mu}{r^2 L^2} - \frac{2L^2 \mu}{r^2}) U$$

$$= \frac{r^2 \omega^2 \rho_0}{2L^2} \frac{d^2 U}{dr^2} + \frac{4r\omega^2 \rho_0}{2L^2} \frac{dU}{dr}$$

$$+ \ (\frac{2\omega^2 \rho_0}{2L^2} - \omega^2 \rho_0) U. \qquad (35)$$

Solutions of this linear differential equation which are regular in the neighbourhood of the singular point are then found by means of the method of Frobenius [Ince, 1956]. These solutions, which satisfy Eqs. (9)-(14), are [see also Crossley, 1975; Takeuchi and Saito, 1971; Jeffreys, 1970]:

$$U = A_1 r^{-l} [1 - \frac{\omega^2 \rho_0}{\mu} \frac{1}{4(-2l+3)} r^2 + ...]$$

$$+ \ A_2 l r^{l-1} [1 - \frac{\omega^2 \rho_0}{\mu} \frac{1}{2(2l+3)} r^2 + ...]$$

$$+ \ A_3 (l+2) r^{l+1} [1 - \frac{\omega^2 \rho_0}{\mu} \frac{1}{4(2l+5)} r^2 + ...]$$

$$+ \ A_4 (-l-1) r^{-l-2} [1 -$$

$$\frac{\omega^2 \rho_0}{\mu} \frac{1}{2(-2l+1)} r^2 + ...] \qquad (36)$$

$$V = -\frac{A_1}{L} r^{-l} [(-l+2) - \frac{\omega^2 \rho_0}{\mu} \frac{(-l+4)}{4(-2l+3)} r^2 + ...]$$

$$- \ \frac{A_2}{L} l r^{l-1} [(l+1) - \frac{\omega^2 \rho_0}{\mu} \frac{(l+3)}{2(2l+3)} r^2 + ...]$$

$$- \ \frac{A_3}{L} (l+2) r^{l+1} [(l+3) -$$

$$\frac{\omega^2 \rho_0}{\mu} \frac{(l+5)}{4(2l+5)} r^2 + ...]$$

$$- \ \frac{A_4}{L} (-l-1) r^{-l-2} [-l -$$

$$\frac{\omega^2 \rho_0}{\mu} \frac{(-l+2)}{2(-2l+1)} r^2 + ...] \qquad (37)$$

$$P = A_1 r^{-l-1} [-2\mu \frac{(l^2 + 3l - 1)}{(l+1)} -$$

$$\omega^2 \rho_0 \frac{(l-2)}{2(2l-3)} r^2 + ...]$$

$$+ \ A_2 l r^{l-2} [2\mu(l-1) - \omega^2 \rho_0 \frac{(l+1)}{(2l+3)} r^2 + ...]$$

$$+ \ A_3 (l+2) r^l [2\mu \frac{(l^2 - l - 3)}{l} -$$

$$\omega^2 \rho_0 \frac{(l+3)}{2(2l+5)} r^2 + ...]$$

$$+ \ A_4 (l+1) r^{-l-3} [2\mu(l+2) +$$

$$\omega^2 \rho_0 \frac{l}{(+2l-1)} r^2 + ...]$$

$$+ \ \rho_0 [A_5 r^l + A_6 r^{-l-1}] + ...$$

$$+ \ \rho_0 g_0 [A_1 r^{-l} [1 - \frac{\omega^2 \rho_0}{\mu} \frac{1}{4(-2l+3)} r^2 + ...]$$

$$+ \ A_2 l r^{l-1} [1 - \frac{\omega^2 \rho_0}{\mu} \frac{1}{2(2l+3)} r^2 + ...]$$

$$+ \ A_3 (l+2) r^{l+1} [1 - \frac{\omega^2 \rho_0}{\mu} \frac{1}{4(2l+5)} r^2 + ...]$$

$$+ \ A_4 (-l-1) r^{-l-2} [1 -$$

$$\frac{\omega^2 \rho_0}{\mu} \frac{1}{2(-2l+1)} r^2 + ...]] \qquad (38)$$

$$Q = \mu \frac{A_1}{L} r^{-l-1} [2(l+1)(-l+1) +$$

$$\frac{\omega^2 \rho_0}{\mu} \frac{(-2l+1)(l^2 - 2l + 2)}{2(4l^2 - 8l + 3)} r^2 + ...]$$

$$+ \ \mu \frac{A_2}{L} l r^{l-2} [2(l+1)(-l+1) +$$

$$\frac{\omega^2 \rho_0}{\mu} \frac{l(l+2)}{(2l+3)} r^2 + ...]$$

$$+ \ \mu \frac{A_3}{L} (l+2) r^l [-2l(l+2) +$$

$$\frac{\omega^2 \rho_0}{\mu} \frac{(l^2 + 4l + 5)}{2(2l+5)} r^2 + ...]$$

$$+ \ \mu \frac{A_4}{L} (l+1) r^{-l-3} [2l(l+2) +$$

$$\frac{\omega^2 \rho_0}{\mu} \frac{(-l-1)(l-1)}{(-2l+1)} r^2 + ...]. \qquad (39)$$

Since a power series of the form

$$\sum a_n z^n$$

has

$$R = [\ \lim_{n \to \infty} | \frac{a_{n+1}}{a_n} |]^{-1}$$

as its radius of convergence, we can see that the developments made in Expressions (36)-(39) are convergent for all values of r ($R = \infty$).

The solutions of Eqs. (13)-(14) take the following form

$$\phi = A_5 r^l + A_6 r^{-l-1}, \qquad (40)$$

$$g = A_5 l r^{l-1} - A_6 (l+1) r^{-l-2}. \qquad (41)$$

We denote by A the integration constants for the inner core and by A''' the integration constants for the mantle.

The boundary conditions imposed in $r = 0$ require that $A_1 = 0 = A_4 = A_6$. If we furthermore impose the boundary conditions in $r = r_s$ and require the continuity of U, P, Q, ϕ and g at the inner core-outer core boundary r_{IOB} and at the core-mantle boundary r_{CMB} we obtain a system of 13 equations in the 13 unknowns:

$$\underbrace{A_2, A_3, A_5,}_{inner\ core} \underbrace{A_1'', A_2'', A_3'', A_4'',}_{outer\ core} \underbrace{A_1''', A_2''', A_3''', A_4''', A_5''', A_6'''}_{mantle}.$$
$$(42)$$

The determinant of the 13x13 matrix formed by the co-efficients of the homogeneous equations must be zero for the solutions to be non-trivial. This condition yields the eigenvalue equation for the first approximation of the eigenfrequencies. When the condition on the determinant is satisfied for a specific value of the frequency, the values of the constants (42) can be determined. Since the system of algebraic equations is homogeneous, one of the constants remains undetermined and is used as normalization factor. We choose A_4''' to be this normalization factor and set $A_4''' = 1$.

Numerical Results

All numerical computations were restricted to small frequencies of the order of the subseismic frequencies. In that case the series of solutions are truncated after the term in ω^2 (or ω_0^2).

For the numerical computation of the eigenfrequencies and the corresponding eigenfunctions we assume that the shear modulus takes the following values

$$\mu_{inner\ core} = 1.4 \times 10^8 \frac{kg}{ms^2}, \qquad (43)$$

$$\mu_{outer\ core} = 0, \qquad (44)$$

$$\mu_{mantle} = 1.9 \times 10^8 \frac{kg}{ms^2}, \qquad (45)$$

and that the mean density in the different regions becomes

$$\rho_{inner\ core} = 12.89 \times 10^3 \frac{kg}{m^3}, \qquad (46)$$

$$\rho_{outer\ core} = 10.89 \times 10^3 \frac{kg}{m^3}, \qquad (47)$$

$$\rho_{mantle} = 4.45 \times 10^3 \frac{kg}{m^3}. \qquad (48)$$

For g_0 we assumed

$$g_{0\ inner\ core} = 3 \frac{m}{s^2}, \qquad (49)$$

$$g_{0\ outer\ core} = 8.3 \frac{m}{s^2}, \qquad (50)$$

$$g_{0\ mantle} = 10 \frac{m}{s^2}. \qquad (51)$$

Taking $l = 3$, the requirements of the continuity at $r_{IOB} = 1217 \times 10^3 m$ and at $r_{CMB} = 3485 \times 10^3 m$, together with the boundary conditions at $r_s = 6371 \times 10^3 m$ then lead e.g. to the small value

$$\omega = 0.00017502 s^{-1}, \qquad (52)$$

which lies within the range of values of frequencies for free oscillations of the core. Eventual discontinuities occurring in the corresponding eigenfunctions are caused by numerical inaccuracies.

The variation of the component U of the Lagrangian displacement field shows relatively large amplitudes in the interior layers of the Earth, thus pointing towards free oscillations of the [outer] core. The fact that we do not find any of the known values for normal modes is related to the very simple Earth model as well as to numerical inaccuracies invoked by Mathematica.

RIGIDITY IN THE OUTER CORE IS SMALL BUT NOT NEGLIGIBLE

In this section, we assume that the small rigidity in the outer core gives rise to boundary layers (abbreviated by BL) near the inner core-outer core boundary and near the core-mantle boundary. The Earth is thus divided in 5 different regions : the inner core with a rigidity $\mu_{ic} = 1.4 \times 10^8 kg/(ms^2)$, the mantle with a rigidity $\mu_m = 1.9 \times 10^8 kg/(ms^2)$, the center of the outer core where the rigidity is zero and the two BL of the outer core near r_{IOB} and r_{CMB} where the rigidity is small (e.g. $10^{-2} kg/(ms^2)$) so that in a first approximation μ can be neglected in the center of the outer core, but the particular solutions removed by that approximation may become dominant in the boundary layers.

In order to study the behaviour of the solutions in the boundary layers, the full system of Eqs. (9)-(14) is used. After transformation of the four first-order differential Eqs. (9)-(12) into the fourth-order differential equation given by Eq. (35), we require that this equation must contain the fourth-order derivative d^4U/dr^4 in order to be of the fourth order and thus to be as rich as possible in solutions. To this end, we examine under which condition the term involving the fourth-order derivative is of the same order of magnitude in the small parameter μ as at least one term of the right-hand member after the introduction of the "stretched" coordinate

$$r^* = \frac{r}{\delta(\mu)} \qquad (53)$$

and the boundary-layer expansions

$$\omega^2 = \sum_{k=0}^{\infty} \nu_k(\mu) \omega^{2(k)}, \qquad (54)$$

$$U^{BL} = \sum_{k=0}^{\infty} \nu_k(\mu) U^{(BL,k)} \qquad (55)$$

(mutatis mutandis for V^{BL}, P^{BL}, Q^{BL}, ϕ^{BL}, g^{BL}).

In Definition (53), the unknown function $\delta(\mu)$ tends to zero with μ. In the boundary-layer expansions (54) and (55), the functions $\nu_k(\mu)$ form an asymptotic sequence that is also to be determined. We then come to the conclusion that, since we may set $\nu_0(\mu) = 1$ without any loss of generality, the dominant term of the left-hand member of Eq. (35) is of the same order of magnitude as the smallest of the dominant terms of the right-hand member of that equation if we take

$$\delta(\mu) = \mu^{1/2}. \qquad (56)$$

Definition (50) of the stretched coordinate r^* then becomes

$$r^* = \frac{r}{\mu^{1/2}}. \qquad (57)$$

The effective thickness of the boundary layer (d_{BL}) is of the order of the value of $\delta(\mu)$ [see Degryse et al., 1992]. More exactly, the highest-order terms of Eq. (35) lead to

$$d_{BL} = (\frac{\mu}{\omega^2 \rho_0})^{1/2}, \qquad (58)$$

so that the value of the boundary-layer thickness is essentially based on the small parameter, being the shear modulus.

By introducing the boundary-layer Expansions (54) and (55) into Eqs. (9)-(14) and considering the powers of r that appear in the coefficients of these equations, we are led to set

$$\nu_k(\mu) = \mu^{k/2}. \qquad (59)$$

We refer to the main part of the outer core outside the boundary layer as the outer region, as is usual in boundary-layer theory. In the outer region, we must construct expansions of the solutions that can be matched with the boundary-layer expansions. We thus distinguish between boundary-layer expansions in the outer core, outer expansions in the outer core, expansions in the inner core and expansions in the mantle.

Since Expansion (54) for the frequency must be used in the whole mass of the Earth, we introduce expansions of the dependent variables of the same type in the outer region of the outer core as well as in the mantle and the inner core. This leads to different systems of differential equations in the subsequent orders of approximation.

Zeroth-Order Approximation

In the mantle and inner core, the zeroth-order differential equations take the following form

$$\frac{dU^{(0)}}{dr} = -\frac{2}{r}U^{(0)} - \frac{L}{r}V^{(0)}, \qquad (60)$$

$$\frac{dV^{(0)}}{dr} = \frac{2L}{r}U^{(0)} + \frac{1}{r}V^{(0)} + \frac{1}{\mu^*}Q^{(0)}, \qquad (61)$$

$$\frac{dP^{(0)}}{dr} = -\frac{L}{r}Q^{(0)} + \frac{12\mu^*}{r^2}U^{(0)} + \frac{6L\mu^*}{r^2}V^{(0)} + \rho_0 g^{(0)}$$
$$- \frac{4}{r}\rho_0 g_0 U^{(0)} - \frac{L}{r}\rho_0 g_0 V^{(0)} + 4\pi G\rho_0^2 U^{(0)}$$
$$- \omega_0^2 \rho_0 U^{(0)}, \qquad (62)$$

$$\frac{dQ^{(0)}}{dr} = -\frac{3}{r}Q^{(0)} + \frac{2L}{r}P^{(0)} + \frac{12L\mu^*}{r^2}U^{(0)}$$
$$+ \frac{6L^2\mu^*}{r^2}V^{(0)} + \frac{2L'^2\mu^*}{r^2}V^{(0)}$$
$$- \frac{2L\rho_0}{r}\phi^{(0)} - \frac{2L}{r}\rho_0 g_0 U^{(0)} - \omega_0^2 \rho_0 V^{(0)}, \qquad (63)$$

$$\frac{d\phi^{(0)}}{dr} = g^{(0)}, \qquad (64)$$

$$\frac{dg^{(0)}}{dr} = -\frac{2}{r}g^{(0)} + \frac{2L^2}{r^2}\phi^{(0)}. \qquad (65)$$

Here, the superscript 0 points at the first term in the series expansions of the variables, ω_0 is the zeroth-order approximation of the frequency and μ^* stands for the values of the rigidity in the inner core and in the mantle. We thus make a distinction between μ^* and the small expansion parameter μ.

Since the equations (60)-(65) have the same form as Eqs. (9)-(14), the solutions are given by Expressions (36)-(41) where we remind that $A_1 = A_4 = A_6 = 0$ because of the boundary conditions, and where we substitute ω^2 by ω_0^2 and μ by μ^*.

In the outer region (superscript o) of the outer core the zeroth-order (superscript 0) differential equations are (see Eqs. (23)-(28)).

$$\frac{dU^{(o;0)}}{dr} = -\frac{2}{r}U^{(o;0)} - \frac{L}{r}V^{(o;0)} \qquad (66)$$

$$Q^{(o;0)} = 0 \qquad (67)$$

$$\frac{dP^{(o;0)}}{dr} = \rho_0 g^{(o;0)} - \frac{4}{r}\rho_0 g_0 U^{(o;0)} - \frac{L}{r}\rho_0 g_0 V^{(o;0)}$$
$$+ 4\pi G\rho_0^2 U^{(o;0)} - \omega_0^2 \rho_0 U^{(o;0)} \qquad (68)$$

$$0 = \frac{2L}{r}P^{(o;0)} - \frac{2L\rho_0}{r}\phi^{(o;0)} - \frac{2L}{r}\rho_0 g_0 U^{(o;0)}$$
$$- \omega_0^2 \rho_0 V^{(o;0)} \qquad (69)$$

$$\frac{d\phi^{(o;0)}}{dr} = g^{(o;0)} \qquad (70)$$

$$\frac{dg^{(o;0)}}{dr} = -\frac{2}{r}g^{(o;0)} + \frac{2L^2}{r^2}\phi^{(o;0)}. \qquad (71)$$

The solutions of this fourth-order system take the form

$$U^{(o;0)} = A_3'' r^{l-1} + A_4'' r^{-l-2} \qquad (72)$$

$$V^{(o;0)} = -A_3'' \frac{(l+1)}{L} r^{l-1} + A_4'' \frac{l}{L} r^{-l-2} \qquad (73)$$

$$P^{(o;0)} = A_1'' \rho_0 r^l + A_2'' \rho_0 r^{-l-1} - A_3'' \frac{\omega_0^2 \rho_0}{l} r^l$$
$$+ A_4'' \frac{\omega_0^2 \rho_0}{(l+1)} r^{-l-1} + A_3'' \rho_0 g_0 r^{l-1}$$

$$+ \quad A_4'' \rho_0 g_0 r^{-l-2} \tag{74}$$

$$Q^{(o;0)} = 0 \tag{75}$$

$$\phi^{(o;0)} = A_1'' r^l + A_2'' r^{-l-1} \tag{76}$$

$$g^{(o;0)} = A_1'' l r^{l-1} - A_2''(l+1) r^{-l-2}. \tag{77}$$

In the boundary layers near r_{IOB} and r_{CMB}, also called the inner regions (superscript i) of the outer core, the introduction of the stretched variable given by Def. (57) leads to the following zeroth-order (superscript 0) differential equations :

$$\frac{d^4 U^{(i;0)}}{dr^{*4}} \quad + \quad \frac{8}{r^*}\frac{d^3 U^{(i;0)}}{dr^{*3}} + \frac{(10 - 2L^2 - 2L'^2)}{r^{*2}}\frac{d^2 U^{(i;0)}}{dr^{*2}}$$

$$+ \quad \omega_0^2 \rho_0 \frac{d^2 U^{(i;0)}}{dr^{*2}} + \frac{(-4 - 4L^2 - 4L'^2)}{r^{*3}}\frac{d U^{(i;0)}}{dr^*}$$

$$+ \quad \frac{4}{r^*}\omega_0^2 \rho_0 \frac{d U^{(i;0)}}{dr^*}$$

$$+ \quad \frac{[4 - 8L^2 + 4L'^2 + (2L^2)^2]}{r^{*4}} U^{(i;0)}$$

$$+ \quad \frac{(2 - 2L^2)}{r^{*2}}\omega_0^2 \rho_0 U^{(i;0)} = 0, \tag{78}$$

$$V^{(i;0)} = -\frac{r^*}{L}[\frac{d U^{(i;0)}}{dr^*} + \frac{2}{r^*}U^{(i;0)}], \tag{79}$$

$$P^{(i;0)} = \rho_0 g_0 U^{(i;0)}, \tag{80}$$

$$Q^{(i;0)} = 0, \tag{81}$$

$$\phi^{(i;0)} = 0, \tag{82}$$

$$\frac{d^2 g^{(i;0)}}{dr^{*2}} + \frac{4}{r^*}\frac{d g^{(i;0)}}{dr^*} + \frac{(2 - 2L^2)}{r^{*2}} g^{(i;0)} = 0. \tag{83}$$

The zeroth-order boundary-layer solutions then become

$$U^{(i;0)} = A_1' r^{*(-l)}[1 - \omega_0^2 \rho_0 \frac{1}{4(-2l+3)} r^{*2} + ...]$$

$$+ \quad A_2' l r^{*(l-1)}[1 - \omega_0^2 \rho_0 \frac{1}{2(2l+3)} r^{*2} + ...]$$

$$+ \quad A_3'(l+2) r^{*(l+1)}[1 - \omega_0^2 \rho_0 \frac{1}{4(2l+5)} r^{*2} + ...]$$

$$+ \quad A_4'(-l-1) r^{*(-l-2)}[1 -$$
$$\omega_0^2 \rho_0 \frac{1}{2(-2l+1)} r^{*2} + ...], \tag{84}$$

$$V^{(i;0)} = -\frac{A_1'}{L} r^{*(-l)}[(-l+2) -$$
$$\omega_0^2 \rho_0 \frac{(-l+4)}{4(-2l+3)} r^{*2} + ...]$$

$$- \quad \frac{A_2'}{L} l r^{*(l-1)}[(l+1) -$$
$$\omega_0^2 \rho_0 \frac{(l+3)}{2(2l+3)} r^{*2} + ...]$$

$$- \quad \frac{A_3'}{L}(l+2) r^{*(l+1)}[(l+3) -$$

$$\omega_0^2 \rho_0 \frac{(l+5)}{4(2l+5)} r^{*2} + ...]$$

$$- \quad \frac{A_4'}{L}(-l-1) r^{*(-l-2)}[-l -$$
$$\omega_0^2 \rho_0 \frac{(-l+2)}{2(-2l+1)} r^{*2} + ...], \tag{85}$$

$$P^{(i;0)} = \rho_0 g_0[A_1' r^{*(-l)}[1 -$$
$$\omega_0^2 \rho_0 \frac{1}{4(-2l+3)} r^{*2} + ...]$$

$$+ \quad A_2' l r^{*(l-1)}[1 - \omega_0^2 \rho_0 \frac{1}{2(2l+3)} r^{*2} + ...]$$

$$+ \quad A_3'(l+2) r^{*(l+1)}[1 -$$
$$\omega_0^2 \rho_0 \frac{1}{4(2l+5)} r^{*2} + ...]$$

$$+ \quad A_4'(-l-1) r^{*(-l-2)}[1 -$$
$$\omega_0^2 \rho_0 \frac{1}{2(-2l+1)} r^{*2} + ...]], \tag{86}$$

$$Q^{(i;0)} = 0, \tag{87}$$

$$\phi^{(i;0)} = 0, \tag{88}$$

$$g^{(i;0)} = A_5' r^{*(l-1)} + A_6' r^{*(-l-2)}. \tag{89}$$

In order to match, at order μ^0, the boundary-layer solutions (84)-(89) and the outer solutions (72)-(77) valid in the outer core in their common domain of validity, we introduce the matching variable [see Van Dyke, 1964; Kevorkian and Cole, 1981]

$$r_m = \frac{r}{m(\mu)}, \tag{90}$$

were $m(\mu)$ is a class of functions in a range

$$m_1(\mu) \ll m(\mu) \ll m_2(\mu) \tag{91}$$

with the two as yet undetermined functions $m_1(\mu)$ and $m_2(\mu)$. We express the solutions in terms of this matching variable and we adopt as a rule for the matching of any of the functions U, V, P, Q, ϕ, and g to order $\mu^{J/2}$ the requirement that, for example,

$$\lim_{\mu \to 0, r_m \text{ fixed}} \frac{1}{\mu^{(J/2)}} \sum_{k=0}^{J} \mu^{k/2}[U^{(i;k)}(\frac{m r_m}{\mu^{1/2}}) - U^{(o;k)}(m r_m)] = 0 \tag{92}$$

for all m satisfying $m_1 \ll m \ll m_2$. In Expression (92), J and k are integers.

When we apply Condition (92) for the matching of the expansions for U to order μ^0, we can see that terms in $A_2' m^{l-1} r_m^{l-1} \mu^{-(l-1)/2}$, $A_3' m^{l+1} r_m^{l+1} \mu^{-(l+1)/2}$, as well as those in $A_4'' m^{-(l+2)} r_m^{-(l+2)}$ become singular since $m/\mu^{1/2} \to \infty$ and $m \to 0$. Therefore, we must set $A_2' = A_3' = A_5' = A_2'' = A_4'' = 0$. We also have to require $l \geq 3$. The term of $U^{(o,0)}$ that has no counterpart in the boundary-layer expansions is of the order m^{l-1}. The contribution of this term to the matching conditions will vanish if $m \ll 1$. The contribution of the unmatched terms of $U^{(i,0)}$ will vanish under Condition (91) imposed on m if $m \gg \mu^{1/2}$. Similar conclusions hold true for the variables V, P, Q, ϕ and g.

The matching of the boundary-layer and outer expansions to order μ^0 then is possible with the non-empty overlap domain determined by

$$\mu^{1/2} \ll m \ll 1. \qquad (93)$$

Composite expansions that are uniformly valid in the outer core to a certain order in the expansion parameter $\mu^{1/2}$ are given by the sum of the boundary-layer expansions and the outer expansions valid in the outer core minus the terms that are common to both expansions [Van Dyke, 1964]. Since our model contains two boundary layers, one near the inner core-outer core boundary and one near the core-mantle boundary, we construct a uniformly valid solution coming from r_{IOB}

$$
\begin{aligned}
U^{(oc1;0)} &= A_1' \mu^{1/2} r^{-l} [1 - \omega_0^2 \rho_0 \frac{1}{4(-2l+3)} \mu^{-1} r^2 ...] \\
&+ A_4' \mu^{(l+2)/2} r^{-l-2}(-l-1)[1 - \\
&\omega_0^2 \rho_0 \frac{1}{2(-2l+1)} \mu^{-1} r^2 ...] + A_3'' r^{l-1}, \qquad (94)
\end{aligned}
$$

$$
\begin{aligned}
V^{(oc1;0)} &= -\frac{A_1'}{L} \mu^{1/2} r^{-l} [((-l+2) - \\
&\omega_0^2 \rho_0 \frac{(-l+4)}{4(-2l+3)} \mu^{-1} r^2 ...] \\
&- \frac{A_4'}{L} \mu^{(l+2)/2} r^{-l-2}(-l-1)[-l - \\
&\omega_0^2 \rho_0 \frac{(-l+2)}{2(-2l+1)} \mu^{-1} r^2 ...] \\
&- A_3'' \frac{(l+1)}{L} r^{l-1}, \qquad (95)
\end{aligned}
$$

$$
\begin{aligned}
P^{(oc1;0)} &= A_1'' \rho_0 r^l - A_3'' \omega_0^2 \frac{\rho_0}{l} r^l + \rho_0 g_0 A_3'' r^{l-1} \\
&+ \rho_0 g_0 \mu^{1/2} A_1' r^{-l} [1 - \omega_0^2 \rho_0 \frac{1}{4(-2l+3)} \mu^{-1} r^2 ...] \\
&+ \rho_0 g_0 (-l-1) A_4' \mu^{(l+2)/2} r^{-l-2} [1 - \\
&\omega_0^2 \rho_0 \frac{1}{2(-2l+1)} \mu^{-1} r^2 ...], \qquad (96)
\end{aligned}
$$

$$Q^{(oc1;0)} = 0, \qquad (97)$$

$$\phi^{(oc1;0)} = A_1'' r^l, \qquad (98)$$

$$g^{(oc1;0)} = A_6' \mu^{(l+2)/2} r^{-l-2} + A_1'' l r^{l-1} \qquad (99)$$

and a uniformly valid solution coming from r_{CMB}

$$
\begin{aligned}
U^{(oc2;0)} &= A_1'^* \mu^{1/2} r^{-l} [1 - \omega_0^2 \rho_0 \frac{1}{4(-2l+3)} \mu^{-1} r^2 ...] \\
&+ A_4'^* \mu^{(l+2)/2} r^{-l-2}(-l-1)[1 - \\
&\omega_0^2 \rho_0 \frac{1}{2(-2l+1)} \mu^{-1} r^2 ...] + A_3'' r^{l-1}, \qquad (100)
\end{aligned}
$$

$$
V^{(oc2;0)} = -\frac{A_1'^*}{L} \mu^{1/2} r^{-l} [((-l+2) -
$$

$$
\begin{aligned}
&\omega_0^2 \rho_0 \frac{(-l+4)}{4(-2l+3)} \mu^{-1} r^2 ...] \\
&- \frac{A_4'^*}{L} \mu^{(l+2)/2} r^{-l-2}(-l-1)[-l - \\
&\omega_0^2 \rho_0 \frac{(-l+2)}{2(-2l+1)} \mu^{-1} r^2 ...] \\
&- A_3'' \frac{(l+1)}{L} r^{l-1}, \qquad (101)
\end{aligned}
$$

$$
\begin{aligned}
P^{(oc2;0)} &= A_1'' \rho_0 r^l - A_3'' \omega_0^2 \frac{\rho_0}{l} r^l + \rho_0 g_0 A_3'' r^{l-1} \\
&+ \rho_0 g_0 \mu^{1/2} A_1'^* r^{-l} [1 - \omega_0^2 \rho_0 \frac{1}{4(-2l+3)} \mu^{-1} r^2 ...] \\
&+ \rho_0 g_0 (-l-1) A_4'^* \mu^{(l+2)/2} r^{-l-2} [1 - \\
&\omega_0^2 \rho_0 \frac{1}{2(-2l+1)} \mu^{-1} r^2 ...], \qquad (102)
\end{aligned}
$$

$$Q^{(oc2;0)} = 0, \qquad (103)$$

$$\phi^{(oc2;0)} = A_1'' r^l, \qquad (104)$$

$$g^{(oc2;0)} = A_6'^* \mu^{(l+2)/2} r^{-l-2} + A_1'' l r^{l-1}. \qquad (105)$$

Both uniformly valid solutions must be joined for U, P, Q, ϕ, g at any point of their common domain of validity in the outer core. The continuity of $U^{(oc1,0)}$ and $U^{(oc2,0)}$ leads to an extra condition in the unknowns A_1', $A_1'^*$, A_4', $A_4'^*$, whereas the continuity of $g^{(oc1,0)}$ and $g^{(oc2,0)}$ leads to

$$A_6' = A_6'^*. \qquad (106)$$

Afterwards, the composite expansions for U, V, Q, P, ϕ and g that are uniformly valid in the outer core must be joined to the corresponding expansions valid in the mantle and the inner core. The requirements of continuity at r_{IOB} and r_{CMB}, together with the extra condition of continuity of $U^{(oc1,0)}$ and $U^{(oc2,0)}$ and with the 3 boundary conditions at the surface lead to a system of 16 algebraic, linear and homogeneous equations in the 16 undetermined constants :

$$
\begin{aligned}
&A_2, A_3, A_5 : inner\ core \\
&A_1', A_4', A_6' : boundary\ layer\ 1\ in\ the\ outer\ core \\
&A_1'', A_3'' : outer\ region\ outer\ core \\
&A_1'^*, A_4'^* : boundary\ layer\ 2\ in\ the\ outer\ core \\
&A_1''', A_2''', A_3''', A_4''', A_5''', A_6''' : mantle. \qquad (107)
\end{aligned}
$$

As in the case $\mu = 0$, the determinant of the 16 × 16 matrix formed by the coefficients of the homogeneous equations must be zero for the solutions to be non-trivial. This condition yields the eigenvalue equation for the zeroth-order approximations ω_0 of the eigenfrequencies.

Numerical Results. The shear modulus of the inner core, respectively mantle, is given by Expression (43), respectively (45). For the mean density, we again take the values (46)-(48). For the outer core, we assume $\mu = 10^{-2} kg/(ms^2)$.

The continuity requirements together with the boundary conditions lead to the determination of the zeroth-order values of the eigenfrequency. One such possible value (for $l = 3$) is

$$\omega_0 = 0.0000984559 s^{-1}, \tag{108}$$

representing a free core oscillation frequency. Again the corresponding eigenfunctions display a rather peculiar behaviour which is probably caused by numerical inaccuracies.

At this moment, we can obtain a numerical value for the boundary-layer thickness defined by Eq. (58). For $\mu = 10^{-2} kg/(ms^2)$, ρ_0 given by Eq. (47) en ω_0 by Eq. (108) we get

$$d_{BL} = 9.74 m. \tag{109}$$

Higher-Order Approximations

Approximation to Order $\mu^{1/2}$. In the mantle and inner core, the equations of order $\mu^{1/2}$ take the form

$$\frac{dU^{(1/2)}}{dr} = -\frac{2}{r}U^{(1/2)} - \frac{L}{r}V^{(1/2)}, \tag{110}$$

$$\frac{dV^{(1/2)}}{dr} = \frac{2L}{r}U^{(1/2)} + \frac{1}{r}V^{(1/2)} + \frac{1}{\mu^*}Q^{(1/2)}, \tag{111}$$

$$\frac{dP^{(1/2)}}{dr} = -\frac{L}{r}Q^{(1/2)} + \frac{12\mu^*}{r^2}U^{(1/2)} + \frac{6L\mu^*}{r^2}V^{(1/2)}$$
$$+ \rho_0 g^{(1/2)} - \frac{4}{r}\rho_0 g_0 U^{(1/2)} - \frac{L}{r}\rho_0 g_0 V^{(1/2)}$$
$$+ 4\pi G \rho_0^2 U^{(1/2)} - \omega_0^2 \rho_0 U^{(1/2)} - \omega_1^2 \rho_0 U^{(0)}, \tag{112}$$

$$\frac{dQ^{(1/2)}}{dr} = -\frac{3}{r}Q^{(1/2)} + \frac{2L}{r}P^{(1/2)} + \frac{12L\mu^*}{r^2}U^{(1/2)}$$
$$+ \frac{6L^2\mu^*}{r^2}V^{(1/2)} + \frac{2L'^2\mu^*}{r^2}V^{(1/2)} - \frac{2L\rho_0}{r}\phi^{(1/2)}$$
$$- \frac{2L}{r}\rho_0 g_0 U^{(1/2)} - \omega_0^2 \rho_0 V^{(1/2)}$$
$$- \omega_1^2 \rho_0 V^{(0)}, \tag{113}$$

$$\frac{d\phi^{(1/2)}}{dr} = g^{(1/2)}, \tag{114}$$

$$\frac{dg^{(1/2)}}{dr} = -\frac{2}{r}g^{(1/2)} + \frac{2L^2}{r^2}\phi^{(1/2)}. \tag{115}$$

The homogeneous system is considered as the part containing variables characterized by the superscript 1/2.

Any general solution of this sixth-order system of differential equations can be expressed as a linear combination of six linearly independent solutions of the homogeneous system and a particular solution of the non-homogeneous system. Since the homogeneous system is similar to the zeroth-order system given by Eqs. (60)-(65), its solution reduces to a sum of multiples (constants B_i instead of A_i) of the six linearly independent zeroth-order solutions. In the inner core, a particular solution of the non-homogeneous system takes the form

$$U_p^{(i;1/2)} = A_2 \frac{\omega_1^2 \rho_0}{\mu^*} \frac{\omega_0^2 \rho_0}{\mu^*} \frac{l}{8(3+2l)(5+2l)} r^{l+3}$$

$$- A_3 \frac{\omega_1^2 \rho_0}{\mu^*} \frac{(l+2)}{4(5+2l)} r^{l+3}$$
$$+ A_3 \frac{\omega_1^2 \rho_0}{\mu^*} \frac{\omega_0^2 \rho_0}{\mu^*} \frac{(l+2)}{12(5+2l)(7+2l)} r^{l+5}$$
$$+ \cdots, \tag{116}$$

$$V_p^{(i;1/2)} = -A_2 \frac{\omega_1^2 \rho_0}{\mu^*} \frac{\omega_0^2 \rho_0}{\mu^*} \frac{l(l+5)}{8(3+2l)(5+2l)L} r^{l+3}$$
$$+ A_3 \frac{\omega_1^2 \rho_0}{\mu^*} \frac{(l+2)(l+5)}{4(5+2l)L} r^{l+3}$$
$$- A_3 \frac{\omega_1^2 \rho_0}{\mu^*} \frac{\omega_0^2 \rho_0}{\mu^*} \frac{(l+2)(l+7)}{12(5+2l)(7+2l)L} r^{l+5}$$
$$+ \cdots, \tag{117}$$

$$P_p^{(i;1/2)} = -A_2 \omega_1^2 \rho_0 r^l$$
$$+ A_2 \omega_1^2 \rho_0 \frac{\omega_0^2 \rho_0}{\mu^*} \frac{2l(l+3)}{8(3+2l)(5+2l)} r^{l+2}$$
$$- A_3 \omega_1^2 \rho_0 \frac{(l+2)(l+3)}{2(5+2l)} r^{l+2}$$
$$+ A_3 \omega_1^2 \rho_0 \frac{\omega_0^2 \rho_0}{\mu^*} \frac{(l+2)(l+5)}{6(5+2l)(7+2l)} r^{l+4}$$
$$+ \rho_0 g_0 [A_2 \frac{\omega_1^2 \rho_0}{\mu^*} \frac{\omega_0^2 \rho_0}{\mu^*} \frac{l}{8(3+2l)(5+2l)} r^{l+3}$$
$$- A_3 \frac{\omega_1^2 \rho_0}{\mu^*} \frac{(l+2)}{4(5+2l)} r^{l+3}$$
$$+ A_3 \frac{\omega_1^2 \rho_0}{\mu^*} \frac{\omega_0^2 \rho_0}{\mu^*} \frac{(l+2)}{12(5+2l)(7+2l)} r^{l+5}]$$
$$+ \cdots, \tag{118}$$

$$Q_p^{(i;1/2)} = -A_2 \omega_1^2 \rho_0 \frac{\omega_0^2 \rho_0}{\mu^*} \frac{2l(5+4l+l^2)}{8(3+2l)(5+2l)L} r^{l+2}$$
$$+ A_3 \omega_1^2 \rho_0 \frac{(l+2)(5+4l+l^2)}{2(5+2l)L} r^{l+2}$$
$$- A_3 \omega_1^2 \rho_0 \frac{\omega_0^2 \rho_0}{\mu^*} \frac{(l+2)(14+6l+l^2)}{6(5+2l)(7+2l)L} r^{l+4}$$
$$+ \cdots. \tag{119}$$

In the mantle, a particular solution of the non-homogeneous system (110)-(115) is

$$U_p^{(m;1/2)} = A_1''' \frac{\omega_1^2 \rho_0}{\mu^*} \frac{1}{4(-3+2l)} r^{-l+2}$$
$$+ A_1''' \frac{\omega_1^2 \rho_0}{\mu^*} \frac{\omega_0^2 \rho_0}{\mu^*} \frac{1}{12(-5+2l)(-3+2l)} r^{-l+4}$$
$$+ A_2''' \frac{\omega_1^2 \rho_0}{\mu^*} \frac{\omega_0^2 \rho_0}{\mu^*} \frac{l}{8(3+2l)(5+2l)} r^{l+3}$$
$$- A_3''' \frac{\omega_1^2 \rho_0}{\mu^*} \frac{(l+2)}{4(5+2l)} r^{l+3}$$
$$+ A_3''' \frac{\omega_1^2 \rho_0}{\mu^*} \frac{\omega_0^2 \rho_0}{\mu^*} \frac{(l+2)}{12(5+2l)(7+2l)} r^{l+5}$$
$$- A_4''' \frac{\omega_1^2 \rho_0}{\mu^*} \frac{\omega_0^2 \rho_0}{\mu^*} \frac{(l+1)}{8(-1+2l)(-3+2l)} r^{-l+2}$$

$$+ \quad ..., \tag{120}$$

$$V_p^{(m;1/2)} = -A_1''' \frac{\omega_1^2 \rho_0}{\mu^*} \frac{(-l+4)}{4L(-3+2l)} r^{-l+2}$$

$$- A_1''' \frac{\omega_1^2 \rho_0}{\mu^*} \frac{\omega_0^2 \rho_0}{\mu^*} \frac{(-l+6)}{12L(-5+2l)(-3+2l)} r^{-l+4}$$

$$- A_2''' \frac{\omega_1^2 \rho_0}{\mu^*} \frac{\omega_0^2 \rho_0}{\mu^*} \frac{l(l+5)}{8L(3+2l)(5+2l)} r^{l+3}$$

$$+ A_3''' \frac{\omega_1^2 \rho_0}{\mu^*} \frac{(l+2)(l+5)}{4L(5+2l)} r^{l+3}$$

$$- A_3''' \frac{\omega_1^2 \rho_0}{\mu^*} \frac{\omega_0^2 \rho_0}{\mu^*} \frac{(l+2)(l+7)}{12L(5+2l)(7+2l)} r^{l+5}$$

$$+ A_4''' \frac{\omega_1^2 \rho_0}{\mu^*} \frac{\omega_0^2 \rho_0}{\mu^*} \frac{(l+1)(-l+4)}{8L(-1+2l)(-3+2l)} r^{-l+2}$$

$$+ \quad ..., \tag{121}$$

$$P_p^{(m;1/2)} = -A_1''' \omega_1^2 \rho_0 \frac{(-2+l)}{2(-3+2l)} r^{-l+1}$$

$$- A_1''' \omega_1^2 \rho_0 \frac{\omega_0^2 \rho_0}{\mu^*} \frac{(-4+l)}{6(-5+2l)(-3+2l)} r^{-l+3}$$

$$- A_2''' \omega_1^2 \rho_0 r^l$$

$$+ A_2''' \omega_1^2 \rho_0 \frac{\omega_0^2 \rho_0}{\mu^*} \frac{2l(l+3)}{8(3+2l)(5+2l)} r^{l+2}$$

$$- A_3''' \omega_1^2 \rho_0 \frac{(l+2)(l+3)}{2(5+2l)} r^{l+2}$$

$$+ A_3''' \omega_1^2 \rho_0 \frac{\omega_0^2 \rho_0}{\mu^*} \frac{(l+2)(l+5)}{6(5+2l)(7+2l)} r^{l+4}$$

$$- A_4''' \omega_1^2 \rho_0 r^{-l-1}$$

$$+ A_4''' \omega_1^2 \rho_0 \frac{\omega_0^2 \rho_0}{\mu^*} \frac{(l+1)(-2+l)}{4(-1+2l)(-3+2l)} r^{-l+1}$$

$$+ \rho_0 g_0 [A_1''' \frac{\omega_1^2 \rho_0}{\mu^*} \frac{1}{4(-3+2l)} r^{-l+2}$$

$$+ A_1''' \frac{\omega_1^2 \rho_0}{\mu^*} \frac{\omega_0^2 \rho_0}{\mu^*} \frac{1}{12(-5+2l)(-3+2l)} r^{-l+4}$$

$$+ A_2''' \frac{\omega_1^2 \rho_0}{\mu^*} \frac{\omega_0^2 \rho_0}{\mu^*} \frac{l}{8(3+2l)(5+2l)} r^{l+3}$$

$$- A_3''' \frac{\omega_1^2 \rho_0}{\mu^*} \frac{(l+2)}{4(5+2l)} r^{l+3}$$

$$+ A_3''' \frac{\omega_1^2 \rho_0}{\mu^*} \frac{\omega_0^2 \rho_0}{\mu^*} \frac{(l+2)}{12(5+2l)(7+2l)} r^{l+5}$$

$$- A_4''' \frac{\omega_1^2 \rho_0}{\mu^*} \frac{\omega_0^2 \rho_0}{\mu^*} \frac{(l+1)}{8(-1+2l)(-3+2l)} r^{-l+2}]$$

$$+ \quad ..., \tag{122}$$

$$Q_p^{(m;1/2)} = -A_1''' \omega_1^2 \rho_0 \frac{(-l+4)(-l+1)}{4L(-3+2l)} r^{-l+1}$$

$$- A_1''' \omega_1^2 \rho_0 \frac{\omega_0^2 \rho_0}{\mu^*} \frac{(-l+6)(-l+3)}{12L(-5+2l)(-3+2l)} r^{-l+3}$$

$$- A_2''' \omega_1^2 \rho_0 \frac{\omega_0^2 \rho_0}{\mu^*} \frac{2l(5+4l+l^2)}{8L(3+2l)(5+2l)} r^{l+2}$$

$$+ A_3''' \omega_1^2 \rho_0 \frac{(l+2)(5+4l+l^2)}{2L(5+2l)} r^{l+2}$$

$$- A_3''' \omega_1^2 \rho_0 \frac{\omega_0^2 \rho_0}{\mu^*} \frac{(l+2)(14+6l+l^2)}{6L(5+2l)(7+2l)} r^{l+4}$$

$$+ A_4''' \omega_1^2 \rho_0 \frac{\omega_0^2 \rho_0}{\mu^*} \frac{(l+1)(2-2l+l^2)}{4L(-1+2l)(-3+2l)} r^{-l+1}$$

$$+ \quad .... \tag{123}$$

The boundary conditions (17)-(22) imposed on the general solution of Eqs. (110)-(115) require that $B_1 = B_4 = B_6 = 0$.

In the outer region of the outer core, the equations of order $\mu^{1/2}$ become

$$\frac{dU^{(o;1/2)}}{dr} = -\frac{2}{r} U^{(o;1/2)} - \frac{L}{r} V^{(o;1/2)}, \tag{124}$$

$$Q^{(o;1/2)} = 0, \tag{125}$$

$$\frac{dP^{(o;1/2)}}{dr} = \rho_0 g^{(o;1/2)} - \frac{4}{r} \rho_0 g_0 U^{(o;1/2)} - \frac{L}{r} \rho_0 g_0 V^{(o;1/2)}$$

$$+ 4\pi G \rho_0^2 U^{(o;1/2)} - \omega_0^2 \rho_0 U^{(o;1/2)}$$

$$- \omega_1^2 \rho_0 U^{(o;0)}, \tag{126}$$

$$0 = \frac{2L}{r} P^{(o;1/2)} - \frac{2L\rho_0}{r} \phi^{(o;1/2)}$$

$$- \frac{2L}{r} \rho_0 g_0 U^{(o;1/2)} - \omega_0^2 \rho_0 V^{(o;1/2)}$$

$$- \omega_1^2 \rho_0 V^{(o;0)}, \tag{127}$$

$$\frac{d\phi^{(o;1/2)}}{dr} = g^{(o;1/2)}, \tag{128}$$

$$\frac{dg^{(o;1/2)}}{dr} = -\frac{2}{r} g^{(o;1/2)} + \frac{2L^2}{r^2} \phi^{(o;1/2)}. \tag{129}$$

A general solution of the fourth-order system of Eqs. (124)-(129) is composed of a linear combination of four linearly independent solutions of the homogeneous system, which are given by a multiple of Solutions (72)-(77) (B_i'' instead of A_i''), and a particular solution of the non-homogeneous system. A particular solution of the non-homogeneous system takes the form

$$U_p^{(o;1/2)} = -A_3'' \frac{\omega_1^2}{\omega_0^2} r^{l-1}, \tag{130}$$

$$V_p^{(o;1/2)} = A_3'' \frac{\omega_1^2}{\omega_0^2} \frac{(l+1)}{L} r^{l-1}, \tag{131}$$

$$P_p^{(o;1/2)} = -\rho_0 g_0 A_3'' \frac{\omega_1^2}{\omega_0^2} r^{l-1}, \tag{132}$$

$$Q_p^{(o;1/2)} = 0, \tag{133}$$

$$\phi_p^{(o;1/2)} = 0, \tag{134}$$

$$g_p^{(o;1/2)} = 0. \tag{135}$$

The differential equations for the boundary layers of the outer core at order $\mu^{1/2}$ are

$$\frac{d^4 U^{(i;1/2)}}{dr^{*4}} + \frac{8}{r^*}\frac{d^3 U^{(i;1/2)}}{dr^{*3}}$$
$$+ \frac{(10 - 2L^2 - 2L'^2)}{r^{*2}}\frac{d^2 U^{(i;1/2)}}{dr^{*2}}$$
$$+ \omega_0^2 \rho_0 \frac{d^2 U^{(i;1/2)}}{dr^{*2}}$$
$$+ \frac{(-4 - 4L^2 - 4L'^2)}{r^{*3}}\frac{dU^{(i;1/2)}}{dr^*}$$
$$+ \frac{4}{r^*}\omega_0^2 \rho_0 \frac{dU^{(i;1/2)}}{dr^*}$$
$$+ \frac{[4 - 8L^2 + 4L'^2 + (2L^2)^2]}{r^{*4}}U^{(i;1/2)}$$
$$+ \frac{(2 - 2L^2)}{r^{*2}}\omega_0^2 \rho_0 U^{(i;1/2)}$$
$$= -\omega_1^2 \rho_0 \frac{d^2 U^{(i;0)}}{dr^{*2}} - \frac{4}{r^*}\omega_1^2 \rho_0 \frac{dU^{(i;0)}}{dr^*}$$
$$- \frac{(2 - 2L^2)}{r^{*2}}\omega_1^2 \rho_0 U^{(i;0)}, \tag{136}$$

$$V^{(i;1/2)} = -\frac{r^*}{L}\left[\frac{dU^{(i;1/2)}}{dr^*} + \frac{2}{r^*}U^{(i;1/2)}\right], \tag{137}$$

$$P^{(i;1/2)} = \frac{r^*}{2L}\left[\frac{dQ^{(i;1/2)}}{dr^*} + \frac{3}{r^*}Q^{(i;1/2)} - \frac{6L^2}{r^{*2}}V^{(i;0)}\right.$$
$$- \frac{12L}{r^{*2}}U^{(i;0)} - \frac{2L'^2}{r^{*2}}V^{(i;0)} + \frac{2L\rho_0}{r^*}\phi^{(i;1/2)}$$
$$\left. + \frac{2L}{r^*}\rho_0 g_0 U^{(i;1/2)} + \omega_0^2 \rho_0 V^{(i;0)}\right], \tag{138}$$

$$Q^{(i;1/2)} = \frac{dV^{(i;0)}}{dr^*} - \frac{V^{(i;0)}}{r^*} - \frac{2L}{r^*}U^{(i;0)}, \tag{139}$$

$$\phi^{(i;1/2)} = \frac{r^{*2}}{2L^2}\left[\frac{dg^{(i;0)}}{dr^*} + \frac{2}{r^*}g^{(i;0)}\right], \tag{140}$$

$$\frac{d^2 g^{(i;1/2)}}{dr^{*2}} + \frac{4}{r^*}\frac{dg^{(i;1/2)}}{dr^*} + \frac{(2 - 2L^2)}{r^{*2}}g^{(i;1/2)} = 0. \tag{141}$$

Again, a general solution of this system is composed of a linear combination of linearly independent solutions of the homogeneous system (proportional to Solutions (84)-(89) with B_i' instead of A_i') and a particular solution of the nonhomogeneous system. A particular solution is given by

$$U_p^{(i;1/2)} = -A_1'\omega_1^2 \rho_0 \frac{1}{4(-2l+3)}r^{*(-l+2)}$$
$$+ A_1'\omega_1^2 \rho_0 \omega_0^2 \rho_0 \frac{1}{12(-2l+3)(-2l+5)}r^{*(-l+4)}$$
$$- A_4'\omega_1^2 \rho_0 \omega_0^2 \rho_0 \frac{(l+1)}{8(-2l+1)(-2l+3)}r^{*(-l+2)}$$
$$+ \cdots, \tag{142}$$

$$V_p^{(i;1/2)} = A_1'\omega_1^2 \rho_0 \frac{(-l+4)}{4L(-2l+3)}r^{*(-l+2)}$$
$$- A_1'\omega_1^2 \rho_0 \omega_0^2 \rho_0 \frac{(-l+6)}{12L(-2l+3)(-2l+5)}r^{*(-l+4)}$$

$$+ A_4'\omega_1^2 \rho_0 \omega_0^2 \rho_0 \frac{(l+1)(-l+4)}{8L(-2l+1)(-2l+3)}r^{*(-l+2)}$$
$$+ \cdots, \tag{143}$$

$$P_p^{(i;1/2)} = -A_1'\frac{2(l^2 + 3l - 1)}{(l+1)}r^{*(-l-1)}$$
$$+ A_1'\omega_0^2 \rho_0 \frac{(l-2)}{2(-2l+3)}r^{*(-l+1)}$$
$$+ A_4'(3l+2)(l+2)r^{*(-l-3)}$$
$$- -A_4'\omega_0^2 \rho_0 \frac{2(l+1)(l-1)}{(-2l+1)}r^{*(-l-1)}$$
$$- A_6'\rho_0 \frac{1}{(l+1)}r^{*(-l-1)}$$
$$+ \rho_0 g_0 [-A_1'\omega_1^2 \rho_0 \frac{1}{4(-2l+3)}r^{*(-l+2)}$$
$$+ A_1'\omega_1^2 \rho_0 \omega_0^2 \rho_0 \frac{1}{12(-2l+3)(-2l+5)}r^{*(-l+4)}$$
$$- A_4'\omega_1^2 \rho_0 \omega_0^2 \rho_0 \frac{(l+1)}{8(-2l+1)(-2l+3)}r^{*(-l+2)}]$$
$$+ \cdots, \tag{144}$$

$$Q_p^{(i;1/2)} = A_1'\frac{2(l+1)(-l+1)}{L}r^{*(-l-1)}$$
$$+ A_1'\omega_0^2 \rho_0 \frac{(l^2 - 2l + 2)}{2L(-2l+3)}r^{*(-l+1)}$$
$$+ A_4'\frac{l(l+1)(l+2)}{L}r^{*(-l-3)}$$
$$+ A_4'\omega_0^2 \rho_0 \frac{(l+1)^2}{L(-2l+1)}r^{*(-l-1)} + \cdots, \tag{145}$$

$$\phi_p^{(i;1/2)} = -A_6'\frac{1}{(l+1)}r^{*(-l-1)}, \tag{146}$$

$$g_p^{(i;1/2)} = 0. \tag{147}$$

We attempt to match the general boundary-layer expansions and the general solutions of the outer region of the outer core to order $\mu^{1/2}$ by applying Condition (92). Similarly as at order μ^0, we avoid singularities in the condition by setting $B_2' = B_3' = B_5' = B_2'' = B_4'' = 0$ and taking $l \geq 3$. If we want the contribution of the dominant term of the outer solutions to vanish in the matching conditions, we must require

$$m \ll \mu^{1/[2(l-1)]}. \tag{148}$$

If we only take one term in the boundary-layer expansions, it vanishes in the matching conditions if

$$\mu^{(l+1)/[2(l+2)]} \ll m. \tag{149}$$

In this case the matching of the boundary-layer and outer expansions to order $\mu^{1/2}$ is possible with the non-empty overlap domain

$$\mu^{(l+1)/[2(l+2)]} \ll m \ll \mu^{1/[2(l-1)]}. \tag{150}$$

This overlap domain is smaller than the overlap domain determined by Condition (93).

If we take into account more terms of the boundary-layer expansions, the overlap domain continues to shrink.

From the matching, we can determine the two composite expansions for U, V, P, Q, ϕ and g that are uniformly valid in the outer core to order $\mu^{1/2}$.

The junction of the composite expansion for U, V, P, Q, ϕ and g that are uniformly valid in the outer core with, firstly, at r_{IOB} the solutions of the inner core and, secondly, at r_{CMB} the solutions of the mantle requires that the functions of order $\mu^{1/2}$ in the expansions be continuous. The continuity of these functions together with the boundary conditions(17)-(22)and with the continuity of $U^{(oc1,1/2)}$ and $U^{(oc2,1/2)}$ at an arbitrary point of the outer core, leads to a system of 16 algebraic, linear and non-homogeneous equations in the 16 undetermined constants

$$
\begin{aligned}
&B_2,\ B_3,\ B_5 : inner\ core \\
&B_1',\ B_4',\ B_6' : boundary\ layer\ 1\ in\ the\ outer\ core \\
&B_1'',\ B_3'' : outer\ region\ outer\ core \\
&B_1'^{*},\ B_4'^{*} : boundary\ layer\ 2\ in\ the\ outer\ core \\
&B_1''',\ B_2''',\ B_3''',\ B_4''',\ B_5''',\ B_6''' : mantle. \quad (151)
\end{aligned}
$$

The determinant of the 16 x 16 matrix formed by the coefficients of the homogeneous system is the determinant whose zeros determine the first approximations ω_0 of the eigenfrequencies and is thus singular. For the non-homogeneous system to be non-contradictory, we have to require the determinant of a 16 x 16 matrix composed of the previous homogeneous system matrix, where one column is replaced by the vector containing the non-homogeneous part of the system, to be zero [Martens and Smeyers, 1982]. The condition on the determinant leads to the expression for the correction term on the frequency.

The value for ω_1 ($l = 3$) is derived on numerical grounds using the same numerical values for the physical quantities involved as in the previous order. As expected, the first order correction term on the frequency is smaller ($\omega_1 \to 0$) than the zeroth-order eigenvalue.

Approximation to Order μ. The equations at order μ differ formally from those of order $\mu^{1/2}$ in the outer region of the outer core and take the form

$$\frac{dU^{(o;1)}}{dr} = -\frac{2}{r}U^{(o;1)} - \frac{L}{r}V^{(o;1)}, \quad (152)$$

$$\frac{dV^{(o;0)}}{dr} = Q^{(o;1)} + \frac{2L}{r}U^{(o;0)} + \frac{1}{r}V^{(o;0)}, \quad (153)$$

$$
\begin{aligned}
\frac{dP^{(o;1)}}{dr} = &-\frac{L}{r}Q^{(o;1)} + \frac{6L^2}{r^2}V^{(o;0)} + \frac{12L}{r^2}U^{(o;0)} \\
&+ \rho_0 g^{(o;1)} - \frac{4}{r}\rho_0 g_0 U^{(o;1)} - \frac{L}{r}\rho_0 g_0 V^{(o;1)} \\
&+ 4\pi G \rho_0^2 U^{(o;1)} - \omega_0^2 \rho_0 U^{(o;1)} - \omega_1^2 \rho_0 U^{(o;1/2)} \\
&- \omega_2^2 \rho_0 U^{(o;0)}, \quad (154)
\end{aligned}
$$

$$
\begin{aligned}
\frac{dQ^{(o;1)}}{dr} = &-\frac{3}{r}Q^{(o;1)} + \frac{2L}{r}P^{(o;1)} - \frac{2L\rho_0}{r}\phi^{(o;1)} \\
&+ \frac{6L^2}{r^2}V^{(o;0)} + \frac{12L}{r^2}U^{(o;0)} + \frac{2L'^2}{r^2}V^{(o;0)} \\
&- \frac{2L}{r}\rho_0 g_0 U^{(o;1)} - \omega_0^2 \rho_0 V^{(o;1)} - \omega_1^2 \rho_0 V^{(o;1/2)} \\
&- \omega_2^2 \rho_0 V^{(o;0)}, \quad (155)
\end{aligned}
$$

$$\frac{d\phi^{(o;1)}}{dr} = g^{(o;1)}, \quad (156)$$

$$\frac{dg^{(o;1)}}{dr} = -\frac{2}{r}g^{(o;1)} + \frac{2L^2}{r^2}\phi^{(o;1)}. \quad (157)$$

This implies that for the first time the tangential stress is different from zero in the outer region of the outer core.

The corrections on the eigenfrequency of this and higher orders can be found by following the procedure we used in the previous section.

CONCLUDING REMARKS

The aim of this work was to present an unusual approach, using boundary-layer theory, to global eigenvalue problems. In order to fully explain the working-method, we have chosen to emphasize the mathematical development in the case of a simple Earth model. The simplicity of the Earth model, which led to unique analytical solutions of the eigenvalue problem, is the cause of the absence of an extended presentation of numerical results. The few results presented here lie within the range of values of frequencies for free core oscillations and can possibly correspond to Slichter modes [see e.g. Smylie et al., 1992]. We have the intention to present, in the future, the same mathematical treatment for more realistic Earth models which will include viscosity, rotation ... We therefore think it is more appropriate to postpone a detailed discussion of numerical results to a next paper.

Acknowledgment. One of the authors (V. D.) is supported by the Belgian National Fund for Scientific Research.

REFERENCES

Backus, G., and Gilbert, F., The rotational splitting of the free oscillations of the Earth, *Proc. Nat. Acad. Sci. U.S.A.*, *47*, 362-371, 1961.

Chandrasekhar, S., Hydrodynamic and Hydromagnetic Stability, Clarendon Press, Oxford, 1961.

Crossley, D.J., The free-oscillation equations at the centre of the Earth, *Geophys. J.R. astr. Soc.*, *41*, 153-163, 1975.

Dahlen, F.A., The normal modes of a rotating elliptical Earth, *Geophys. J.R. astr. Soc.*, *16*, 329-367, 1968.

Dahlen, F.A., On the static deformation of the Earth model with a fluid core, *Geophys. J.R. astr. Soc.*, *36*, 461-485, 1974.

Degryse, K., Noels, A., Gabriel, M., Waelkens, C., and Smeyers, P., Vibrational stability of an evolved 5 $M_\odot$ star towards higher-order g-modes, *Astron. Astrophys. (in press)*, 1992.

Ince, E.L., Ordinary Differential Equations, Dover Publications, 1956.

Jeffreys, H., The Earth, Cambridge University Press, 1970.

Kevorkian, J., and Cole, J.D., Perturbation Methods in Applied Mathematics, Springer-Verlag, New York, 1981.

Martens, L., and Smeyers, P., On the linear adiabatic oscillations of a uniformly and synchronously rotating component of a binary, *Astron. Astrophys.*, *106*, 317-326, 1982.

Mochizuki, E., Effects of core rigidity on free oscillations of the Earth: Reexamination of analytic solutions, *J. Phys. Earth*, *38*, 251-260, 1990.

Phinney, R.A., and Burridge, R., Representation of the elastic-gravitational excitation of a spherical Earth model by generalized spherical harmonics, *Geophys. J.R. astr. Soc., 34*, 451-487, 1973.

Smith, M.L., The scalar equations of infinitesimal elastic-gravitational motion for a rotating, slightly elliptical Earth, *Geophys. J.R. astr. Soc., 37*, 491-526, 1974.

Smylie, D.E., Xianhua Jiang, Brennan, B.J., and Kachishige Sato, Numerical calculation of modes of oscillation of the Earth's core, *Geophys. J. Int., 108*, 465-490, 1992.

Takeuchi, H., and Saito, R., Seismic surface waves, in *Methods of Computational Physics, 11*, pp. 217-298, Academic Press, New York, 1971.

Van Dyke, M., Perturbation Methods in Fluid Mechanics, Academic Press, Yew York, 1964.

K. Degryse and V. Dehant, Koninklijke Sterrenwacht van België, Ringlaan, 3, B-1180 Brussels, Belgium

Numerical Calculation of Modes of Oscillation
of the
Earth's Core

D. E. SMYLIE AND XIANHUA JIANG

Department of Earth and Atmospheric Science, York University, North York, Ontario, M3J 1P3, Canada

Abstract The small oscillation equations for the Earth's fluid outer core are considered for periods long compared to purely acoustic periods. Acoustic periods are less than ten minutes, and acoustic effects on the displacement field are shown to scale as the square of the ratio of the longest acoustic period to the period in question. Thus, for periods in the range of three hours and longer, acoustic effects are negligible compared to the effects of transport through the initial pressure and gravity fields. Applied uniformly to both the governing equations and boundary conditions, the resulting subseismic approximation leads to a fully Hermitean system and a variational principle which forms the basis of the numerical calculation of long period core modes. Coriolis acceleration is paramount at periods of hours and longer, and specially adapted finite element support functions are used to implement the variational calculation to avoid the poor convergence of conventional vector spherical harmonic expansions due to tight Coriolis coupling. The subseismic approximation to the boundary conditions yields effective Love numbers for the shell and inner core allowing their elasto-gravitational response to be incorporated. Our theoretical and numerical methods have led to the identification of the translational triplet of the inner core in superconducting gravimeter spectra.

Introduction

The very interesting dynamics of the Earth's fluid outer core at periods long compared to conventional seismic periods presents a number of computational challenges. On the one hand, computational stability demands (Holton, 1979, p.175) that we suppress short period acoustic effects describing seismic radiation, and on the other, tight Coriolis coupling renders the usual vector spherical harmonic expansions of the displacement and gravity fields poorly convergent (Johnson and Smylie, 1977). Analytic solutions of the problem of rotating fluids in rigid containers from the classical literature cannot be extended to realistic Earth models, and provide only qualitative insight into some aspects of the geophysical setting (Poincaré, 1885; Bryan, 1884; Kudlick, 1966; Greenspan, 1969).

In this paper, we outline[1] theoretical and numerical methods we have developed and implemented for the accurate calculation of core modes, and we illustrate the application of our methods with the example of the recent identification (Smylie, 1992) of the translational triplet of the solid inner core in the spectra of superconducting gravimeter data.

Long Period Equations and Boundary Conditions

The small oscillation equation of motion in the displacement vector **u** is

$$-\omega^2 \mathbf{u} + 2i\omega \mathbf{\Omega} \times \mathbf{u} = -\nabla\chi - (1-\eta)\mathbf{g}_0 \nabla \cdot \mathbf{u}, \quad (1)$$

where

$$\chi = \frac{p_1}{\rho_0} - V_1, \quad (2)$$

Dynamics of Earth's Deep Interior and Earth Rotation
Geophysical Monograph 72, IUGG Volume 12
Published in 1993 by the International Union of Geodesy and Geophysics and the American Geophysical Union.

[1]A full journal article account of this work is given by Smylie, Jiang, Brennan and Sato (1992).

and ω is the oscillation angular frequency, $\mathbf{\Omega}$ is the constant vector angular rotation rate of the Earth reference frame, η is the stratification parameter, $\mathbf{g}_0$ is unperturbed gravity, p_1 is the flow pressure perturbation, ρ_0 is the unperturbed density, and V_1 is the decrease in gravitational potential.

Conservation of mass and the barotropic equation of state resulting from the still accurately isentropic deformations give the flow pressure perturbation as

$$p_1 = -\alpha^2 \rho_0 \nabla \cdot \mathbf{u} - \rho_0 \mathbf{u} \cdot \mathbf{g}_0 \qquad (3)$$

with α denoting the P-wave speed.

In the linearized, small oscillations treatment followed here, there is no contribution to the flow pressure perturbation from the non-linear dynamic flow pressure, and in the absence of acoustic effects, it derives entirely from rotational forces. In comparison to the second term on the right of equation (3) arising from transport through the initial static pressure field, the flow pressure perturbation scales as the ratio $\Omega^2 L/g_0$ for motions of characteristic length L. Even for length scales of 10^3 km, this amounts to only 1/1500. To the same order, we may then take

$$|p_1| \ll |\rho_0 \mathbf{u} \cdot \mathbf{g}_0| \qquad (4)$$

or

$$|p_1/\rho_0| \ll |\mathbf{u} \cdot \mathbf{g}_0|. \qquad (5)$$

In this approximation, the dilatation-compression is controlled entirely by transport through the pre-existing static pressure field arising from self-gravitation and equation (3) becomes

$$\nabla \cdot \mathbf{u} = -\frac{1}{\alpha^2} \mathbf{u} \cdot \mathbf{g}_0. \qquad (6)$$

Of course for motions with negligible radial extent, (6) reduces to the usual assumption of incompressible flow and (4) and (5) express the usual neglect of flow pressure perturbations on the density for periods beyond the acoustic period range. Acoustic effects on the density scale as $\omega^2 L^2/\alpha^2$ (Batchelor, 1967, p.169). Their neglect therefore involves a 'subseismic' approximation, applicable below regular seismic frequencies, which improves as the square of the ratio of the period to the longest acoustic period. For periods of several hours and greater, the approximation is less than 1/1000. The Hermitean nature of the problem allows construction of a variational principle and the approximation to eigenfrequencies is proportional to the square of the error in the eigenfunctions. Thus, we expect eigenfrequencies to be accurate to no worse than one part in 10^6.

Through the introduction of a dimensionless scalar decompression factor f (Smylie, Szeto and Sato, 1990), an equivalent solenoidal vector field $f\mathbf{u}$ can be found allowing equation (6) to be written

$$\nabla \cdot (f\mathbf{u}) = 0. \qquad (7)$$

The decompression factor is a very benign function which varies monotonically from unity at the core-mantle boundary to a value of about 1.23 at the inner core boundary for most Earth models.

Using equation (6), the equation of motion (1) transforms to

$$-\omega^2 \mathbf{u} + 2i\omega \mathbf{\Omega} \times \mathbf{u} = -\nabla \chi - \omega_V^2 (\mathbf{u} \cdot \hat{\mathbf{s}}) \hat{\mathbf{s}}, \qquad (8)$$

where $\hat{\mathbf{s}}$ is the unit vector in the orthometric direction and ω_V^2 is the signed square of the Väisälä angular frequency given by

$$\omega_V^2 = (\eta - 1) \left(\frac{g_0}{\alpha} \right)^2. \qquad (9)$$

Equation (8) is purely algebraic in $\mathbf{u}$ and can be solved to allow recovery of the displacement field from the generalized displacement potential χ via

$$\mathbf{u} = \mathcal{T} \nabla \chi. \qquad (10)$$

$\mathcal{T}$ is a second order Hermitean tensor.

Substitution of expression (10) for the displacement field into the subseismic relation (7) yields the second order scalar subseismic wave equation

$$\sigma^2 \nabla^2 \chi - \left(\hat{\mathbf{k}} \cdot \nabla \chi \right)^2 - \frac{A}{B} \mathbf{C} \cdot \nabla \chi - \mathbf{C}^* \cdot \nabla \left(\frac{\mathbf{C} \cdot \nabla \chi}{B} \right) = 0. \quad (11)$$

$$A = -\frac{g_0^2}{\alpha^2 N^2} \sigma^2 \left(\sigma^2 - 1 \right) + \sigma^2 \left(4\pi G \rho_0 - 2\Omega^2 \right) - g_0 \left(\hat{\mathbf{k}} \cdot \nabla \right) \hat{\mathbf{k}} \cdot \hat{\mathbf{s}} \quad (12)$$

$$B = -g_0^2 \left[\frac{1}{N^2} \sigma^2 \left(\sigma^2 - 1 \right) + \left(\hat{\mathbf{k}} \cdot \hat{\mathbf{s}} \right)^2 - \sigma^2 \right], \qquad (13)$$

$$\mathbf{C} = -g_0 \left[\left(\hat{\mathbf{k}} \cdot \hat{\mathbf{s}} \right) \hat{\mathbf{k}} + i\sigma \hat{\mathbf{k}} \times \hat{\mathbf{s}} - \sigma^2 \hat{\mathbf{s}} \right]. \qquad (14)$$

G is the universal constant of gravitation, $\hat{\mathbf{k}}$ is a unit vector aligned with the rotation axis and $\mathbf{\Omega} = \hat{\mathbf{k}}\Omega$, $\sigma = \omega/2\Omega$. $N^2 = \omega_V^2/4\Omega^2$ is the dimensionless square of ω_V.

The approximations (4) and (5), leading to the subseismic relation (6), must be uniformly applied to the conditions at the fluid-solid interfaces of the outer core to obtain the subseismic boundary conditions. On the fluid sides of the deformed interfaces, the normal stress is $-p_1 - \rho_0 \mathbf{u} \cdot \mathbf{g}_0$ balanced on the solid sides by $\rho_0 \alpha^2 \nabla \cdot \mathbf{u}$. In view of relation (4), in the subseismic approximation, normal stress on the fluid sides is $-\rho_0 \mathbf{u} \cdot \mathbf{g}_0$. Similarly, the gravity potential perturbation on the fluid sides of the deformed boundaries, $-V_1 - \mathbf{u} \cdot \mathbf{g}_0 = -p_1/\rho_0 - \mathbf{u} \cdot \mathbf{g}_0 + \chi$ is approximated by $-\mathbf{u} \cdot \mathbf{g}_0 + \chi$ according to relation (5).

The subseismic boundary conditions permit the description of the elasto-gravitational deformation of the shell and inner core in terms of sequences of internal load Love numbers. For core-mantle boundary radius b, inner core radius a, the outward normal components of the displacement vector $\mathbf{u}_B$ on the solid sides of the boundaries are expanded as

$$\mathbf{u}_B \cdot \hat{\mathbf{n}} = \mathbf{u}_B \cdot \hat{\mathbf{s}} = e^{im\phi} \sum_{n=m}^{\infty} y_{1_n}(b) P_n^m(\cos\theta), \quad (15)$$

on the core-mantle boundary, and as

$$\mathbf{u}_B \cdot \hat{\mathbf{n}} = -\mathbf{u}_B \cdot \hat{\mathbf{s}} = -e^{im\phi} \sum_{n=m}^{\infty} y_{1_n}(a) P_n^m(\cos\theta) \quad (16)$$

on the inner core surface. The colatitude is denoted by θ, the east longitude by ϕ and P_n^m are Legendre functions. The generalized displacement potential χ on the fluid sides of the boundaries is similarly expanded as

$$\chi = e^{im\phi} \sum_{n=m}^{\infty} \chi_n(r_0) P_n^m(\cos\theta), \quad (17)$$

Table 1. Internal load Love numbers for the shell and inner core computed for the Earth model CORE11.

Degree n	h_n^S	h_n^I
0	0.1347	0.0
1	0.1149	-25.8426
2	0.2079	-0.1113
3	0.1519	-0.0679
4	0.1033	-0.0505
5	0.0753	-0.0406
6	0.0593	-0.0341
7	0.0494	-0.0295
8	0.0427	-0.0259
9	0.0378	-0.0232
10	0.0340	-0.0210
11	0.0309	-0.0192
12	0.0284	-0.0176
13	0.0262	-0.0163
14	0.0244	-0.0152
15	0.0228	-0.0142
16	0.0214	-0.0134
17	0.0202	-0.0126
18	0.0191	-0.0119
19	0.0181	-0.0113
20	0.0172	-0.0108
21	0.0164	-0.0103
22	0.0157	-0.0098
23	0.0150	-0.0094

where r_0 is the mean radius. After all other elasto-gravitational boundary conditions are satisfied, continuity of normal displacement is assured by the equations (Smylie, Szeto and Sato, 1990)

$$y_{1_n}(b) = \frac{h_n^S}{g_0(b)} \chi_n(b), \quad (18)$$

and

$$y_{1_n}(a) = \frac{h_n^I}{g_0(a)} \chi_n(a). \quad (19)$$

The dimensionless coefficients h_n^S and h_n^I are internal load Love numbers describing the complete elasto-gravitational deformational response of the shell and inner core respectively, independent of fluid outer core properties. Their values are listed in Table 1 for Earth model CORE11 of Widmer, Masters and Gilbert (1988).

A functional for the solution of the subseismic wave equation (11), or its equivalent expression (7) in terms of the displacement field, is

$$\mathcal{F} = -\int f \nabla \chi^* \cdot \mathcal{T} \cdot \nabla \chi dV + \int f \chi^* \mathbf{u} \cdot \hat{\mathbf{n}} dS. \quad (20)$$

Relations (18) and (19) allow expansion of the surface integral in the functional in terms of the internal load Love numbers, giving

$$\int f \chi^* \mathbf{u} \cdot \hat{\mathbf{n}} dS = \frac{\pi b}{\Omega^2} \Sigma_S = \frac{4\pi b^2}{g_0(b)} \sum_{n=m}^{\infty} \frac{1}{2n+1} \frac{(n+m)!}{(n-m)!} \cdot$$

$$\left[h_n^S \chi_n(b) \chi_n^*(b) - f(a) \frac{a^2}{b^2} \frac{g_0(b)}{g_0(a)} h_n^I \chi_n(a) \chi_n^*(a) \right]. \quad (21)$$

At short periods, such as those involved in the translational modes of the solid inner core, the Love number h_1^I has a frequency dependence due entirely to inertial effects given by

$$\frac{1}{h_1^I} = \alpha\sigma\left(\sigma - \delta_m^1\right) - \gamma,$$

where α (different from the P-velocity) and γ are constants. For these modes, we then remove the term in h_1^I from the summation (21) and denote the remainder of the sum by a superscript prime in later usage.

The variation of the functional is

$$\delta\mathcal{F} = \int \left[\delta\chi \nabla \cdot (f\mathbf{u}^*) + \delta\chi^* \nabla \cdot (f\mathbf{u}) \right] dV$$

$$- \int f \left[\delta\chi (\mathbf{u}^* - \mathbf{u}_B^*) \cdot \hat{\mathbf{n}} + \delta\chi^* (\mathbf{u} - \mathbf{u}_B) \cdot \hat{\mathbf{n}} \right] dS. \quad (22)$$

Stationarity of the functional therefore ensures solution of the subseismic equation (7), and continuity of normal displacements at the interfaces as 'natural' boundary conditions. All of the other boundary conditions are incorporated as 'essential' boundary conditions in the calculation of the Love numbers h_n^S and h_n^I for the shell and inner core.

Numerical Implementation and Results

The numerical implementation of the variational calculation is considerably simplified by a number of general symmetries that the core oscillation problem possesses (Smylie and Rochester, 1986).

Modes of a given azimuthal number m are completely independent of other modes, the tight Coriolis, ellipticity and centrifugal coupling present in vector spherical harmonic representations of the problem all being across zonal number. Explicit account of azimuthal dependence is therefore not required and the problem is reduced to two dimensions.

Reversing the signs of both m and σ results in the same governing equations and boundary conditions, allowing the restriction of the azimuthal number to zero and the positive integers. The axisymmetric motions are standing waves characterized by σ^2, although in general they have all three components of displacement present. Non-axisymmetric modes with σ positive are retrograde and the pattern of displacements drifts westward at the rate σ/m. Non-axisymmetric modes with negative σ are prograde and their displacement pattern drifts eastward at the rate $-\sigma/m$.

Modes which are even in the equatorial plane are decoupled from modes odd in the equatorial plane. We develop special local support functions for the representation of core modes, which are either globally even or globally odd in the equatorial plane, in order to suppress spurious numerical effects. Further, the most general domain of the problem is reduced to the quarter annulus of the meridional plane by the parity separation of modes.

Finally, all core mode solutions are known to have the factor $sin^m\theta$ in their solutions and we eliminate the possibility of branch points by the change of variable

$$\chi = \left(1 - z^2\right)^{m/2}\Phi \tag{23}$$

with z used as a shorthand for $cos\theta$. An appropriately scaled version of the functional (20), suitable for numerical implementation, is then

$$\mathcal{F}_R = \sigma^2 \int_0^1 \int_{a/b}^1 f \left(1 - z^2\right)^m \mathbf{V}^{\mathbf{T}} \times$$
$$\begin{pmatrix} m^2\left(1 + z^2\right)/\left(1 - z^2\right) & 0 & -mz \\ 0 & 1 & 0 \\ -mz & 0 & \left(1 - z^2\right) \end{pmatrix} \mathbf{V} dr dz$$
$$- \int_0^1 \int_{a/b}^1 f \left(1 - z^2\right)^m \mathbf{V}^{\mathbf{T}} \begin{pmatrix} mz \\ -z \\ z^2 - 1 \end{pmatrix} \otimes$$

$$\left(\; mz, \; -z, \; z^2 - 1 \; \right) \mathbf{V} dr dz$$
$$- \int_0^1 \int_{a/b}^1 f \frac{\left(1 - z^2\right)^m}{\zeta_1^2 - z^2} \mathbf{V}^{\mathbf{T}} \begin{pmatrix} m\left(\sigma + z^2\right) \\ \sigma^2 - z^2 \\ z\left(z^2 - 1\right) \end{pmatrix} \otimes \tag{24}$$
$$\left(\; m\left(\sigma + z^2\right), \; \sigma^2 - z^2, \; z\left(z^2 - 1\right) \; \right) \mathbf{V} dr dz$$
$$+ m\sigma \int_0^1 \int_{a/b}^1 f \left(1 - z^2\right)^m \mathbf{V}^{\mathbf{T}}$$
$$\begin{pmatrix} 2mz^2/\left(1 - z^2\right) & 1 & -z \\ 1 & 0 & 0 \\ -z & 0 & 0 \end{pmatrix} \mathbf{V} dr dz - \sigma^2 \left(\sigma^2 - 1\right) \Sigma_S$$

with

$$\zeta_1^2 = \sigma^2 \left(1 - \frac{\sigma^2 - 1}{N^2}\right). \tag{25}$$

The scalar field Φ and its partial derivatives Φ_r, Φ_z are described by the vector $\mathbf{V}^{\mathbf{T}} = \left(\Phi, r\Phi_r, \Phi_z\right)$. The dimensionless radius r is measured in units of the mean radius r_0 and the symbol $\otimes$ denotes the outer vector product. In the r, z plane the domain of the problem is just the rectangle shown in Figure 1.

When the non-neutral stratification is weak and the period is neither large nor near one half sidereal day, the parameter ζ_1^2 defined by (25) is large and we can use the expansion

$$\frac{1}{\zeta_1^2 - z^2} = -\frac{N^2}{\sigma^2\left(\sigma^2 - 1\right)} \left[1 + \frac{N^2}{\sigma^2\left(\sigma^2 - 1\right)} \left(\sigma^2 - z^2\right) + \cdots\right], \tag{26}$$

valid for

$$\left| \frac{N^2}{\sigma^2 - 1} \right| \ll \sigma^2. \tag{27}$$

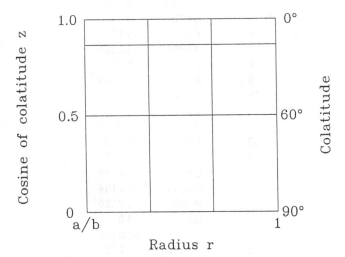

Figure 1. Finite element grid in the (r, z) plane where r is scaled by the outer core radius and $z = cos\theta$ with θ the colatitude. In the 3×3 element grid shown here, values of z are chosen for equi-spaced values of θ.

At the boundaries of the outer fluid core, normal displacement and stress, gravitational potential perturbation and its gradient are all continuous. All but continuity of normal displacement are incorporated into the calculation of the internal load Love numbers h_n^S, h_n^I listed in Table 1. The normal displacement is

$$\mathbf{u} \cdot \hat{\mathbf{n}} = \frac{\hat{\mathbf{n}} \cdot \hat{\mathbf{s}} (1-z^2)^{m/2}}{4\Omega^2 N^2 r_0 (z^2 - \zeta_1^2)} \cdot$$

$$\left(m(\sigma + z^2), \ (\sigma^2 - z^2), \ z(z^2 - 1) \right) \mathbf{V}. \quad (28)$$

Polynomial support functions in r and z are used to match the values of Φ, Φ_r, Φ_z and Φ_{rz} at each vertex of the rectangular finite elements shown in Figure 1. In radius, the required support functions are the canonical piecewise Hermite cubic splines, while in z, for modes even in the equatorial plane, they are cubics in z^2, and for modes odd in the equatorial plane they are z times cubics in z^2. Along the equator, the support functions in z are degenerate and become quadratics in z^2 in the even case and z times quadratics in z^2 in the odd case. The degeneracy arises because Φ_z vanishes on the equator in the even case and Φ itself vanishes there in the odd case, and there is therefore one less parameter to be matched for each parity.

The expansion coefficients for the vector $\mathbf{V}$ become the nodal values of Φ, Φ_r, Φ_z and Φ_{rz}, and performing the integrations in the functional (24) and invoking the range and summation conventions, it takes the bilinear, symmetric form

$$\mathcal{F}_R = x_i a_{ij} x_j, \quad a_{ij} = a_{ij} \quad (29)$$

where x_i is a vector made up of all the expansion coefficients.

Continuity of normal displacement at the boundaries is imposed by calculating the radial coefficients y_{1n} there from expression (28) by spherical harmonic expansion, and applying conditions (18) and (19) to the radial coefficients χ_n of similar expansions of χ at the boundaries. A system of homogeneous linear algebraic equations

$$x_p'' = c_{pk} x_k' \quad (30)$$

results with constrained variables x_p'' expressed in terms of unconstrained variables x_k'. Elimination of the constrained variables yields a functional

$$\mathcal{F}_R = x_k' a_{kl} x_l' + 2 x_k' a_{kp} c_{pl} x_l' + x_k' c_{pk} a_{pq} c_{ql} x_l' \quad (31)$$

in the constrained variables alone. Stationarity of the functional $\mathcal{F}_R$ leads to a linear system in the unconstrained variables with a still symmetric modified coefficient matrix

$$a_{kl} + a_{kp} c_{pl} + a_{lp} c_{pk} + c_{pk} a_{pq} c_{ql}. \quad (32)$$

We are then presented with the eigenvalue-eigenvector problem

$$A(\sigma_i) \mathbf{x}_i = 0, \quad (33)$$

where the matrix A is a λ-matrix of degree eight expressible as

$$A(\sigma) = A_8 \sigma^8 + A_7 \sigma^7 + \cdots + A_1 \sigma + A_0. \quad (34)$$

A specially adapted inverse iteration scheme is used to recover the eigenvector $\mathbf{x}_i$ once the eigenfrequency σ_i is found by a stable determinant routine.

Having found a solution for the generalized displacement potential χ, the components of the displacement vector field in spherical polar coordinates u_r, u_θ, u_ϕ can be calculated using equation (10). They are

$$ru_r = F^m \left[m \left(\sigma + z^2 \right) \Phi + \left(\sigma^2 - z^2 \right) r \Phi_r - \left(1 - z^2 \right) \Phi_z \right] e^{im\phi},$$

$$\sigma^2 r \sin\theta u_\theta = F^m \left[\left\{ \sigma \left(z^2 - \zeta_2^2 \right) + \left(1 - \zeta_2^2 \right) \right\} m\sigma z \Phi + z \left(1 - z^2 \right) \sigma^2 r \Phi_r - \left(z^2 - \zeta_2^2 \right) \left(1 - z^2 \right) \sigma^2 \Phi_z \right] e^{im\phi}, \quad (35)$$

$$\sigma r \sin\theta u_\phi = F^m \left[m \left(\sigma + z^2 \right) \left(1 - \zeta_2^2 \right) \Phi + \sigma^2 \left(1 - z^2 \right) r \Phi_r - \left(1 - \zeta_2^2 \right) z \left(1 - z^2 \right) \Phi_z \right] e^{i(m\phi + \pi/2)},$$

where F^m is the common factor

$$F^m = \frac{(1 - z^2)^{m/2}}{4\Omega^2 \sigma^2 (\sigma^2 - 1)} \quad (36)$$

and ζ_2^2 is the parameter

$$\zeta_2^2 = 1 - \sigma^2 + N^2 = \frac{N^2}{\sigma^2} \zeta_1^2. \quad (37)$$

Although a complete catalogue of modes has yet to be completed, we illustrate our methods with the spatially simple oscillation with 15.8 hr period shown in Figure 2. It is an axisymmetric mode in the intertidal band computed with the finite element grid of Figure 1. A near neutrally stable density profile has been used with $N^2 = 10^{-3}$ and the first twelve odd degree ($n = 1, 3, \cdots, 23$) Love numbers have been used in constraining continuity in the normal component of $\mathbf{u}$.

Detection and Identification of Core Modes

The periods of core modes are extremely sensitive to the density profile. While this is an advantage in the deep interior where conventional seismic methods are least sen-

Generalized Displacement Potential

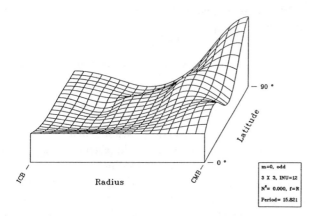

Displacement Vector Field

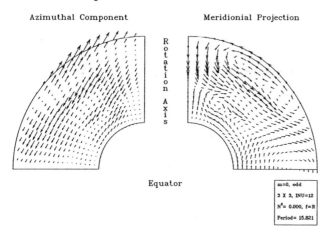

Figure 2. Retrograde axisymmetric core mode in the intertidal band at 15.821 hr period. Absence of tidal lines in this band enhances the possibility of observing such modes.

sitive, it makes direct identification of core modes in observed spectra difficult. In this Section, we propose a detection and identification algorithm based on very general rotational splitting properties, and we illustrate its application with the recently discovered translational triplet of the solid inner core found in the spectra of four European superconducting gravimeter records (see Smylie, Hinderer, Richter, Ducarme and Mansinha (1992) elsewhere in this Volume).

The equation of motion (8) for a particular eigenfrequency-eigenfunction pair $\omega_j, \mathbf{u}_j$ is multiplied scalarly by $f\mathbf{u}_j$. The resulting equation added to its conjugate and integrated throughout the outer fluid core, after using the properties of the scalar triple product, the divergence theorem and relations (18), (19) and (21), can be put in the form

$$\omega_j^2 - 2\omega_j S = E + F \tag{38}$$

with S defined by

$$\left[2 \int f\boldsymbol{\Omega} \cdot (\mathbf{u}_{j_R} \times \mathbf{u}_{j_I})\, d\mathcal{V} + \delta_m^1 \alpha \frac{g_0(a)}{4\Omega} \int f\left(\mathbf{u}_T \cdot \hat{\mathbf{s}}\right)^2 d\mathcal{S} \right] / I, \tag{39}$$

EI being the expansion (21) of the surface integral, and F is defined by

$$F = \left[\int f\omega_V^2 \left| \mathbf{u}_j \cdot \hat{\mathbf{s}}^2 \right| d\mathcal{V} + \gamma g_0(a) \int f\left(\mathbf{u}_T \cdot \hat{\mathbf{s}}\right)^2 d\mathcal{S} \right] / I \tag{40}$$

where the modal intensity is

$$I = \left[\int f \left| \mathbf{u}_j \right|^2 d\mathcal{V} + \alpha \frac{g_0(a)}{4\Omega^2} \int f\left(\mathbf{u}_T \cdot \hat{\mathbf{s}}\right)^2 d\mathcal{S} \right]. \tag{41}$$

The real part, $\mathbf{u}_{j_R}$, and the imaginary part, $\mathbf{u}_{j_I}$, of the displacement field $\mathbf{u}_j$, without loss of generality, can be taken respectively as the meridional and azimuthal components given by (35), since the azimuthal component leads the meridional component by 90° in phase.

We can write the spin term as

$$S = g\Omega. \tag{42}$$

The dimensionless geometrical factor g for the displacement field is defined as

$$g = \left[2 \int f u_R u_\phi d\mathcal{V} + \delta_m^1 \alpha \frac{g_0(a)}{4\Omega^2} \int f\left(\mathbf{u}_T \cdot \hat{\mathbf{s}}\right)^2 d\mathcal{S} \right] / I. \tag{43}$$

The geometrical factor is not directly dependent on Earth model but reflects only the spatial structure of the displacement field for a given mode.

The quantity E in equation (38) represents the elasto-gravitational potential energy stored in the deformation of the shell and inner core while F represents the internal energy generated through work done against a non-neutrally stratified density profile. The latter term is negative if the profile is unstably stratified. Both E and F directly reflect Earth model properties. A non-rotating period T_0 for the oscillation has the square

$$T_0^2 = \frac{4\pi^2}{E + F}. \tag{44}$$

For a mode with period T_j, rotational splitting is described by the simple quadratic equation

$$\left(\frac{T_j}{T_0}\right)^2 + 2g\left(\frac{T_j}{T_0}\right)\frac{T_0}{T_S} - 1 = 0 \tag{45}$$

with $T_S = 2\pi/\Omega$ as the length of the sidereal day. It is then possible to delineate the loci of modal periods as functions of their non-rotating values as illustrated in Figure 3. For a stably stratified core, the prograde periods are depressed by rotation and fall on one branch of the hy-

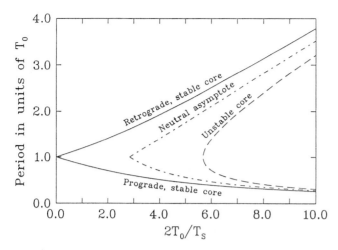

Figure 3. Possible core mode periods in units of non-rotating period shown as functions of non-rotating period in units of the semi-sidereal day. All curves but the neutral asymptote are branches of hyperbolas. This rotational splitting diagram forms the basis of a detection and identification technique for core modes.

perbola (45), while retrograde periods are increased and fall on the other branch of the hyperbola. For an outer core sufficiently unstably stratified to just balance the potential energy stored in deforming the shell and inner core, the prograde branch becomes a rectangular hyperbola and the retrograde branch degenerates to the neutral asymp-

Displacement Vector Field

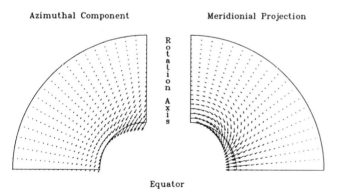

Retrograde equatorial mode

Figure 4. Displacement fields of the three translational modes of the solid inner core. The azimuthal components are shown in perspective. The retrograde equatorial mode is illustrated in the top diagram. The axial mode is illustrated in the middle diagram. The prograde equatorial mode is illustrated in the lower diagram. In each case the azimuthal component leads the meridional component by $90°$.

Displacement Vector Field

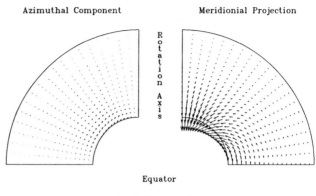

Axial mode

Displacement Vector Field

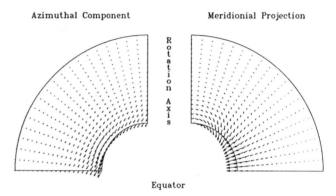

Prograde equatorial mode

Fig. 4. (continued)

totic line. For an outer core so severely unstable that $E + F < 0$, if T_0 is interpreted as the magnitude of the purely imaginary square root of T_0^2, then real oscillation periods still exist for T_0 greater than $1/|g|$ sidereal days.

Now consider the application of the properties of rotational splitting just outlined to the detection and identification of core modes. The weak dependence of the geometrical factor g on Earth model is well-illustrated by the example of the three translational modes of the inner core with displacement fields shown in Figure 4. In Table 2 we list the parameters of the splitting equation (45) determined by iteration for Earth models CORE11 and 1066A (Gilbert and Dziewonski, 1975) as well as for the observed values of the translational periods.

Table 2[2].   Splitting parameters of equation (45) for the three translational periods (T_R, T_C, and T_P) of the inner

[2]These require future correction for inner core inertial effects.

core. The corresponding values of the geometrical factor are given as g_R, g_C, and g_P. Results are shown for Earth models CORE11 and 1066A as well as for the observed resonances.

	T_0(hr)	g_R	g_C	g_P
CORE11	3.5314	-0.3520	0.0496	-0.3713
1066A	2.7141	-0.3524	0.0384	-0.3670
Observed	3.7985	-0.3495	0.0513	-0.3700

For a given Earth model, the geometrical factor g and the non-rotating period T_0 then specify the locus of acceptable predicted periods. As a check, the forecast period for another known Earth model is found and compared with calculation. Candidate observed periods can then be compared with the forecast based on rotational splitting. In Figure 5 we show the results of using the computed periods of Earth model CORE11 to forecast those for Earth models 1066A and PREM (Dziewonski and Anderson, 1981), and to predict the locations of the observed periods. The respective relative errors in the forecast locations of the three observed periods are 0.040%, 0.027% and 0.020%.

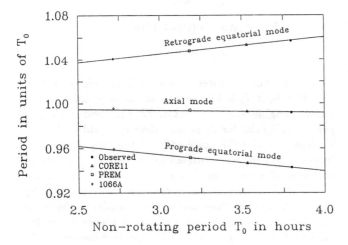

Figure 5. Rotational splitting of translational triplet periods as a function of the period in the absence of rotation. The curves are branches of the hyperbolas shown in Figure 3 and are generated using the parameters of Earth model CORE11 given in Table 2. Computed periods for Earth model 1066A and PREM, and the observed periods, are all accurately forecast. Relative errors in predicting locations of the observed resonances are 0.040%, 0.027% and 0.020%.

Discussion

A solid computational basis has now been established for the calculation of long period core oscillations in realistic models of Earth's deep interior. Together with the simple detection and identification algorithm developed in this paper using rotational splitting properties, our methods have been successfully applied to find the three translational resonances in the spectra of superconducting gravimeter records. In turn, the translational modes have given very accurate values for inner core density and have provided direct estimates of outer core viscosity and an upper bound for magnetic field strength there (Smylie, 1992). The identification of other long period core modes, particularly in the band between the diurnal and semi-diurnal tides, would further expand the range of this important new tool for study of Earth's deep interior.

Acknowledgements

This work has been supported by the Natural Sciences and Engineering Research Council of Canada. The Tri-University Meson Facility (TRIUMF) in Vancouver, Canada has generously donated their outstanding graphics and analysis package. D. E. S. tient à remercier les collègues de l'I.P.G. Strasbourg et Paris pour leur hospitalité durant son année sabbatique en France, en particulier Jacques Hinderer, Roland Schlich et Jean-Louis Le Mouël.

References

Batchelor, G. K., *An Introduction to Fluid Dynamics*, Cambridge University Press, Cambridge, 1967.

Bryan, G. H., The waves on a rotating liquid spheroid of finite ellipticity, *Phil. Trans. R. Soc. Lond.* A, **180**, 187-219, 1889.

Dziewonski, A. M. and D. L. Anderson, Preliminary reference Earth model, *Phys. Earth Planet. Int.*, **25**, 297-356, 1981.

Gilbert, F. and Dziewonski, A.M., An application of normal mode theory to the retrieval of structural parameters and source mechanisms from seismic spectra, *Phil. Trans. R. Soc. Lond.* A, **278**, 187-269, 1975.

Greenspan, H.P., *The Theory of Rotating Fluids*, Cambridge University Press, Cambridge, 1969.

Holton, J. R., *An Introduction to Dynamic Meteorology*, second edition, Academic Press, New York, 1979.

Johnson, I. M. and Smylie, D. E., A variational approach to whole-Earth dynamics, *Geophys. J. R. astr. Soc.*, **50**, 35-54, 1977.

Kudlick, M. D., On transient motions in a contained, rotating fluid, *Ph.D. thesis*, M. I. T., 1966.

Poincaré, H., Sur l'équilibre d'une masse fluide animée d'un mouvement de rotation, *Acta Math.*, **7**, 259-380, 1885.

Smylie, D. E., The inner core translational triplet and the density near Earth's center, *Science*, **255**, 1678-1682, 1992.

Smylie, D. E., Jiang, Xianhua, Brennan, B. J., and Kachishige Sato, Numerical calculation of modes of oscillation of the Earth's core, *Geophys. J. Int.*, **108**, 465-490, 1992.

Smylie, D.E., Hinderer, J., Richter, B., Ducarme, B. and L. Mansinha, A comparative analysis of superconducting gravimeter data, (this issue) 1992.

Smylie, D. E., Szeto, A. M. K. and Kachishige Sato, Elastic boundary conditions in long-period core oscillations, *Geophys. J. Int.*, **100**, 183-192, 1990.

Smylie, D. E. and M. G. Rochester, Long period core dynamics, *Earth Rotation: Solved and Unsolved Problems*, A. Cazenave, ed., D. Reidel, Amsterdam, 297-324, 1986.

Widmer, R., Masters, G. and F. Gilbert, The spherical Earth revisited, *17th International Conference on Mathematical Geophysics*, Blanes, Spain, I. U. G. G., June 1988.

D.E. Smylie and Xianhua Jiang, Department of Earth and Atmospheric Sciences, York University, North York, Ontario, M3J 1P3, Canada.

A Comparative Analysis of Superconducting Gravimeter Data

D. E. SMYLIE

Department of Earth and Atmospheric Science, York University, North York, Ontario M3J 1P3, Canada

JACQUES HINDERER

Institut de Physique du Globe, 67084 Strasbourg Cedex, France

BERND RICHTER

Institut für Angewandte Geodasie, D-6000 Frankfurt am Main 70, Germany

BERNARD DUCARME

Observatoire Royal de Belgique, B-1180 Bruxelles, Belgium

AND

LALU MANSINHA

Department of Geophysics, University of Western Ontario, London Ontario N6A 5B7, Canada

Abstract   Four long records of tidal residuals from
European superconducting gravimeters are analyzed and
compared. Power spectral densities are estimated from
the four records and a product power spectrum is computed to bring out common features. Resonances with
the same rotational splitting as the three translational
modes of Earth's solid inner core are found in the product spectrum. A shorter record from the Canadian superconducting gravimeter installation is also examined. Our
analysis serves as a pilot study of the effectiveness of a
global network of superconducting gravimeter observatories proposed for the international Global Geodynamics
Project.

Introduction

There are now as many as eleven superconducting
gravimeters in operation world-wide. The principal ob-
jectives of this study are to examine four of the longest
records available in a comparative analysis, and to search
for geodynamical signals in the frequency spectral domain.

Our analysis subjects the four tidal residual series to
a uniform power spectral density estimation procedure,
and then we introduce a novel product spectrum of the
four records to emphasize signals seen in them all and to
suppress uncorrelated noise. The product spectrum is a
straightforward generalization of the cross-spectrum fre-
quently used for the detection of coherent signals between
two records.

The first of the four tidal residual series used in our
study is the old Brussels record running from June 2, 1982
to October 14, 1986. It is the longest continuous record
available, consisting of 38,304 hourly values. The second
record was taken at the Bad Homburg observatory near
Frankfurt, Germany and the tidal residual series it gives
runs from March 22, 1986 to December 27, 1988 for 24,272
hourly values. Record three is from Strasbourg and covers
the period from October 1, 1987 to March 12, 1991 yielding 30,177 hourly values of tidal residuals. The fourth
long record we consider is the new Brussels series beginning July 22, 1987 and running to December 30, 1989,
giving 21,432 hourly tidal residuals.

Dynamics of Earth's Deep Interior and Earth Rotation
Geophysical Monograph 72, IUGG Volume 12
Published in 1993 by the International Union of Geodesy and Geophysics
and the American Geophysical Union.

It turns out that the four longest available records are all from European stations in close geographical proximity, and we therefore include for interest an analysis of data from the new Canadian installation at Cantley, Quebec, although the record is too short (8,665 hourly values of tidal residuals) to be included in the common product spectrum.

Our analysis is the first to use superconducting gravimeter data from several stations and may therefore be regarded as a pilot study of what might be achieved by the international Global Geodynamics Project adopted by SEDI (Study of the Earth's Deep Interior) at the August, 1991 IUGG Assembly in Vienna (Aldridge et al, 1991). An unexpected outcome has been the identification of a triplet of resonances with the same spectral splitting (to within parts in 10^4) as the three translational modes of the solid inner core (see Smylie and Jiang (1992) this volume).

Spectral Analysis and the Product Spectrum

The removal of the tidal signal from gravimeter measurements is a specialized task which has evolved over the course of development of earth tide research. Similarly, the corrections for the effects of the atmosphere, both its direct attraction and loading, and for ocean tides, require special consideration. The practice in making these adjustments varies from one observatory to another according to local conditions, and to some extent, local traditions. Since our primary purpose here is to examine the residual records for geodynamical signals, these corrections have been left entirely to the observers familiar with local site and instrumental characteristics.

Some adjustments of the data sets were made prior to analysis. The Bad Homburg data is subject to a very severe drift near the beginning of the record. Accordingly, a drift function $a + be^{-ct}$ is subtracted with $a = -104.087\mu gals$, $b = 110.339\mu gals$ and $c = 1.536321 \times 10^{-4}hr^{-1}$. In the case of the Strasbourg series, as a check on the classical least squares fitting procedure for removing the tidal signal, we began with 5 minute samples including tides and then removed them using a synthetic tide generated with the G-Wave programme of James B. Merriam. This circumvents completely the possibility that signal might be absorbed in the least squares fitting procedure commonly used in the removal of tides. Decimation to hourly samples was then accomplished with a 480 point (40 hour) low pass filter with a Parzen window roll-off beginning at 0.4 cycles per hour and cutting off at the Nyquist frequency of 0.5 cycles per hour. A linear drift function $c + dt$ was then subtracted with $c = -68.7751\mu gals$ and $d = 3.147954 \times 10^{-3}\mu gals \cdot hr^{-1}$. The four residual series subjected to analysis are shown in Figure 1.

Our choice of method for spectral analysis is the windowed, overlapping segment method of P. D. Welch (Welch, 1967) which has become a nearly universal method of balancing the trade-off between reliability and resolution. The resolution is controlled by the length M of the time domain window $w(t)$, a real function, even about its centre. A sample spectral estimator for the M-length segment is then

$$\tilde{S}_{gg}(f_k) = \frac{|H_k|^2}{I} \qquad (1)$$

where $h(t) = w(t)g(t)$ is the windowed version of the data $g(t)$ on the segment. The Finite Fourier transform of $h(t)$ is H_k and

$$I = \int_{-M/2}^{M/2} w^2(t)\,dt. \qquad (2)$$

The original record $g(t)$ is assumed to be band limited between the Nyquist frequencies and to be of indefinite extent. Expression (1) estimates the power spectral density at intervals of $1/M$ on the frequency axis between the Nyquist frequencies.

The final spectral density estimate $\bar{\bar{S}}_{gg}(f)$ is taken as the average of the estimates (1) over κ overlapping segments of the complete record of length T. The variance of the final spectral estimate is found to be

$$var\left\{\bar{\bar{S}}_{gg}(f)\right\} = \frac{S_{gg}^2(f)}{\kappa}\left[1 + 2\sum_{m=1}^{\kappa-1}\frac{\kappa - m}{\kappa}\rho(m)\right] \qquad (3)$$

with $S_{gg}(f)$ as the true spectral density and $\rho(m) = J_m^2/I^2$ where

$$J_m = \int_{-M/2}^{M/2} w(t)w(t - mM)\,dt \qquad (4)$$

is the integral of the product of windows separated by m segments.

An unfortunate confusion of nomenclature arises in the Welch paper where a triangular window is referred to as a Parzen window. Normally, the triangular window is called a Bartlett window and a Parzen window refers to the piecewise cubic resulting from the convolution of two Bartlett windows. Welch's choice of 50% overlap of segments and his estimate of variance, which have spread broadly through the literature and into mathematical subroutine libraries, apply only to the Bartlett window.

For the Parzen window, which we have chosen in our analysis because of its superior side lobe behaviour, an overlap of 75% is near the optimum and we then find by direct integration $I = 302/1120$, $J_1 = 397J_2/40$, $J_2 = 3/224$ and $J_3 = J_2/120$, all given in units of M. The variance for a Parzen window with 75% overlap is found from expression (3) to be

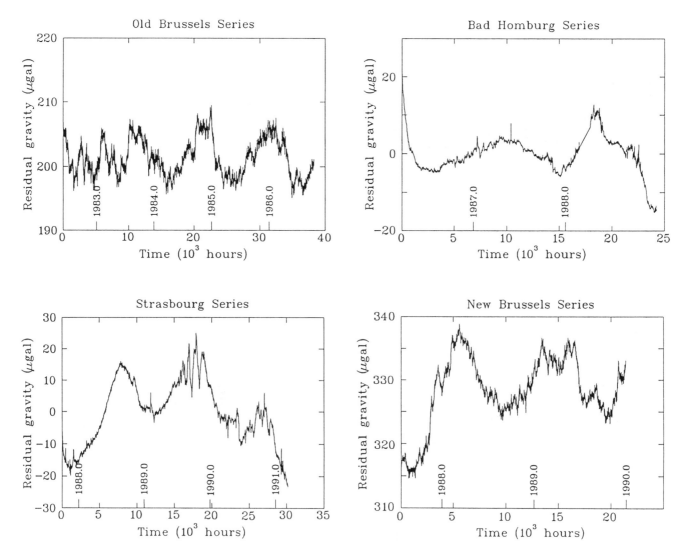

Figure 1. The four European tidal residual series. (A) (upper left) The old Brussels series running from June 2, 1982, 1 hr and running to October 14, 1986, 24 hr for a total of 38,304 hourly values. (B) (upper right) Bad Homburg tidal residuals after subtraction of the drift function $a + be^{-ct}$ with t in hours starting from $t = 1$ and with $a = -104.087\mu gals$, $b = 110.339\mu gals$ and $c = 1.536321 \times 10^{-4} hr^{-1}$. The complete record consists of 24,272 hourly points running from March 22, 1986, 0 hr to December 27, 1988, 7 hr. One gap has been filled by straight-line interpolation. (C) (lower left) The Strasbourg data series formed by the subtraction of a synthetic tide, based on the G-Wave programme of James B. Merriam, from 5 minute samples. Decimation to hourly samples is done using a 480 point filter with roll-off beginning at 0.4 cph and continuing to the cut off at the Nyquist frequency of 0.5 cph. The decimated data begins at October 1, 1987, 20 hr and ends March 12, 1991, 4 hr for a total of 30,177 hourly samples. A drift function $c + dt$ with t in hours beginning at $t = 1$ and with $c = -68.7751\mu gals$, $d = 3.147954 \times 10^{-3}\mu gals \cdot hr^{-1}$ is subtracted. (D) (lower right) New Brussels tidal residuals series beginning July 22, 1987, 1 hr and running to December 30, 1989, 24 hr for a total of 21,432 hourly readings.

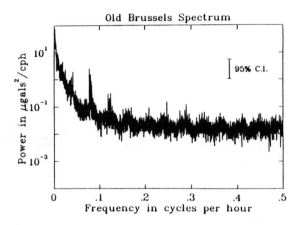

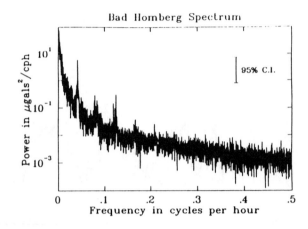

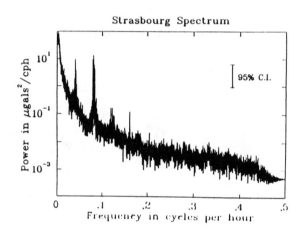

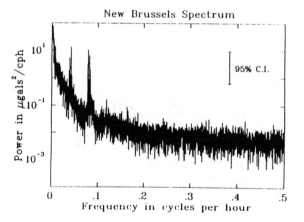

Figure 2. (A) (upper left) Old Brussels spectrum using the first 36,000 hourly values of the record. A 12,000 hr Parzen window is employed with 75% overlap giving nine segments with 12.54 equivalent degrees of freedom. (B) (upper right) Bad Homburg spectrum using the first 24,000 hourly values of the record. A 12,000 hr Parzen window is employed with 75% overlap giving five segments with 7.19 equivalent degrees of freedom. (C) (lower left) Strasbourg spectrum using the first 30,000 hourly values of the record. A 12,000 hr Parzen window is employed with 75% overlap giving seven segments with 9.86 equivalent degrees of freedom. (D) (lower right) New Brussels spectrum using the first 21,000 hourly values of the record. A 12,000 hr Parzen window is employed with 75% overlap giving four segments with 5.85 equivalent degrees of freedom.

$$var\left\{\bar{\bar{S}}_{gg}(f)\right\} = \frac{S_{gg}^2(f)}{\kappa}\left[1.490960 - \frac{0.495895}{\kappa}\right]. \quad (5)$$

With 75% overlap, the number of segments is

$$\kappa = \left(4\frac{T}{M} - 3\right) \quad (6)$$

giving

$$\nu = 2\kappa / \left[1.490960 - \frac{0.495895}{\kappa}\right] \quad (7)$$

equivalent degrees of freedom. The final spectral density estimate is χ_ν^2 distributed if the pairs of individual harmonic amplitudes at each discrete frequency are regarded as resulting from a Gaussian process.

Thus, for long records the variance inflation by the factor 1.49 arising from the 75% overlap is more than offset by the fourfold increase in effective record length.

The width between half-power points of the Parzen window in the frequency domain is approximately $1.82/M$. We have chosen a window width of 12,000 hrs which allows frequency resolution in the range of 10^{-4} while yielding reasonable segment numbers for good reliability. The first 36,000 hours of the old Brussels record are used producing 9 segments with 12.5 equivalent degrees of freedom. The first 24,000 hours of the Bad Homburg series is used giving 5 segments with 7.2 equivalent degrees of freedom. For the Strasbourg record we utilize the first 30,000 hourly values broken down into 7 segments with 9.9 equivalent degrees of freedom. The first 21,000 hours of the new Brussels series produce 4 segments with 5.9 equivalent degrees of freedom. The power spectral densities for the four records are shown in Figure 2. The 95% confidence interval, parallel transportable on the logarithmic scale is shown directly on each plot.

Each of the spectral densities in Figure 2 is $\chi^2_{\nu_i}$ distributed with probability density function

$$f_i(x_i) = \frac{1}{2^{\nu_i/2}\Gamma(\nu_i/2)}x^{\nu_i/2-1}e^{-x/2}. \qquad (8)$$

If we form the product of the four spectral densities, the cumulative distribution function for the newly created product spectrum is

$$F(z) = \int_0^\infty f_1(x_1)\int_0^\infty f_2(x_2)\int_0^\infty f_3(x_3)\cdot \qquad (9)$$

$$P(\nu_4/2, z/2x_1x_2x_3)\,dx_3dx_2dx_1, \qquad (10)$$

where

$$P(a,y) = \frac{1}{\Gamma(a)}\int_0^y t^{a-1}e^{-t}dt \qquad (11)$$

is the incomplete gamma function. After transforming each of the integrals to the interval $(0,1)$, expression (9) can be evaluated numerically and iteration for the appropriate percentiles of the cumulative distribution permits the calculation of confidence intervals for the product spectrum.

The band between the diurnal and semi-diurnal tides, the so-called intertidal band, and the short period subtidal band are of the most interest in searching for geodynamical signals. The product spectra in these two bands are shown in Figure 3. Modes associated through their rotational splitting with the translational oscillations of the solid inner core are indicated directly on the subtidal band plot.

Finally, we include a spectral analysis of tidal residuals from the new Cantley, Canada superconducting gravime-

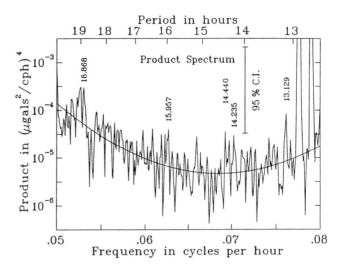

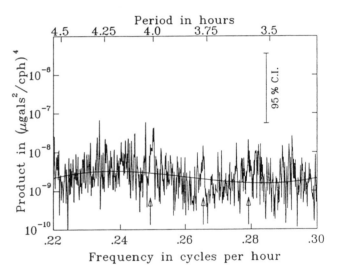

Figure 3. (A) (upper plot) Product spectrum of the four superconducting gravimeter records in the intertidal band. A parabolic reference noise level is fitted excluding the two tidal lines at the right. The periods of the five peaks with the largest values above the reference noise are shown directly on the diagram. The peak at 13.129 hr is likely a tidal residual line. (B) (lower plot) Product spectrum in the subtidal band. A sinusoid is shown fitted across the whole spectrum to provide a reference noise level. The locations of resonances with the same rotational splitting as the triplet of inner core oscillations are indicated by arrows.

ter installation. The record runs from February 23, 1990 to February 18, 1991 for 8,665 hourly values. This record is too short to be included in the product spectrum of the four longer European records. We use the first 8,659 hours of the record and a Parzen window length of 4,948 hours to produce 4 segments with 75% overlap giving 5.9 equivalent degrees of freedom. The residual data series and the power density spectrum are shown in Figure 4. The noise level for this instrument in the subtidal band

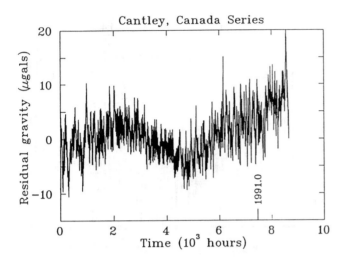

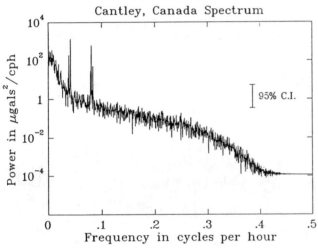

Figure 4. (A) (upper plot) Cantley, Canada tidal residuals running from February 23, 1990, 0 hr to February 18, 1991, 23 hr giving a total of 8,665 hourly values. (B) (lower plot) Cantley, Canada spectrum using the first 8,659 hourly values of the record. A 4,948 hr Parzen window is used with 75% overlap giving four segments with 5.85 equivalent degrees of freedom.

seems somewhat higher than for the four European installations and there is evidence of a roll off beginning at 0.3 cycles per hour which may be more severe than required to prevent aliasing.

Discussion

While our analysis may be regarded as a preliminary study, it illustrates very well the power of combining records from several stations. A very important next step will be to combine records from stations with a broader global distribution. This will add a new dimension to the effort of detecting geodynamical signals in what has become an exciting new field in terrestrial spectroscopy.

Acknowledgements

The authors are grateful to the agencies supporting the very highly specialized superconducting gravimeter installations which make the observations used in our analysis. The Tri-University Meson Facility (TRIUMF) in Vancouver, Canada has generously donated their outstanding graphics and analysis package.

References

Aldridge, K. D., Crossley, D. J., Mansinha, L. and D. E. Smylie, GGP:The global geodynamics project, *Cahiers du Centre Européen de Géodynamique et de Séismologie*, **3**, 169-196, 1991.

Smylie, D. E. and Xianhua Jiang, Numerical calculation of modes of oscillation of the Earth's core, (this volume) 1992.

Welch, P. D., The use of the Fast Fourier Transform for the estimation of power spectra: a method based on time averaging over short, modified periodograms, *IEEE Trans. Audio Electroacous.*, **AU-15**, 70-74, 1967.

D. E. Smylie, Department of Earth and Atmospheric Science, York University, North York, Ontario M3J 1P3, Canada.

Jacques Hinderer, Institut de Physique du Globe, 67084 Strasbourg Cedex, France.

Bernd Richter, Institut fur Angewandte Geodasie, D-6000 Frankfurt am Main 70, Germany.

Bernard Ducarme, Observatoire Royal de Belgique, B-1180 Bruxelles, Belgium.

Lalu Mansinha, Department of Geophysics, University of Western Ontario, London Ontario N6A 5B7, Canada.

Time-Dependent Flow at the Core Surface and Conservation of Angular Momentum in the Coupled Core-Mantle System

ANDREW JACKSON

Department of Earth Sciences, Oxford University, Parks Road, Oxford, U.K.

JEREMY BLOXHAM

Department of Earth and Planetary Sciences, Harvard University, Cambridge, Massachusetts, U.S.A.

DAVID GUBBINS

Department of Earth Sciences, Leeds University, Leeds, U.K.

We present the results of calculations to determine fluid velocities at the core-mantle boundary from magnetic data collected over the period 1840–1990. The calculations adopt the frozen-flux and tangentially geostrophic approximations which neglect firstly the contribution of magnetic diffusion to the observed secular variation, and secondly the contribution of Lorentz forces in the horizontal force balance at the top of the core. We show that in general the flows are able to fit observatory data to a reasonable degree of accuracy. Of particular interest is the time-dependency of the flows: steady flows must be discounted on such a long timescale because changes in the length-of-day on a decade timescale are believed to indicate core-mantle coupling and changes of angular momentum of the core. We use a recent theory which predicts the response of the core to topographic and gravitational torques to extrapolate our core surface motions throughout the core and give an estimate of the changes of angular momentum of the core. The resulting predictions of the changes in the angular momentum of the mantle are highly encouraging, indicating that the time variations in the core flow are of the correct magnitude to conserve angular momentum in the core-mantle system.

INTRODUCTION

The last decade has seen renewed interest in tackling the problem of understanding the secular variation (SV) of the earth's magnetic field in terms of fluid flow at the core surface. Much attention has been given to both the underlying physics and the mathematical problem of constructing solutions for the flow. In addition, the fidelity of models of the magnetic field which are designed to describe the evolution of the field in time and space either at the earth's surface [e.g. Langel et al., 1986] or at the core-mantle boundary (CMB) [e.g. Bloxham, 1987; Bloxham and Jackson, 1989; Bloxham and Jackson, *Time-dependent mapping of the magnetic field at the core-mantle boundary*, submitted to *J. Geophys. Res.*, referred to hereinafter as BJ92] has been steadily increasing. The calculations presented in this paper are, to

Dynamics of Earth's Deep Interior and Earth Rotation
Geophysical Monograph 72, IUGG Volume 12
Published in 1993 by the International Union of Geodesy and Geophysics and the American Geophysical Union.

a certain extent, a result of this combination of theoretical advances and improvement in the quality of data available for probing the fluid flow at the core surface.

Ever since the work of Roberts and Scott [1965], one of the basic tenets of core-flow calculations has been the use of the frozen-flux approximation, namely the approximation which ascribes all of the observed SV to that produced by advection of existing field **B** by a fluid flow **v**, and the possibility of magnetic diffusion is ignored. The calculations presented here are performed using this same approximation; however it should be remembered that this device is indeed an approximation which is expected to break down over long timescales and over the smallest spatial scales. Indeed, the very existence of a dynamo (in the geophysical context) rests on a balance between diffusion and advection over long timescales. Observational evidence for a breakdown of frozen-flux on a century timescale has been presented previously by Bloxham and Gubbins [1985] and most recently by Bloxham et al. [1989]; the strongest evidence comes from the monotonic growth of a reverse flux patch in the South

Atlantic during the 20th century. Although theories allowing for diffusion have been developed (Voorhies, personal communication), the frozen-flux hypothesis is taken as a zeroth-order approximation here.

Roberts and Scott [1965] recognised that solutions for fluid velocities **v** based solely on the radial component of the induction equation would be nonunique; this nonuniqueness was explored further by Backus [1968]. Most of the progress over the last ten years has been directed towards investigating the effect on the issue of nonuniqueness of plausible constraints on admissible types of fluid velocities. An in-depth review is given by Bloxham and Jackson [1991]. Of the methods recently proposed, the one adopted here is that of tangential geostrophy developed independently by Hills [1979] and LeMouël [1984]. This hypothesis requires that the contribution of the Lorentz force to the radial vorticity balance be sufficiently small that it can be neglected, and that the horizontal force balance is almost entirely between pressure and Coriolis forces. It is well-known that even a poorly-conducting mantle allows a possible large contribution to the Lorentz force from coupling of toroidal and poloidal magnetic fields [e.g. LeMouël, 1984]; the size of the contribution from poloidal fields alone is much smaller than that of the Coriolis force. Specifically, for a toroidal field **B**$_T$ which vanishes at the CMB we require

$$\frac{\partial \mathbf{B}_T}{\partial r} \ll 10^3 \text{ nT m}^{-1} \tag{1}$$

for geostrophy to be a reasonable approximation.

In principle it is possible to test the validity of the hypothesis of frozen-flux and tangential geostrophy against actual magnetic data (or a time-varying field model) since certain integral constraints must be satisfied at each point in time. This is discussed in the next section.

TESTING FROZEN-FLUX AND TANGENTIAL GEOSTROPHY

In this section we ask whether magnetic data can be used to resolve if the frozen-flux and tangentially geostrophic theories are a viable combination for the earth on a decade timescale. Under the frozen-flux hypothesis, the radial induction equation reads

$$\dot{B}_r = -\nabla_{\mathrm{h}} \cdot (\mathbf{v} B_r) \tag{2}$$

where B_r and $\dot{B}_r$ are the radial magnetic field at the CMB and its rate of change with time respectively, **v** is a fluid velocity, and ∇_{h} is the horizontal gradient: $\nabla_{\mathrm{h}} = \nabla - \hat{\mathbf{r}}\nabla$. The assumption of tangential geostrophy leads to a constraint on the fluid velocity of the form

$$\nabla_{\mathrm{h}} \cdot (\mathbf{v}\cos\theta) = 0 \tag{3}$$

where θ is colatitude [LeMouël et al., 1985]. As Backus [1968] pointed out, it follows from (2) that for the frozen-flux hypothesis to hold we require that certain integrals of the radial magnetic field B_r must be invariant with time, namely

$$\frac{d}{dt}\int_S B_r \, dS = 0 \tag{4}$$

where S is a patch whose boundary ∂S is a curve on which $B_r = 0$. These conditions form a finite set for non-trivial fields. Provided these frozen-flux conditions hold, and if the core flow is tangentially geostrophic, then additional conditions apply, namely

$$\frac{d}{dt}\int_{S_\psi} B_r \, dS = 0 \tag{5}$$

where S_ψ is a patch whose boundary ∂S_ψ is a curve on which $\psi = B_r/\cos\theta = $ constant (see e.g. Jackson [1989]; Bloxham and Jackson [1991]). These latter conditions form an infinite set since there are an infinite number of patches bounded by curves on which ψ is constant. Note also that both conditions (4) and (5) are nonlinear in the field because the boundary of the patch is determined by the field B_r itself. In fact one of the conditions (5) is linear in the field: for $\psi = \pm\infty$ the boundary of the patch is the geographical equator irrespective of the model of B_r. Constancy of the flux through the patch bounded by the geographical equator follows from the fact that the flux is frozen and fluid particles cannot cross the equator [e.g. Benton, 1985]. Let us call this flux out of the northern geographical hemisphere $\mathcal{N}$: then

$$\mathcal{N} = \int_\Omega G_\psi B_r \, d\Omega \tag{6}$$

where Ω is the whole CMB and G_ψ is a prediction kernel (a boxcar with value unity over the northern hemisphere and value zero over the southern hemisphere). Estimating the numerical value of $\mathcal{N}$ at different times gives the prospect of testing whether the dual frozen-flux and tangential geostrophy hypothesis is valid. This has been done previously by Benton et al. [1987], who conclude that the constraint is reasonably satisfied. However, to test the hypothesis requires some estimate of error on the value of $\mathcal{N}$. Unless the prediction kernel appropriate for synthesizing $\mathcal{N}$ can be written as a linear combination of the data kernels associated with a particular dataset, nothing in principle can be said about $\mathcal{N}$, unless an *a priori* bound is held on the model [Backus and Gilbert, 1968; Backus, 1970].

One such *a priori* bound relevant to this problem is the heat flow bound, which requires that the minimum Ohmic heating in the core associated with the poloidal magnetic field does not exceed that observed to be flowing through the earth's surface, namely 3×10^{13} W. With an appropriate choice of Hilbert space, the norm can be used to bound the size of the uncertainty in the prediction by using Backus linear inference [e.g. Backus, 1989]. Although we will not present the theory or results here [Jackson, unpublished manuscript, 1990], it is unfortunately the case that the heat flow bound is sufficiently slack, and the prediction kernel sufficiently difficult to synthesize that the associated error in the prediction is much larger than the changes in the value of $\mathcal{N}$ over a few decades. It appears that although models of the field such as the one used in this analysis do not conserve (6), it would be possible to construct models which do not exhibit this property whilst still satisfying the data and the heat-flow bound. As a result, neither geostrophy nor frozen-flux can be dismissed on the basis of this single (admittedly special) prediction.

METHOD

We now describe our method for calculating tangentially geostrophic flows. Our formalism is essentially the same as that used by almost all other authors and we only outline the method. All the flows were calculated using the time-dependent field model for the period 1840–1990 of BJ92, which is an expansion of the field at the CMB spatially in spherical harmonics and temporally in B-splines.

The starting point is the spectral expansion for the velocity

$$\mathbf{v} = \mathbf{v}_t + \mathbf{v}_p = \nabla \wedge (T\mathbf{r}) + \nabla_h(rS) \qquad (7)$$

where the toroidal and poloidal vectors in a system with spherical coordinates (r, θ, ϕ) are

$$\mathbf{v}_t = \left(0, \frac{1}{\sin\theta}\frac{\partial T}{\partial \phi}, -\frac{\partial T}{\partial \theta}\right) \quad \mathbf{v}_p = \left(0, \frac{\partial S}{\partial \theta}, \frac{1}{\sin\theta}\frac{\partial S}{\partial \phi}\right) \quad (8)$$

We expand the potentials T and S in spherical harmonics

$$T(\theta, \phi) = \sum_{l,m} t_l^m Y_l^m(\theta, \phi)$$

$$S(\theta, \phi) = \sum_{l,m} s_l^m Y_l^m(\theta, \phi) \qquad (9)$$

where Y_l^m is a real Schmidt quasi-normalised spherical harmonic. Clearly

$$\mathbf{v}_t = \sum_{l,m} t_l^m \mathbf{T}_l^m \quad ; \quad \mathbf{v}_p = \sum_{l,m} s_l^m \mathbf{S}_l^m \qquad (10)$$

where the basis vectors $\mathbf{T}_l^m$ and $\mathbf{S}_l^m$ are given by

$$\mathbf{S}_l^m = r\nabla_h Y_l^m \qquad (11)$$

$$\mathbf{T}_l^m = \nabla \wedge (Y_l^m \mathbf{r}) \qquad (12)$$

We expand the geomagnetic field in the usual way:

$$B_r(c, \theta, \phi) = \sum_{l,m} (l+1) \left(\frac{a}{c}\right)^{l+2}$$

$$\times \left[g_l^m \cos m\phi + h_l^m \sin m\phi\right] P_l^m(\cos\theta) \qquad (13)$$

where $P_l^m(\cos\theta)$ is an associated Legendre function and a and c are the earth and core radii respectively; a similar expansion is used for $\dot{B}_r$. Substituting into the radial induction equation, equation (2), we find that under this parametrisation it can be written in the form

$$\dot{\mathbf{b}} = A\mathbf{m} = \left[E \vdots G\right] \begin{pmatrix} \mathbf{t} \\ \cdots \\ \mathbf{s} \end{pmatrix} \qquad (14)$$

Here $\dot{\mathbf{b}}$ is a vector of SV coefficients $\{\dot{g}_l^m; \dot{h}_l^m\}$, $\mathbf{m}$ is a model vector comprised of the vectors $\mathbf{t}$ and $\mathbf{s}$, vectors of toroidal and poloidal coefficients respectively, and G and E contain elements which are related to the Gaunt and Elsasser integrals respectively [e.g. Gibson and Roberts, 1969]. Explicit expressions for the elements of G and E can be found in various places, for example Whaler [1986], Bloxham [1988] or Jackson and Bloxham [1991]. The Gaunt and Elsasser integrals have been evaluated exactly using the transform method of Lloyd and Gubbins [1990]. The elements of E and G are non-zero only for certain choices of l and m — these are the selection rules of Bullard and Gellman [1954].

The geostrophic constraint (3) can be implemented in a number of ways. Backus and LeMouël [1986] show that velocities which satisfy (3) can be written in terms of a new basis, the geostrophic basis, whose members are linear combinations of the poloidal and toroidal vectors (11) and (12). Gire and LeMouël [1990] implement this basis in the calculation of tangentially geostrophic flows compatible with secular variation. LeMouël et al. [1985] write the constraint (3) in the form

$$G\mathbf{m} = 0 \qquad (15)$$

and then implement the constraint approximately using a penalty method. It is equally convenient to implement the constraint exactly using orthogonal transformations such as the QR decomposition [Golub and Van Loan, 1983 p. 411]. Care must be taken because the matrix G is not of full rank; if the flow is truncated at degree L_v, then the matrix G has rank $L_v(L_v + 4)$ [LeMouël et al., 1985]. After implementing the constraint, the number of free parameters (before implementing any regularisation) becomes $2L_v(L_v + 2) - L_v(L_v + 4) = L_v^2$.

To implement (9) numerically it is necessary to fix some truncation level L_v (chosen here as 14) before which point we ensure the spectral velocity expansion has converged. To provide convergence we apply a regularisation condition. Noting that the horizontal divergence D and radial vorticity V can be written as

$$D = \nabla_h \cdot \mathbf{v} \quad ; \quad V = (\hat{\mathbf{r}} \wedge \nabla_h) \cdot \mathbf{v} \qquad (16)$$

we penalise spatial gradients in these quantities, a regularisation procedure common in geophysics [e.g. Constable et al., 1987]. We minimise the quantity

$$\frac{c^4}{4\pi} \int (\nabla_h D)^2 + (\nabla_h V)^2 \, d\Omega$$

$$= \sum_l \frac{l^3(l+1)^3}{2l+1} \sum_m \left[(t_l^m)^2 + (s_l^m)^2\right] = \mathbf{m}^T N\mathbf{m} \qquad (17)$$

a norm used previously by Bloxham [1988]. As usual, a solution is sought which minimises the square of the error $\mathbf{e}$ between the observations and the predictions, subject to the regularisation: we minimise Φ where

$$\Phi = \mathbf{e}^T C_e^{-1} \mathbf{e} + \lambda \mathbf{m}^T N\mathbf{m} \qquad (18)$$

and $\mathbf{e} = \mathbf{Am} - \dot{\mathbf{b}}$, $\mathbf{C_e}$ is the estimated covariance matrix of the data, and λ is used to find an appropriate balance between smoothness of the solution and fit to the data.

The errors in each of the secular variation coefficients $\{\dot{g}_l^m; \dot{h}_l^m\}$, which enter the weight matrix $\mathbf{C_e}$, require a little discussion. Recall that our data are actually estimates of SV from the time-dependent model of BJ92; they are linear predictions made from the complete model vector which we will call $\mathbf{M}$, so for example, $\dot{\mathbf{b}} = \mathbf{S}(t_i)\mathbf{M}$ at one particular epoch t_i. In principle, it is possible to calculate the covariance matrix for $\dot{\mathbf{b}}$ (which we call $\mathbf{C_e}$ here) from the covariance matrix for $\mathbf{M}$, $\mathbf{C_M}$. This proves to be rather inconvenient, because of the size of $\mathbf{C_M}$: it occupies almost 100 M words of memory. For this reason we have used just the diagonal elements of $\mathbf{C_M}$ to calculate the diagonal elements of $\mathbf{C_e}$. One further approximation is made: the errors for each harmonic of degree l were found by averaging over all the errors in the harmonics within that degree from the covariance matrix. When this procedure was compared to the average errors per degree found when the full matrix $\mathbf{C_M}$ was used, the difference was slight. However this process does assign a correct variability in the size of errors with time, reflecting the quality and distribution of data which originally contributed to the time-dependent model. An example of the variability in the error estimate on a single coefficient with time can be found in BJ92. Some previous authors have fit SV harmonics either "at the earth's surface" or "at the CMB" — using

the proper error estimates on the coefficients eliminates this problem, since the two are the same [e.g. Whaler and Clarke 1988]. Minimising the squared misfit between predicted and observed SV power at the earth's surface [e.g. LeMouël et al., 1985] or its radial component [Voorhies, 1986] has been implemented by previous authors; the former assigns an error of $1/\sqrt{l+1}$ to each SV Gauss coefficient of degree l. The reason that fitting at the earth's surface works reasonably is that the errors in B_r are much closer to being white at the earth's surface than at the CMB, because this is where the source of noise is. Indeed, similar results to those presented herein are obtained if the SV power is fit at the earth's surface rather than using the *a posteriori* error estimates from the spline model.

RESULTS

Velocities were calculated at 2.5 year intervals over the period 1840–1990. All the models presented here were calculated using the same value for the regularising parameter, λ, namely the value $\lambda = 2 \times 10^{-2}$, a value chosen so that a satisfactory fit to the data is achieved. As a result, the number of degrees of freedom in each model and the norm of each model (equation (17)) varies with time. However, this procedure seems reasonable if the models are to be allowed to vary depending on the information content of the data. Figure 1 shows example velocity fields for 3 epochs in the 20th century. The morphology of the flows are very similar

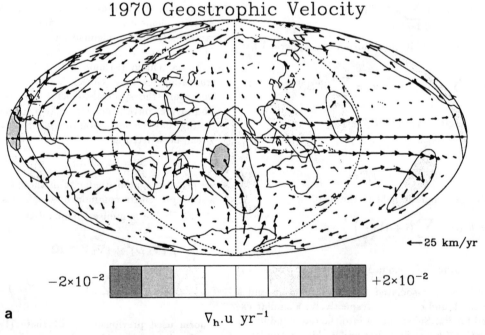

Fig. 1. Geostrophic velocities for three epochs. The vectors show the speed and direction of the flow near the core surface and the grey scale the intensity of the horizontal divergence (upwelling and downwelling) of the flow. The sign of the horizontal divergence can be determined from the flow vectors: flow towards the equator corresponds to positive divergence (upwelling) and flow away from the equator to negative divergence (downwelling). (a) 1970. (b) 1930. (c) 1890.

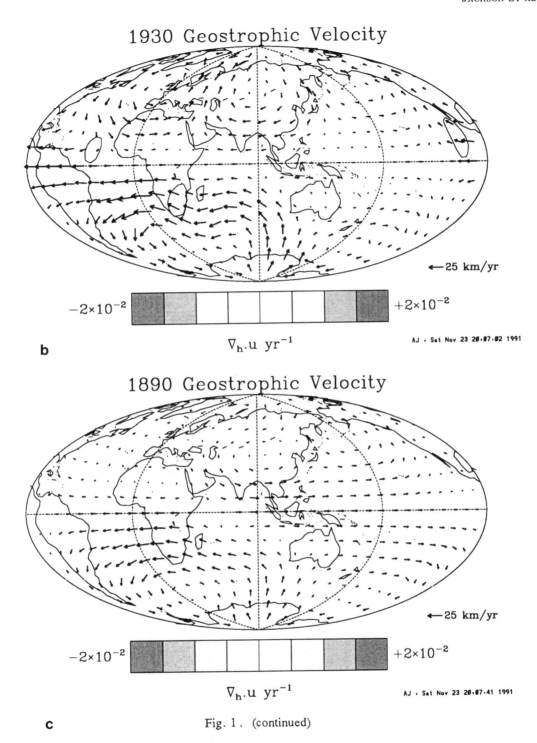

Fig. 1. (continued)

to tangentially geostrophic velocities published previously (see e.g. Bloxham and Jackson [1991] for intercomparisons of published flows) and we do not dwell on their description. In Figure 2 we show the fit of the secular variation predicted by the flows to that observed at sample permanent observatories. The fit is superior for the northern hemisphere observatories, probably because our estimates of the actual secular variation are most accurate there and possibly because some of the signal for the southern hemisphere observatories may be generated diffusively in the core; indeed, the field model used for the velocity inversions does not conserve the necessary frozen-flux integrals in the southern hemisphere.

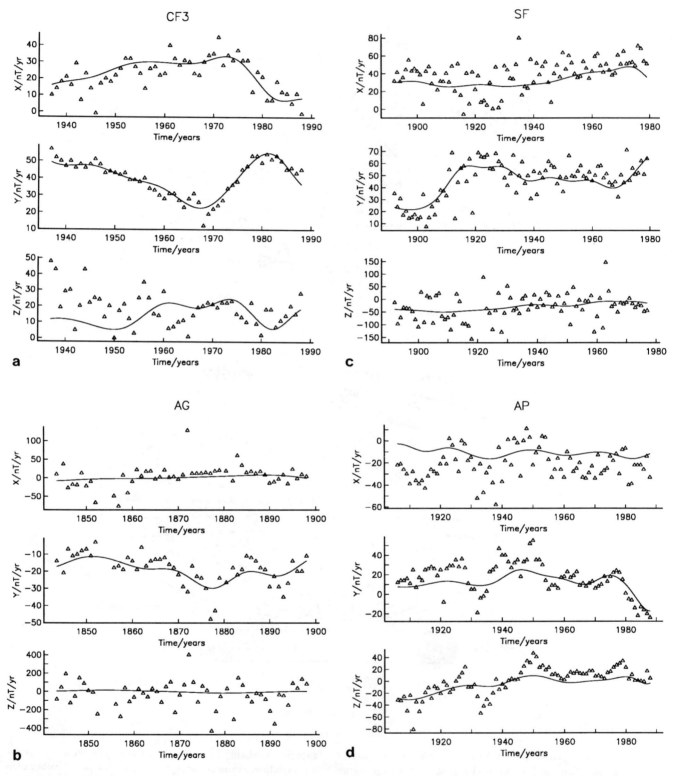

Fig. 2. Comparison of secular variation recorded at selected observatories and that predicted by the time-dependent tangentially geostrophic flows calculated in this paper. The observatory estimates of SV are calculated from first-differences of annual mean observations. Note the low noise level in Y, the easterly component. (a) Chambon-la-Forêt (CF3), France. (b) Agincourt (AG), Canada. (c) San Fernando (SF), Spain. (d) Apia (AP), W. Samoa.

Nevertheless, we believe that the fit is generally reasonable, although there remains some unexplained signal. The high frequency residuals between the model and the observatory values are mostly due to external fields; note the low levels of noise in the Y component of field, a consequence of the geometry of the external ring current. The X and Z components are not fitted as well as the Y component, an effect which appears to be due to the imposition of the geostrophic constraint because it is not seen in unconstrained flows. We note that the secular variation shows a sufficiently large degree of time dependency that it is impossible to attain such a good degree of fit to the data with a steady flow: although steady flows do attain some variability in predicted secular variation with time, because of the non-linearity of the steady forward problem, they cannot achieve the same level of complexity of secular variation over a 150 year timespan as that seen in observatory records.

We now ask if these flows can be used to infer anything about the flows occurring within the core, and in particular whether they can account for observed changes in the angular momentum of the mantle, thought to be indicative of angular momentum transfer to the core. This section is necessarily somewhat speculative, but the ideas are sufficiently interesting to warrant testing. The theory behind the arguments which follow was laid out in an important paper by Jault and LeMouël [1989] (hereinafter referred to as JLM), and results were presented by Jault et al. [1988] (hereinafter referred to as JGLM) for the period 1969–1985, and preliminary results for the 20th century were given by Jackson [1989]. JLM are concerned with the response of the core to torques which can act between the core and mantle. These torques can be in the form of viscous coupling at the CMB (generally accepted to be too small), electromagnetic coupling (which JLM believe to be discounted by recent observations), gravitational coupling (due to the action of an aspherical geoid acting on density inhomogeneities in the core) or topographic coupling due to dynamical pressure variations acting on an aspherical CMB, a suggestion originally due to Hide [1969]. The system considered by JLM is one of an inviscid, rotating fluid carrying a magnetic field. They show that the result of both topographic and gravitational coupling between the core and mantle is to excite motions in cylindrical annuli of the type

$$v_G = v_\phi(s)\hat{\phi} \qquad (19)$$

where s is the distance perpendicular to the rotation axis and $\hat{\phi}$ is the unit vector in the easterly direction. We follow Gubbins [1991] and refer to these motions as the geostrophic degeneracy. Figure 3 shows schematically the form of these flows. The importance of these flows was shown many years ago by Taylor [1963] when considering an inviscid, rotating fluid permeated by a magnetic field and in a container with axisymmetric, insulating boundaries. In this case there can be no coupling between the fluid and its container; therefore Taylor showed that the fluid can undergo slow (non-inertial) motions only if the magnetic field satisfies a certain con-

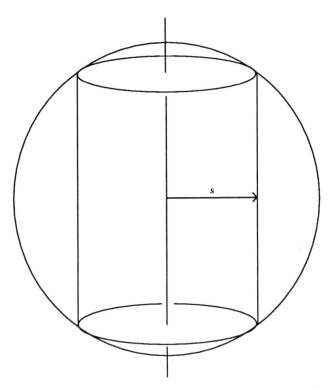

Fig. 3. Schematic representation of zonal flows v_G generated by topographic and gravitational torques according to Jault and LeMouël's theory. The flow is in the form of coaxial cylinders.

straint, namely that the integrated azimuthal component of the Lorentz torque over the surface of every coaxial cylinder vanishes. If the constraint is not met then inertial axisymmetric flows of the form (19) will be excited until the constraint is satisfied. The analysis of JLM begins with a state where the Taylor constraint is met, and finds the axisymmetric flows (19) to be driven by the gravitational and topographic torques. The inductive effect of these flows must be ignored however — the flows are able to create toroidal magnetic field from poloidal field by shearing action, a process known as the ω-effect in dynamo theory. As very little is known regarding the internal magnetic field morphology in the core, these effects must be ignored to a first approximation; this may be adequate if the size of $\partial v_G/\partial s$ is sufficiently small.

There are two ways of using the analysis of JLM, which can be regarded as dynamic and kinematic approaches. The dynamic approach involves the calculation of the forces responsible for the torques which create the changes of angular momentum between core and mantle. This approach has successfully been applied to understanding the secular increase in the length-of-day due to tidal torques; calculated torques agree extremely well with those predicted from the changes in the angular momentum of the moon seen by lunar laser ranging, and corroborate the measurements of the lengthening of the length-of-day deduced from historical observations [Christodoulis et al., 1988]. In the case of the core, calculation of the topographic torque requires accurate models of CMB topography and dynamical fluid pressures derived from a flow model; this approach has been investigated by Jault and LeMouël [1990]. As JLM note, the prospect of calculating the additional contribution from the gravitational torque appears to be very bleak because of the difficulties in mapping density heterogeneities within the core.

The second approach, already used in length-of-day studies when the corresponding dynamic approach is infeasible, is based on the conservation of angular momentum. For example, meteorological data are being used to routinely predict changes in atmospheric angular momentum, which have excellent correlation with short-term changes in length-of-day; the corresponding calculation of mountain torques and viscous torques is much more difficult. In the atmospheric context the kinematic calculation is rather robust because of the high quality of data and the fact that the dynamics of the atmosphere are reflected to a high degree of accuracy by the general circulation models. Performing a similar calculation for the core is rather more difficult.

The kinematic calculation for the core relies on the fact that the surficial expression of the geostrophic degeneracy (19) is in the form of zonal toroidal motions. The component of angular momentum along the rotation axis, J_z, can be written as

$$J_z = \hat{\mathbf{z}} \cdot \int_V \mathbf{r} \wedge \mathbf{v} \, \rho \, dV = \int_V s v_\phi \rho \, dV \qquad (20)$$

where ρ is the core density, $\mathbf{r}$ is the position vector, and V is the volume of the core, and thus the angular momentum only depends on the azimuthal component of flow. We find an approximation to the total angular momentum by ignoring both the presence of the inner core and the small radial variation of ρ, and integrating the angular momentum of the flow throughout the core; the volume element for a cylinder of radius s and width ds is $dV = 2\pi s 2\sqrt{c^2 - s^2}$, so

$$J_z = 4\pi c^4 \rho \int_0^1 x^2 \sqrt{1 - x^2} v_\phi(x) dx \qquad (21)$$

where $x = \cos\theta$, θ is colatitude and c is the core radius. When a flow of the form (19) is represented in terms of the expansion (9) on the boundary, then the angular momentum that it carries is contained in the values of just two modes t_1^0 and t_3^0; the axial angular momentum J_z is [JGLM; Jault, 1990]

$$J_z = \frac{8\pi c^4 \rho}{15} \left(t_1^0 + \frac{12}{7} t_3^0 \right) \qquad (22)$$

This simple result occurs because all the modes except two carry no angular momentum — they are orthogonal to the kernel of integration in (21). Note that solid body rotation eastward is given by $v_\phi(s) = s\omega$, which is represented by $t_1^0 = c\omega, t_n^0 = 0 \; \forall n > 1$ under the expansion (9). Then

$$J_z = \frac{8\pi}{15} c^4 \rho c\omega = I_c \omega \qquad (23)$$

with I_c being the usual moment of inertia of a sphere of mass M:

$$I_c = \frac{2}{5} M c^2 \qquad (24)$$

Following JGLM, equation (22) is conveniently written in the form

$$J_z = I_c \omega_c \qquad (25)$$

where $c\omega_c = t_1^0 + (12/7) t_3^0$. This is the angular momentum as measured in the mantle rest frame; the total z-component of angular momentum of the core in an inertial frame (J_{core}) is

$$J_{\text{core}} = J_z + I_c \Omega_m \qquad (26)$$

where I is a principal moment of inertia, Ω is the earth's rotation rate, and the subscripts c and m stand for core and mantle respectively. Then the total angular momentum of the earth is

$$\begin{aligned} J_{\text{tot}} &= J_{\text{core}} + I_m \Omega_m \\ &= \Omega_m (I_c + I_m) + J_z \end{aligned} \qquad (27)$$

which must remain constant in the absence of any external torque (i.e. after correction for the tidal torque). Thus we expect that changes in Ω_m should be reflected in opposite changes in J_z. Changes in angular momentum of the mantle are measured as changes in the length-of-day, δT. Assuming I_c and I_m are constant in time, then since

$$\frac{\delta\Omega_m}{\Omega_m} = -\frac{\delta T}{T} \qquad (28)$$

where T is the length-of-day, we find

$$\delta T = -\frac{T^2}{2\pi} \delta\Omega_m = \frac{T^2}{2\pi} \left[\frac{I_c}{I_c + I_m} \right] \delta\omega_c \qquad (29)$$

where ω_c is the equivalent rotation from (25). In our calculations we have taken $I_m = 7.2 \times 10^{37}$ kg m^2 and $I_c = 0.85 \times 10^{37}$ kg m^2 [Gubbins and Roberts, 1987, page 62]. These numerical values lead to a useful relation between δT and the coefficients t_1^0, t_3^0: under the Schmidt quasi-normalisation of spherical harmonics which we have used throughout, we find

$$\delta T/\text{ms} = 1.138 \left(\delta t_1^0 + \frac{12}{7} \delta t_3^0 \right) / \text{km yr}^{-1} \qquad (30)$$

where $\delta T/\text{ms}$ is the change in the length-of-day measured in milliseconds and $\delta t_1^0/\text{km yr}^{-1}$ is the change in t_1^0 measured in kilometres/year.

The first calculations to make use of the kinematic analysis were presented by JGLM. They noted that the toroidal zonal motions for the 15 years after 1970 showed a intriguing symmetry about the equator; this symmetry is also found in this study and Figure 4a shows an example of it. JGLM suggested that this type of symmetry could occur if the zonal

Model 1970.0

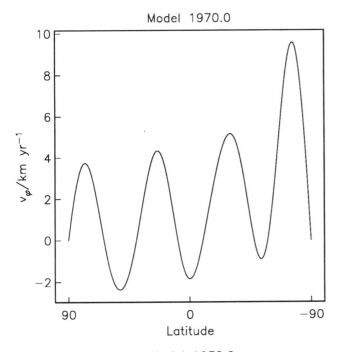

Model 1930.0

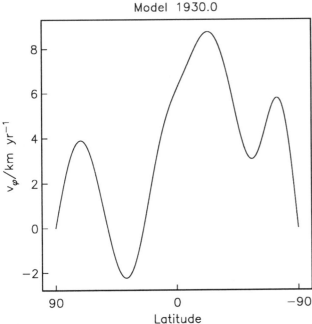

Fig. 4. Zonal toroidal part of velocity versus latitude for (a) 1970 (b) 1930. Note the symmetry in 1970.

lation between the angular momentum of the core from (22) and the length-of-day data. The calculations reported here have not imposed any symmetry on the motions *a priori*. Theory suggests that the annular motions are a response on the inertial timescale; there are changes in the core flow on the convective timescale in addition, including flows which do not have the symmetry possessed by v_G. Therefore it is an implicit assumption that the angular momentum changes associated with the background convection (and therefore the even toroidal zonal modes) are small.

Figure 5 shows the changes in the two zonal coefficients on which the angular momentum calculation depends. The coefficient t_1^0, which represents the solid body drift rate (see above) is in general negative, representing westward flow. At the end of the 19th century t_1^0 can be seen to drop to approximately zero before increasing again. The length-of-day can be deduced quite accurately from astronomical observations taken throughout the last two centuries. These astronomical measurements clearly show a difference between the dynamical timescale (as kept by the earth) and a universal timescale amounting to over 50 seconds in the last 100 years (see Morrison [1979] for details). From this it is deduced that the excess length-of-day over a standard day of 86 400 seconds varies with an amplitude of several milliseconds. These values of the length-of-day effectively measure the mantle's rotation speed, and therefore its angular momentum. Figure 6 shows values for the length-of-day from astronomical observations, taken from McCarthy and Babcock [1986], which draws heavily on the work of Morrison [1979]. These data are augmented by recent Very Long Baseline Interferometry data distributed by subcommission IRIS (International Radio Interferometric Surveying). A correction of 1.4 milliseconds per century has been made to ac-

Degree 1 and Degree 3 Toroidal Zonal Coefficients

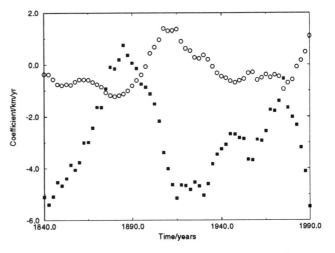

Fig. 5. Variation of the coefficients t_1^0 (solid squares) and t_3^0 (open circles) with time. These coefficients carry the angular momentum of the flow shown in Figure 3.

flow was of the form of the geostrophic degeneracy (19). We find that the tendency for symmetry does not hold as we move back in time, which is in agreement with JGLM (see Figure 4b). JGLM's calculations were made on the assumption that all of the toroidal zonal flow was in the mode v_G by imposing symmetry about the equator on the calculations; this requires $t_{2n} = 0 \ \forall n \geq 1$. They found remarkable corre-

count for the linear trend in the data seen during the last few centuries and thought to be attributable to tidal friction [Stephenson and Morrison, 1984].

In Figure 6 we also show predictions of the length-of-day calculated using the values for t_1^0 and t_3^0 from Figure 5 and equation (30); note that these predictions are made entirely from geomagnetic data. Because only changes in the combination of the coefficients t_1^0, t_3^0 given in (30) are relevant, there is a degree of freedom in choosing the baseline from which to measure these changes from; in Figure 6 this baseline has been chosen to bring the predictions and observations into reasonable agreement. We find an intriguing correlation, which has the correct amplitude and similar variations with time, although the fit clearly degrades back in time. It is instructive to compare Figures 5 and 6, recalling the simple relationship between t_1^0, t_3^0 and the excess length-of-day. We see that the temporal changes in the body rotation t_1^0 alone would lead to an overestimate of the length-of-day changes. This is because in general there is a compensating change in t_3^0 of the opposite sign, so that for example, the large changes in t_1^0 from 1970–1990 do not lead to such dramatically large changes in the length-of-day. In Figure 6 it appears that t_1^0 and t_3^0 are approximately anticorrelated with a repeat time of roughly 80 years. This may be part of a torsional oscillation envisaged by Braginsky [1970]; coaxial cylinders are coupled by the poloidal magnetic field which acts rather like elastic strings reducing their relative displacements.

Our results are evidence in support of the earlier conclusions of JGLM that the time-variations in the flow are of the correct size to balance angular momentum changes. Indeed, this line of evidence is one of the only ways of indicating that the core velocities which are derived are of a physically realistic size.

CONCLUSIONS AND DISCUSSION

We have derived time-dependent core velocities for the period 1840–1990. We use results of the theory of Jault and LeMouël [1989] to extrapolate the zonal toroidal part of these motions throughout the core. The time dependent part of this flow is responsible for carrying angular momentum changes in the core, and the concommitent changes in the mantle's angular velocity can be predicted. We find reasonable correlation between the predictions and historical observations of the length-of-day. Although this extrapolation is somewhat speculative and is difficult to prove firmly since we have no way of probing motions beyond those at the core surface, the theory has a firm physical basis and makes predictions which are highly encouraging. There remains the possibility that the correlation between the observed and predicted length-of-day records is completely fortuitous. The problem of non-uniqueness remains, because whilst part of the flow can be found uniquely, a global measure of flow such as t_1^0 necessarily depends on parts of the flow which cannot be found uniquely, and therefore these measures may be heavily dependent on the regularisation employed. In addition, the dynamics of the inner core have been neglected and the effects of diffusion are implicitly ignored in this study, despite evidence in the field models of non-conservation of frozen-flux. These results also have bearing on the question of the conductivity of the deep mantle. The first order effect of the mantle on the transmission of magnetic signals to the earth's surface is to delay these signals by a time τ and to smooth them [Runcorn, 1955; Backus, 1983]; we would therefore expect that the geodetic estimates of length-of-day would lead the geomagnetic estimates, because the latter would be attributed to the time that the data were collected, whereas the mantle velocity and the core motions causing the secular variation changed at a time of the order of τ previously. Current estimates of the delay time range from less than 3 years [LeMouël et al., 1982; Courtillot et al., 1984] to 13 years [Backus, 1983], and although the correlation of Figure 6 is not sufficiently strong to make firm statements, we might hope that this approach might be refined and have bearing on the question of mantle conductivity in the future. As JLM have stressed, it is now equally important to perform the dynamic calculation of the topographic torques which act at the CMB. Current calculations lead to torques which are too large by two orders of magnitude [Jault and LeMouël, 1990], unless a special property of orthogonality between the pressure field and CMB topography is invoked. Testing these ideas further will require highly accurate models of the structure of the lower mantle and the topography of the core-mantle boundary.

Acknowledgements. We are grateful for support from the NERC (UK), NASA, NSF, the Packard Foundation, and the Royal Society of London. We thank Dominique Jault and Jean-Louis LeMouël for useful discussions.

Observed and Predicted Excess Length of Day

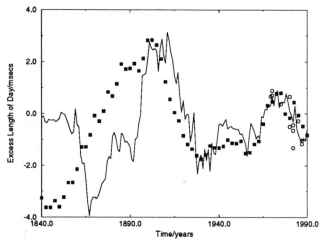

Fig. 6. Predicted excess length-of-day (over a standard day of 86 400 seconds) from this study using the theory of JLM for the period 1840–1990 (squares), compared to the observations (solid line) of McCarthy and Babcock (1986), corrected for secular trend with a value of 1.4 ms per century. Also shown are the results of JGLM (circles).

REFERENCES

Backus, G.E., Kinematics of geomagnetic secular variation in a perfectly conducting core, *Phil. Trans. R. Soc. Lond., A263*, 239–266, 1968.

Backus, G.E., Inference from inadequate and inaccurate data, I, *Proc. Natn. Acad. Sci. U.S.A., 65*, 1–7, 1970.

Backus, G.E., Application of mantle filter theory to the magnetic jerk of 1969, *Geophys. J. R. Astr. Soc., 74*, 713–746, 1983.

Backus, G.E., Confidence set inference with a prior quadratic bound, *Geophys. J., 97*, 119–150, 1989.

Backus, G.E., and F. Gilbert, The resolving power of gross earth data, *Geophys. J. R. Astr. Soc., 16*, 169–205, 1968.

Backus, G.E., and J.-L. LeMouël, The region on the core-mantle boundary where a geostrophic velocity field can be determined from frozen-flux magnetic data, *Geophys. J. R. Astr. Soc., 85*, 617–628, 1986.

Benton, E.R., On the coupling of fluid dynamics and electromagnetism at the top of the earth's core, *Geophys. Astrophys. Fluid Dynam., 33*, 315–330, 1985.

Benton, E.R., R.H. Estes, and R.A. Langel, Geomagnetic field modeling incorporating constraints from frozen-flux electromagnetism, *Phys. Earth Planet. Inter., 48*, 241–264, 1987.

Bloxham, J., Simultaneous stochastic inversion for geomagnetic main field and secular variation 1. A large-scale inverse problem, *J. Geophys. Res., 92*, 11597–11608, 1987.

Bloxham, J., The determination of fluid flow at the core surface from geomagnetic observations, in *Mathematical Geophysics, A Survey of Recent Developments in Seismology and Geodynamics*, edited by N.J. Vlaar, G. Nolet, M.J.R. Wortel, and S.A.P.L. Cloetingh, Reidel, Dordrecht, 1988.

Bloxham, J., and D. Gubbins, The secular variation of earth's magnetic field, *Nature, 317*, 777–781, 1985.

Bloxham, J., and A. Jackson, Simultaneous stochastic inversion for geomagnetic main field and secular variation 2. 1820–1980, *J. Geophys. Res., 94*, 15,753–15,769, 1989.

Bloxham, J., and A. Jackson, Fluid flow near the surface of earth's outer core, *Rev. Geophys., 29*, 97–120, 1991.

Bloxham, J., D. Gubbins, and A. Jackson, Geomagnetic secular variation, *Phil. Trans. R. Soc. Lond., A329*, 415–502, 1989.

Braginsky, S.I, Torsional magnetohydrodynamic vibrations in the Earth's core and variations in day length, *Geomagn. Aeron. 10*, 3–12, (Engl. Trans. 1–8), 1970.

Bullard, E.C., and H. Gellman, Homogeneous dynamos and terrestrial magnetism,*Phil. Trans.R.Soc.Lond.A247*,213–278,1954.

Christodoulis, D.C., D.E. Smith, R.G. Williamson and S.M. Klosko, Observed tidal braking in the Earth/Moon/Sun system, *J. Geophys. Res., 93*, 6216–6236, 1988.

Constable, S.C., R.L. Parker, and C.G. Constable, Occams inversion: A practical algorithm for generating smooth models from electromagnetic sounding data, *Geophysics, 52*, 289–300, 1987.

Courtillot, V., J.-L. LeMouël and J. Ducruix, On Backus' mantle filter theory and the 1969 geomagnetic impulse, *Geophys. J. R. Astr. Soc., 78*, 619–625, 1984.

Gibson, R.D., and P.H. Roberts, The Bullard-Gellman dynamo, in *Applications of Modern Physics to the Earth and Planetary Interiors*, edited by S.K. Runcorn, pp. 577–602, Wiley-Interscience, New York, 1969.

Gire, C., and J.-L. LeMouël, Tangentially geostrophic flow at the core-mantle boundary compatible with the observed geomagnetic secular variation: The large-scale component of the flow, *Phys. Earth Planet. Inter., 59*, 259–287, 1990.

Golub, G.H., and C.F. Van Loan, *Matrix Computations*, Johns Hopkins University Press, Baltimore, 1983.

Gubbins, D., and P.H. Roberts, Magnetohydrodynamics of the earth's core, in *Geomagnetism*, edited by J.A. Jacobs, vol. 2, pp. 1–183, Academic Press, London, 1987.

Gubbins, D., Dynamics of the secular variation, *Phys. Earth Planet. Inter., 68*, 170–172, 1991.

Hide, R., Interaction between the Earth's liquid core and solid mantle, *Nature, 222*, 1055–1056, 1969.

Hills, R.G., Convection in the earth's mantle due to viscous shear at the core-mantle interface and due to large-scale buoyancy, *PhD thesis*, New Mexico State University, Las Cruces, New Mexico, 1979.

Jackson, A., The earth's magnetic field at the core-mantle boundary, *PhD thesis*, University of Cambridge, Cambridge, England, 1989.

Jackson, A., and J. Bloxham, Mapping the fluid flow and shear near the core surface using the radial and horizontal components of the magnetic field,*Geophys. J. Int.,105*, 199–212,1991.

Jault, D., Variation séculaire du champ geomagnetique et fluctuations de la longeur du jour, *PhD thesis*, University of Paris VII, Paris, 1990.

Jault, D., and J.-L. LeMouël, The topographic torque associated with a tangentially geostrophic motion at the core surface and inferences on the flow inside the core, *Geophys. Astrophys. Fluid Dynam., 48*, 273–296, 1989.

Jault, D., and J.-L. LeMouël, Core-mantle boundary shape: Constraints inferred from the pressure torque acting between the core and the mantle, *Geophys. J. Int., 101*, 233–241, 1990.

Jault, D., C. Gire, and J.-L. LeMouël, Westward drift, core motions and exchanges of angular momentum between core and mantle, *Nature, 333*, 353–356, 1988.

Langel, R.A., D.J. Kerridge, D.R. Barraclough, and S.R.C. Malin, Geomagnetic temporal change: 1903–1982, a spline representation, *J. Geomagn. Geoelectr., 38*, 573–597, 1986.

LeMouël, J.-L., Outer core geostrophic flow and secular variation of earth's magnetic field, *Nature, 311*, 734–735, 1984.

LeMouël, J.-L., J. Ducruix, and C. Duyen, The worldwide character of the 1969–1970 impulse of the secular acceleration rate, *Phys. Earth Planet. Inter., 28*, 337–350, 1982.

LeMouël, J.-L., C. Gire, and T. Madden, Motions at the core surface in geostrophic approximation, *Phys. Earth Planet. Inter., 39*, 270–287, 1985.

Lloyd, D., and D. Gubbins, Toroidal fluid motion at the top of earth's core, *Geophys. J. Int., 100*, 455–467, 1990.

McCarthy, D.D., and A.K. Babcock, The length of day since 1656, *Phys. Earth Planet. Inter., 44*, 281–292 1986.

Morrison, L. V., Re-determination of the decade fluctuations in the rotation rate of the earth in the period 1861–1978, *Geophys. J. R. Astr. Soc., 58*, 349–360, 1979.

Roberts, P.H., and S. Scott, On the analysis of the secular variation. 1. A hydromagnetic constraint: Theory, *J. Geomagn. Geoelectr., 17*, 137–151, 1965.

Runcorn, S.K., The electrical conductivity of the earth's mantle, *Trans. Am. Geophys. Un., 36*, 191–198, 1955.

Stephenson, F.R. and L.V. Morrison, Long-term changes in the rotation of the Earth: 700 B.C. to A.D. 1980, *Philos. Trans. R. Soc. Lond., 313*, 47–70, 1984.

Taylor, J. B., The magneto-hydrodynamics of a rotating fluid and the earth's dynamo problem, *Proc. R. Soc., A274*, 274–283, 1963.

Voorhies, C.V., Steady surficial core motions: An alternate method, *Geophys. Res. Lett., 13*, 1537–1540, 1986.

Whaler, K.A., Geomagnetic evidence for fluid upwelling at the core-mantle boundary, *Geophys. J. R. Astr. Soc., 86*, 563–588, 1986.

Whaler, K.A., and S.O. Clarke, A steady velocity field at the top of the earth's core in the frozen-flux approximation, *Geophys. J., 94*, 143–155, 1988.

Andrew Jackson, Department of Earth Sciences, Oxford University, Parks Road, Oxford, OX1 3PR, U.K.

Jeremy Bloxham, Department of Earth and Planetary Sciences, Harvard University, Cambridge, Massachusetts 02138, U.S.A.

David Gubbins, Department of Earth Sciences, Leeds University, Leeds, U.K.

Angular Momentum Transfer between the Earth's Core and Mantle

RAYMOND HIDE

The Robert Hooke Institute, The Observatory, Clarendon Laboratory, Parks Road, Oxford OX1 3PU, England U.K.

Irregular fluctuations in the magnitude and direction of the rotation of the solid Earth (mantle, crust and cryosphere) are produced by torques arising from motions in the overlying hydrosphere and atmosphere and in the underlying liquid metallic core. Hydrospheric and atmospheric torques are largely responsible for fluctuations on interannual, seasonal and intraseasonal timescales, but it is necessary to invoke angular momentum transfer between the core and mantle in order to account for irregular decadal fluctuations in the Earth's rotation. The elucidation of concomitant torques at the core-mantle boundary will be inseparable from advances in geophysical studies of the structure, composition and dynamics of the upper reaches of the core and the lower reaches of the mantle. Irregular planetary-scale topographic features of the core-mantle boundary of about 0.5km in vertical amplitude are strongly implied by the data, but other interpretations are, of course, possible.

INTRODUCTION

Recent improvements in geodetic data and practical meteorology have advanced research on fluctuations in the Earth's rotation. The interpretation of these fluctuations is inextricably linked with studies of the dynamics of the Earth-moon system and dynamical processes in the liquid metallic core of the Earth (where the geomagnetic field originates), other parts of the Earth's interior, and the hydrosphere and atmosphere. Investigations of the most rapid fluctuations bear on meteorological studies of interannual, seasonal, and intraseasonal variations in the general circulation of the atmosphere and the response of the oceans to such variations. Fluctuations occurring on decadal time scales are best regarded as manifestations of angular momentum transfer between the core and mantle and they have implications for the topography of the core-mantle boundary and the electrical, magnetic, and other properties of the core and lower mantle.

Define $\vec{\Omega} = \vec{\Omega}(t)$ (where t denotes time) as the angular velocity with which a geographical reference frame based on observations fixed to the Earth's crust rotates relative to an inertial frame based on the fixed stars. Precise determinations of $\vec{\Omega}$ by geodesists and astronomers reveal temporal fluctuations over time scales ranging from days to centuries and longer. The fluctuations must be manifestations of (i)

Dynamics of Earth's Deep Interior and Earth Rotation
Geophysical Monograph 72, IUGG Volume 12
Published in 1993 by the International Union of Geodesy and Geophysics
and the American Geophysical Union.

changes in the moment of inertia of the solid Earth due to redistribution of matter within it and (ii) the action of net torques resulting from applied stresses. Contributions to the former comprise (i) periodic tidal deformations of the solid Earth produced by the gravitational action of the moon, sun, and other astronomical bodies, (ii) nonperiodic deformations due to surface stresses produced by fluctuating fluid flow in the underlying liquid metallic core and the overlying hydrosphere and atmosphere, and (iii) nonperiodic mass redistribution associated with earthquakes, the melting of ice, and, on geological time scales, mantle convection and movements of tectonic plates.

One important contribution to the applied torques acting on the solid Earth arises from the gravitational action of the sun and moon on the Earth's tidal bulge, which owing to tidal friction within the oceans and other parts of the Earth, is oriented at an angle to the line joining the centres of the interacting bodies. The Earth's rotation speed is thus decelerated at a rate corresponding to a gradual lengthening of the day

$$\Lambda(t) \equiv 2\pi / | \vec{\Omega}(t) | \tag{1}$$

by about 1.4ms per century. As a result, the moon's orbit expands at a detectable rate of 3.7 ± 0.2cm per year. Stronger torques (by a factor of more than 50), fluctuating irregularly in magnitude and sign on time scales of several years and upwards, are thought to be produced by core motions. It has been argued on general quantitive grounds that such motions provide the most likely source for the excitation of irregular fluctuations in $\bar{\Omega}(t)$ on decadal time scales. Decadal changes in the length of the day (LOD) range in amplitude to 5ms; concomitant displacements of the pole of the rotation axis of the solid Earth relative to the geographical axis amount to a few meters. [By far the strongest torques to which the solid Earth is subjected are those generated by atmospheric motions, but they act mainly on sub-decadal time scales and produce LOD changes no larger than about 1ms.]

SPECTRUM OF IRREGULAR LOD FLUCTUATIONS

The quantity of interest in the study of irregular LOD fluctuations is

$$\Lambda^*(t) \equiv \Lambda(t) - \Lambda_0 - \Lambda_1(t) - \Lambda_2(t) \qquad (2)$$

rather than the full $\Lambda(t)$ time series. Here $\Lambda_0 = 86400$s, $\Lambda_1(t)$ is the long-term increase in $\Lambda(t)$ at 1.4×10^{-3}s per century associated with tidal friction (see above) and $\Lambda_2(t)$ is a sum of strictly periodic terms with amplitudes of up to 0.5×10^{-3}s at known tidal frequencies. Changes in the moment of inertia that are associated with $\Lambda_2(t)$ are attributable to calculable distortions in the figure of the Earth produced by the gravitational action of the moon and sun.

The residual time series $\Lambda^*(t)$, representing irregular fluctuations on time scales ranging from decades to days, is conveniently expressed as the sum of four terms:

$$\Lambda^*(t) \equiv \Lambda_\alpha(t) + \Lambda_\beta(t) + \Lambda_\gamma(t) + \Lambda_\delta(t) \qquad (3)$$

denoting respectively the decadal, interannual, seasonal, and intraseasonal components. They are largely attributable to processes that effect the exchange on various time scales of angular momentum between the solid Earth and the fluid regions with which it interacts at its upper and lower surfaces. The interactions are produced by normal and tangential hydrodynamical stresses acting at these interfaces that give rise to a net axial torque $\hat{\Gamma}^*(t)$. It is convenient to introduce the hypothetical equivalent axial torque

$$\Gamma(t) \equiv -2\pi C \dot{\Lambda}^*(t)/\Lambda_0^2 \qquad (4)$$

where $\dot{\Lambda}^* \equiv d\Lambda^*/dt$ and $C = 7.04 \times 10^{37}$ kg m^2 is the principal moment of inertia of the solid Earth (mantle, crust, and cryosphere) about the polar axis, which amounts to 0.9 times that of the whole Earth (to which the solid inner core, liquid outer core, hydrosphere, and atmosphere contribute $7 \times 10^{-4}C, 10^{-1}C, 3 \times 10^{-4}C$, and $10^{-6}C$, respectively). The actual axial torque $\hat{\Gamma}^*(t)$ acting on the solid Earth would be exactly the same as $\Gamma^*(t)$ in the hypothetical case when it is possible to neglect effects due to (i) departures from perfect rigidity of the solid Earth and (ii) motions in other parts of the Earth induced by fluctuations in the rotation of the solid Earth.

In practice, $\Gamma^*(t)$ differs from $\hat{\Gamma}^*(t)$ by no more than about 10%. Analogously to Eq. 3, $\Gamma^*(t)$ can be expressed as the sum of spectral components

$$\Gamma^*(t) = \Gamma_\alpha(t) + \Gamma_\beta(t) + \Gamma_\delta(t) + \Gamma_\gamma(t) \qquad (5)$$

where $\Gamma_a(t) \equiv -2\pi C \dot{\Lambda}_\alpha(t)/\Lambda_0^2$, and so forth.

The seasonal and intraseasonal $[\Lambda_\gamma(t)$ and$\Lambda_\delta(t)]$ components of $\Lambda^*(t)$ are largely due to angular momentum exchange between the atmosphere and the solid Earth, and there is strong evidence that meteorological excitation is also the main cause of interannual fluctuations $[\Lambda_\beta(t)]$. However, the decadal variations $\Lambda_\alpha(t)$ are too large in amplitude to be largely attributable to angular momentum exchange between the atmosphere (or the hydrosphere or both) and the solid Earth, but a case can be made that motions in the Earth's liquid core provide their main source of excitation. Polar motion on decadal time scales is expected to be another manifestation of angular momentum transfer between the core and mantle, the implied equatorial components of the effective fluctuating torque involved being nearly 100 times bigger in amplitude than the axial component.

DECADAL FLUCTUATIONS

Observations of the main geomagnetic field indicate that irregular magnetohydrodynamic flow occurs in the Earth's liquid metallic core with typical speeds of about 3×10^{-4} m s^{-1}; associated horizontal pressure variations on a planetary scale are about 3×10^2N m^{-2}. As we have already noted, it is generally accepted on general quantitative grounds that fluctuations in these motions give rise to angular momentum exchange between the liquid core and overlying "solid" mantle and that the observed decadal variation $\Lambda_\alpha(t)$ is largely a manifestation of that exchange.

The hydrodynamical stresses at the CMB comprise (i) tangential stresses produced by viscous forces in the thin Ekman-Hartman boundary layer just below the core-mantle interface and also by Lorentz forces associated with the interaction of electric currents with the magnetic field there and (ii) normal stresses produced largely by dynamical pressure forces acting on interface topography (that is, departures in shape from that of a sphere). Insufficient knowledge of (a) core motions, (b) the viscosity and electrical conductivity of the core, (c) the electrical conductivity of the lower mantle and the magnetic field there, and (d) the topography of the CMB, make it impossible to determine the torque acting at the CMB with much certainty. But progress is now accelerating as geophysicists intensify their studies of the structure, composition and dynamics of the Earth's deep interior.

Although not completely certain, viscous coupling is likely negligible, so most investigations are concerned with electromagnetic coupling, first proposed by Bullard, or topographic coupling, first proposed by Hide when the adequacy of electromagnetic coupling had first been called into question by various investigators. The strength of electromagnetic coupling depends critically on the assumed electrical conductivity of the lower mantle and the assumed intensity of the magnetic field (poloidal and toroidal) at the core-mantle interface, and it would be zero in the case of a perfectly insulating mantle. The strength of topographic coupling depends on the assumed amplitude of the topographic, but not in a monotonic way owing to subtle effects related to the intensity and structure of the magnetic field at the core-mantle interface and elsewhere within the core and also to horizontal density variaitons that are proportionately tiny but dynamically significant.

An essential step in the estimation of both electromagnetic and topographic torques from geophysical observations is the determination of the geomagnetic field and its time rate of change at the CMB. This is done by downward extrapolation of the main geomagnetic field and its secular changes as measured at and near the Earth's surface, a procedure which places great demands on attempts to improve geomagnetic observations and methods of error analysis. from determinations of the geomagnetic field near the CMB it is possible under plausible assumptions to determine the horizontal velocity and pressure fields in the outer core. These pressure fields can be combined with various models of the topography of the core-mantle interface (based on gravity data, seismic tomography, and various assumptions about the rheology of the mantle and the structure of the so-called D'' layer at the base of the mantle) to produce hypothetical values of the topographic torque at the core-mantle interface. One can find hypothetical values of the electromagnetic torque by calculating the Lorentz forces in the lower mantle under various assuptions concerning the electrical conductivity and magnetic field fluctuations there.

Acceptable models of the Earth's deep interior must be such that the total hypothetical axial torque (topographic plus electromagnetic) agrees satisfactorily with $\Gamma_\alpha(t)$ time series and the corresponding equatorial component of the torque must agree with determinations of polar motion on decadal time scales when allowance has been made for other effects. Of particular interest are the cases when one or the other type of coupling dominates, even though further studies might show that both are comparable in importance. The study of core-mantle coupling in connection with the interpretation of decadal variations in the Earth's rotation bears directly in the problem of accounting for the main geomagnetic field in terms of self-exciting magnetohydrodynamic dynamo action in the core.

Dynamo models can be classified in terms of two characteristic features. The first is the ratio of the strengths of the toroidal and poloidal magnetic fields in the core, which in strong field dynamos is much greater that unity and in weak field dynamos is of the order of unity. The second is the extent to which dyamo action is concentrated

in the upper reaches of the core or extends throughout the whole body of the core. Topographic coupling might account for the decadal changes in the Earth's rotation if Lorentz forces associated with strong toroidal magnetic fields are restricted to the deep interior of the core. More extensive and refined calculations that either weaken the case for significant topographic coupling or consistently indicate excessive topographic coupling would constitute evidence in favor of strong electromagnetic torques at the CMB, produced by dynamo action concentrated just below the CMB.

On the basis of the hypothesis that electromagnetic and viscous coupling are negligible in comparison with topographic coupling, the torque $\vec{L}$ applied by the core to the solid Earth is given by

$$\vec{L}(t) = c^2 \int_0^{2\pi} \int_0^\pi (\vec{r} \times h \nabla_s p_s) \sin\theta d\theta d\lambda \qquad (6)$$

where θ denotes co-latitude, λ longitude, and the (nearly spherical) shape of the CMB is the locus of points where the distance from the Earth's center of mass is equal to $c + h(\theta, \lambda)$, c being the mean radius of the CMB (3480km) and the poorly known topography $h(\theta, \lambda)$ is such that $| h | \ll c$. The quantity p_s is the dynamic pressure associated with core motions $\vec{u}_s$ in the free stream just below the viscous boundary layer at the CMB and the operator

$$\nabla_s \equiv c^{-1}(\vec{\theta}\partial/\partial\theta, \vec{\lambda}cosec\theta\partial/\partial\lambda) \qquad (7)$$

if $\vec{\theta}$ and $\vec{\lambda}$ are unit vectors in the directions of increasing θ and λ, respectively. If we assume that there is geostrophic balance at that level, the term $\nabla_s p_s$ can be replaced by $-2\bar{\rho}\vec{\Omega} \times \vec{u}_s$ (where $\bar{\rho}$ is the horizontally averaged value of the density in the upper reaches of the core), giving for the axial component of $\vec{L}(t)$ the expression

$$-2\Omega_0\bar{\rho} \int_0^{2\pi} \int_0^\pi h(\theta, \lambda)v_s \sin^2\theta \cos\theta d\theta d\lambda \qquad (8)$$

where v_s is the θ component of $\vec{u}_s$. The axial component of $\vec{L}$ depends on λ-dependent features of h. These features are not known with any degree of confidence, in spite of efforts made over several decades to infer $h(\theta, \lambda)$ from the pattern of long-wave anomalies of the Earth's gravitational field under various assumptions about the structure and rheology of the solid Earth, as in more recent studies, from the results of seismic tomography. The quantity $\vec{u}_s(t)$ is obtained from determination of the geomagnetics secular variation at the CMB using "Alfven's frozen magnetic flux"hypothesis and the geostrophic relationship.

Rough dynamical arguments show that even with a high degree of cancelling of positively and negatively directed contributions to the axial component of $\vec{L}$ the presence of λ-dependent variations in h of no more than 0.5km in effective amplitude should suffice to account for the magnitude of

$\Gamma_\alpha(t)$ inferred through Eq.8 more detailed calculations based on Eq.8 confirm this result. Moreover, these findings are consistent with an estimate of 0.5km for the excess ellipticity of the CMB inferred from determinations of the amplitude and phase of the "free-core nutation" and also from models of irregular CMB topography based on seismic data that allow for mechanical and chemical variations in the so called D'' layer at the base of the mantle.

The foregoing arguments have to be modified slightly if between the fluid metallic core and the overlying mantle there exist either a continuous layer or pools of poorly conducting fluid. The effective topography in Eqs. 6 and 8 could then be significantly less than the actual topography of the CMB, a situation that might appeal to those seismologists who advocate models of irregular CMB topography with amplitudes of several kilometers. Evidence against such high irregular topography and the related need to invoke a poorly-conducting fluid layer comes from considerations of the equatorial components of $\vec{L}$ (see Eq.6). An estimate of the magnitude of these components can be found by inserting for h in Eq.6 the expression corresponding to the "equilibrium" equatorial bulge of the CMB and comparing the estimate with that implied by the observed polar motion on decadal time scales. Detailed studies are needed to enable firm conclusions to be drawn.

CONCLUDING REMARKS

Other speakers in this symposium will discuss encouraging developments in the study of the angular momentum budget between the core and mantle, which involves the introduction of assumptions concerning the (unknown) relationship between surface motions in the core and flow at depth (see Jackson and Bloxham 1992). In the present paper we have concentrated on the torques acting at the CMB. A complete discussion of angular momentum transfer between the core and mantle would involve considerations of both aspects of the problem, as in studies of the angular momentum budget between the atmosphere and the solid Earth. This aspect of Earth rotation studies has seen very rapid progress in the past decade, thanks to the availability of meteorological data at all levels in the atmosphere covering the whole globe (see Hide and Dickey 1991).

No attempt is made here to give an extensive list of the literature relevant to the ideas discussed. The present paper is based on parts of a recent review article by Hide and Dickey (1991), where an extensive bibliography can be found. For many useful references to work on the Earth's deep interior, see also Aldridge et al [1990], Bloxham and Jackson [1991], Gubbins [1991], Hide [1989], Jault and Le Mouël [1989,1991], Knittle and Jeanloz [1991], and Voorhies [1991].

REFERENCES

Aldridge, K.D., et al., Core-mantle interactions, *Surveys in Geophys*, 11, 329-353, 1990.

Bloxham, J. and Jackson, A., Fluid flow near the surface of the Earth's core, *Rev. Geophys.*, 29, 97-120, 1991.

Gubbins, D., Convection in the Earth's core and mantle, *Quart. Jown. Roy. Astron. Soc.* 32, 69-84, 1991.

Hide, R., Fluctuations in the Earth's rotation and the topography of the core-mantle interface, *Philos. Trans. Roy Soc. London*, 351-363, 1989.

Hide, R. and Dickey, J.O., Earth's variable rotation, *Science*, 253, 629-637, 1991.

Jackson, A. and Bloxham, J., Angular momentum of the core during the last two centuries, in proceedings of IUGG Symposium on Dynamics of the Earth's interior and Earth's rotation, edited by J-L. Mouël (in the press).

Jault, D. and Le Mouël, J-L., The topographic torque associated with tangentially-geostrophic motion at the core surface and inferences on flow inside the core, Geophys. Astrophys. Fluid Dyn., 48, 273-296, 1989.

Jault, D. and Le Mouël, J-L., Exchange of angular momentum between core and mantle, *J. Geomag. Geoelect.*, 43, 111-129, 1991.

Knittle, E. and Jeanloz, R., Earth's core-mantle boundary: results of experiments at high pressures and temperatures, *Science*. 251, 1438-1443, 1991.

Loper, D.E., The nature and consequences of thermal interactions twixt core and mantle, *J. Geomag. Geoselect.*, 43, 79-91, 1991.

Voorhies, C.V., Coupling an inviscid core to an electrically-insulating mantle, *J. Geomag. Geoselect.*, 43, 131-156, 1991.

Raymond Hide, The Robert Hooke Institute, The Observatory, Clarendon Laboratory, Parks Road, Oxford OX1 3PU, England U.K.

Geomagnetic Estimates of Steady Surficial Core Flow and Flux Diffusion: Unexpected Geodynamo Experiments

COERTE V. VOORHIES

Geodynamics Branch, Code 921/NASA, Goddard Space Flight Center, Greenbelt, MD 20771, USA.

The hypothesis that motional induction in a sense opposes magnetic flux diffusion is arguably central to core geodynamo theory. Evidence substantiating this hypothesis was found while investigating steady induction effects in geomagnetism.

In fitting geomagnetic secular variation, if the effects of a 10% change in flow near the core surface (suggested on the basis of Earth rotation data) are not modeled, one can expect residuals of about 10%. If the geomagnetic effects of mantle conductivity, core asphericity and core resistivity are not modeled, one can expect residuals of about 7%. In the source-free mantle/frozen-flux core magnetic earth model, a spatially smooth steady flow leaves weighted residuals of 7% with respect to the Definitive Geomagnetic Reference Field models (DGRFs 1945-1980). Weighted residuals ≤ 1.1% are tolerable, so allowance for these special effects is reasonable and could boost the core flow change signal-to-noise ratio needed to resolve time-dependent flow.

To make an allowance for core resistivity the DGRFs were fitted in terms of both steady core surface flow and steady magnetic flux diffusion. This approach has a simple dynamo-theoretic interpretation: order zero closure of the mean field radial induction equation and retention of only the leading terms in the time series expansions for the mean flow and mean flux diffusion. The experiment unexpectedly resulted in an operational dynamo: core secular variation due to motional induction is negatively correlated with that due to magnetic flux diffusion.

1. INTRODUCTION

Core geodynamo theory would have motional induction sustain Earth's main magnetic field against the long term decay caused by magnetic flux diffusion and Ohmic dissipation [see, e.g., Moffat, 1978]; geodynamo experiments would test this theory against observation. The variety of geomagnetically relevant geophysical processes leads to a formidable mathematical description of the core geodynamo. At the very least this description includes a set of coupled non-linear partial differential equations describing magnetic induction, transport of mass, momentum and energy densities, and an equation of state within Earth's core [see, e.g., Jacobs, 1987]. An attempt to solve such equations can be imagined; yet requisite initial conditions, boundary values, and key physical properties are extremely uncertain [see, e.g., Voorhies, 1984]. It thus seems difficult to make specific theoretical predictions regarding observable quantities, notably the geomagnetic field. Moreover, such a prediction would seem difficult to test due to the rich variety of geomagnetically relevant geophysical processes contributing to the measured field.

Dynamics of Earth's Deep Interior and Earth Rotation
Geophysical Monograph 72, IUGG Volume 12

Motional induction is essential to the operation of a dynamo. Sometime and somewhere within a core geodynamo, motional induction must not only oppose, but overcome, magnetic diffusion. Recent magnetic field measurements might not absolutely require motional induction to currently oppose magnetic diffusion near the top of Earth's core, but I found unexpected evidence of such geodynamo action while investigating steady induction effects in geomagnetism. Here I try to describe this finding and a few closely related results.

2. IMPLICATIONS OF DECADE FLUCTUATIONS IN THE LENGTH OF THE DAY FOR GEOMAGNETIC ESTIMATION OF CORE SURFACE FLOW

Several interpretations of geomagnetic secular variation (SV) have used the source-free mantle/frozen-flux core (SFM/FFC) magnetic earth model—wherein an effectively rigid, impenetrable, electrical insulator of uniform magnetic permeability mantles an effectively spherical, inviscid, perfect fluid conductor in anelastic flow [see, e.g., Benton, 1979; Hide & Malin, 1981; Voorhies, 1984, 1986a, 1991]. Among these are a few of the published efforts to estimate the fluid motion near the top of the core reviewed by Bloxham & Jackson [1991]. Some attempts to explicate the SV indicated by the Definitive Geomagnetic Reference Field models (DGRFs) [IAGA, 1988] further suppose surficially steady core flow from 1945 to 1980. The implied non-linear inverse steady motional induction problem was solved iteratively via the method of damped weighted least squares. The resulting estimates of fluid flow near the core surface, such as shown in Figure 3, indicate speeds of about 10 km/yr. This also holds when the geostrophic radial vorticity balance [Backus & LeMouël, 1986; Benton, 1985; LeMouël, 1984; Voorhies, 1991] is imposed, as shown in Figure 7. The root mean square (rms) speeds of these steady flows are 8.78 km/yr (Fig. 3) and 12.85 km/yr (Fig. 7).

Length of day (LOD) data show decade fluctuations of about ±1.5 milliseconds per decade (ms/d) [see, e.g., Stephenson & Morrison, 1984]. Such changes are often attributed to an exchange of angular momentum between Earth's mantle and core: decreased LOD indicates mantle spin up and core spin down [see, e.g., Hide, 1989; Hide & Dickey, 1991; Jault & LeMouël, 1989, 1990, 1991; Voorhies, 1991]. A very simple calculation (see Appendix A) uses such decade fluctuations in the LOD to suggest corresponding fluctuations in core surface flow of about ±1 km/yr–a mere 10% of the estimated rms steady surficial flow speed. Such small changes have long seemed unimportant; they still seem small compared with differences between Figures 3 and 7.

Bloxham & Jackson [1991] cite decadal changes in the LOD and core-mantle coupling as providing rather strong evidence against steady flow. Yet the angular momentum of the core is an integral quantity which can change appreciably even if the flow at the top of the core does not, so decadal changes in the LOD provide rather weak evidence against steady surficial core flow. Fortunately, the possibility that fluid motion near the top of the core (albeit below the viscous sub-layer) is exactly steady seems about as remote as the possibility that the mean flow obtained by averaging over a few decades is exactly zero.

The possibility that dynamical processes occuring well within the core maintain a persistent flow near its surface is, in contrast, not easily dismissed; moreover, a posteriori geodynamical arguments showing that core surface flow can be influenced by long lived thermal, topographic, and/or compositional anomalies in the deep mantle support further consideration of statistically steady core surface flow. Bloxham & Gubbins [1987] advanced thermal interactions, Jault & LeMouël [1990] proposed topographic locking, and I suggest "Lorentz linkage" to lateral heterogeneities in the electrical conductivity of the deep mantle as mechanisms tending to oppose change in the flow near the core surface. The utility and accuracy of supposing statistically steady core surface flow thus remain subjects of investigation. In the interim, one can construct arguments which use LOD data to reckon the magnitude of core flow change.

In fitting secular geomagnetic change, if the effects of a 10% change in flow are not modeled, one can expect residuals of about 10%. Yet the elementary SFM/FFC model omits many geomagnetic effects. If such effects are not modeled, a few calculations indicate root sum square residuals of about 7% (about 6% from core resistivity, 3% from core asphericity, and 2% from mantle conductivity). The normalized rms weighted residual relative to the DGRFs is 6.9% for Figure 3 and 13.9% for Figure 7. These residuals are judged significant; indeed, weighted residuals of 1.1% or less are needed. Estimation of core surface

flow change thus seems reasonable–particularly if allowance for such effects boosts the core flow change signal-to-noise ratio (i.e., the ratio of flow change signal to other signal). To make some allowance for non-zero core resistivity the DGRFs were fitted in terms of both steady core surface flow and steady magnetic flux diffusion.

3. THEORY: A SIMPLIFIED MEAN FIELD INDUCTION EQUATION

Let $\mathbf{r}$ represent position in spherical polar coordinates (r, θ, ϕ) and let t denote time. With magnetic flux density $\mathbf{B}(\mathbf{r},t)$, fluid velocity $\mathbf{V}(\mathbf{r},t)$, and with uniform magnetic diffusivity η, the mean field equation considered is an ensemble average [Krause & Radler, 1980]. An overbar indicates the mean: $\mathbf{B} = \bar{\mathbf{B}} + \mathbf{b}$, $\mathbf{V} = \bar{\mathbf{V}} + \mathbf{v}$, and

$$\partial_t\bar{\mathbf{B}} = \nabla\times(\bar{\mathbf{V}}\times\bar{\mathbf{B}}) + \eta\nabla^2\bar{\mathbf{B}} + \nabla\times\overline{(\mathbf{v}\times\mathbf{b})} \qquad (1)$$

Equation (1) describes secular variation as the sum of the effects of motional induction, magnetic diffusion, and correlated fluctuations. An average over a short time interval $\delta\tau$ may be used instead–provided $\delta\mathbf{B}/\delta\tau$ is not distinguished from $\partial_t\bar{\mathbf{B}}$ [Voorhies, 1986a, p12,447].

Unlike first order closure via the α-effect, the lowest order closure omits $\hat{\mathbf{r}}\bullet\nabla\times\overline{(\mathbf{v}\times\mathbf{b})}$. With current density $\mathbf{J}(\mathbf{r},t)$ and conductivity σ the radial component of (1) is, omitting overbars, then

$$\partial_t B_r + \nabla_s\bullet(B_r\mathbf{V}_s - V_r\mathbf{B}_s) = -\hat{\mathbf{r}}\bullet\nabla\times\sigma^{-1}\mathbf{J}$$

$$= \eta r^{-1}\nabla^2 rB_r \equiv d_r \qquad (2)$$

where $\mathbf{B}_s \equiv \mathbf{B} - \hat{\mathbf{r}}B_r$ and $\mathbf{V}_s \equiv v\hat{\theta} + w\hat{\phi}$.

Consider positions $\mathbf{r} = \mathbf{c}$ just below the sub-layer separating a nearly spherical, low viscosity core in anelastic flow from a relatively rigid, source-free mantle–where $V_r(\mathbf{c},t) \approx 0$ and $J_r(\mathbf{c},t) \approx 0$. Also consider a long time interval $(\Delta t \gg \delta\tau)$ during which B_r and $\partial_t B_r$ at $\mathbf{c}$ change with time. The leading terms in the slow time series expansions for $\mathbf{V}_s(\mathbf{c},t)$ and $d_r(\mathbf{c},t)$ are steady $\mathbf{V}_s^*(\mathbf{c})$ and $d_r^*(\mathbf{c})$. Only these leading terms are retained here, so (2) at $\mathbf{c}$ becomes

$$\partial_t B_r(\mathbf{c},t) + \nabla_s\bullet[B_r(\mathbf{c},t)\mathbf{V}_s^*(\mathbf{c})] \approx d_r^*(\mathbf{c}) \qquad (3)$$

Comparison of (3) with (2) or (1) confirms that only $\mathbf{V}_s$ and d_r at $\mathbf{c}$ are treated as (statistically) steady. Indeed, at $\mathbf{r} = \mathbf{c}$ steady $d_r \equiv -\hat{\mathbf{r}}\bullet\nabla\times\sigma^{-1}J_s$ implies neither steady B_r, J_s, nor $\nabla_s\bullet J_s$. The slow time series expansions need not be truncated deeper in the core; the α-effect on $\partial_t\mathbf{B}_s$ is not barred; and any laterally homogeneous, isotropic α-effect at $\mathbf{c}$ would leave (2) and (3) unchanged (as $\hat{\mathbf{r}}\bullet\nabla\times\alpha\mathbf{B} = \alpha\mu J_r \approx 0$ for uniform magnetic magnetic permeability $\mu = 1/\sigma\eta$). Given (3) as an equality, time-varying $B_r(\mathbf{c},t)$ can uniquely determine both $\mathbf{V}_s^*(\mathbf{c})$ and $d_r^*(\mathbf{c})$ [Voorhies, 1987a]. This "steady radial induction theorem" can be viewed as a corollary to the steady motions theorem [Voorhies & Backus 1985]; alternately, it may be viewed as a special case of the "steady induction theorem" [Voorhies, 1990]. Of course, lacking either true or measured values for $B_r(c,t)$ at any (c,t), one can but estimate $V_s^*(c)$ and $d_r^*(c)$ from available geomagnetic information.

4. DATA, DGRFS, AND APPROACH

Standard models of observatory and survey data represent the geomagnetic field $B(r,t)$ in terms of Gauss coefficients for spherical harmonic expansions of the scalar potential $V(r,t)$: $B = -\nabla V$. The International Association of Geomagnetism and Aeronomy adopted the DGRF from such models [IAGA, 1988]. Candidate models were typically fitted to tens of thousands of data via weighted least squares. The statistical nature of the data fitted and independent observations indicate that the DGRFs represent a few hundred thousand geomagnetic observations quite well.

The DGRFs include the 120 internal Gauss coefficients through spherical harmonic degree and order 10 at 5 year intervals from 1945 to 1980. Unlike a referee, I argue that the DGRFs are well suited to core studies. They may even represent the slowly varying, broad scale part of the core field rather better than do original measurements because narrow scale crustal and rapidly changing external fields remain largely unmodeled. Moreover, a predicted field which matches the DGRFs will also fit the data quite well–provided predicted higher degree field structure does not greatly exceed several nT at Earth's surface. Although derivation of the pre-1965 DGRF candidate models employed both an a priori field model and a priori information about the core field, the latter is also empirical in nature and was studiously confined to the construction of correlated weight matrices [Langel, Barraclough, Kerridge, Golovkov, Sabaka, & Estes, 1988]. These candidate models, and their covariances, are thus free of certain side-effects [elucidated by Langel, 1987 and Backus, 1988] which result from direct assignation of an a priori covariance to core field model parameters.

For each of the 7 pre-1980 DGRFs, a full covariance matrix was derived from either the candidate models (epochs 1965-75) or the information matrix provided by R. Langel & T. Sabaka (1945-60). The inverses of these covariances are the heavier, generalized weight matrices summarized by Voorhies [1991b]. Each DGRF defines a point in a 120 dimensional subspace. Each full covariance corresponds to an error ellipsoid around each point. The experiments attempt quantitative explication of the indicated secular change. I take the 1980 MAGSAT DGRF (with unmodeled coefficients set to zero) as an 'initial' condition and seek a theoretical prediction which, when projected onto the subspace, gives a trajectory passing near each such error ellipsoid. Misfit is measured in units of σ^*: a trajectory grazing each of the 7 error ellipsoids would give a tolerable residual of $1.00\sigma^*$.

Study of systematic errors introduced with simplifying suppositions suggested larger residuals (about $3.3\sigma^*$); yet prior information on core surface flow and flux diffusion, core asphericity, and deep mantle conductivity was judged insufficient to sustain either Gauss-Markov estimation, stochastic inversion, or Bayesian inference [McLeod, 1986, Backus, 1988]. I used the method of damped weighted least squares.

5. EXPERIMENTAL METHOD

The initial geomagnetic condition near Earth's surface has the spherical harmonic expansion: $B_r(a,t_0) = \Sigma_i\, g_i(t_0)S_i(\theta,\phi)$, where $|a| = 6371.2$ km, DGRFs at epoch t_k are represented by $g_i(t_k)$ with $i \leq 120$ and $0 \leq k \leq 7$, and the S_i notation is that of Voorhies [1986b]. Downward continuation via vacuum extrapolation gives $B_r(c,t_0)$, so any $V_s^*(c)$ and $d_r^*(c)$ allow numerical solution of (3) for the corresponding $\partial_t B_r(c,t_0)$. Numerical integration of (3) through time gives the prediction $B_{rp}(c,t\neq t_0)$. Spherical harmonic analysis and upward continuation of this prediction implies $B_{rp}(a,t) = \Sigma_i\, \gamma_i(a,t)S_i(\theta,\phi)$. The $\gamma_i(t)$ with $i \leq 120$ are the projection of the prediction onto the subspace containing the DGRFs; the residuals are $g_i(t_k) - \gamma_i(t_k)$.

Estimates of both $V_s^*(c)$ and $d_r^*(c)$ were derived by solving the weighted, constrained, damped, non-linear, inverse steady diffusive motional induction problem. In summary, I try to minimize the sum, Δ^2, of: the square weighted residual relative to the DGRFs (denoted by Δ_r^2); an optional geostrophic radial vorticity constraint $(\lambda_g\Delta_g^2)$, flow damping of mean square radial vorticity and upwelling $(\lambda_d\Delta_d^2)$; plus diffusion damping of the mean square radial SV at a due to radial flux diffusion at c $(\lambda_\eta\Delta_\eta^2)$. Denote the average of Q over a sphere by $<Q>$, weight matrix elements by $\Omega_{ij}(t)$, radial vorticity by $\omega_r = \hat{r}\cdot\nabla\times V_s^*$, downwelling by $\partial_r u = -V_s\cdot V_s^*$, and the steady radial SV implied by vacuum upward continuation of $d_r^*(c)$ by $d_r^*(a)$. Then

$$\Delta^2 \equiv \Delta_r^2 + \lambda_g\Delta_g^2 + \lambda_d\Delta_d^2 + \lambda_\eta\Delta_\eta^2$$

$$= \sum_{k=1}^{7}\{\ \sum_{i=1}^{120}\sum_{j=1}^{120}\ [g_i(t_k) - \gamma_i(t_k)]\Omega_{ij}(t_k)[g_j(t_k) - \gamma_j(t_k)]\ \}$$
$$+ \lambda_g<[\partial_r u(c)\cos\theta + v(c)\sin\theta/c]^2>$$
$$+ \lambda_d<[\omega_r(c)]^2 + [\partial_r u(c)]^2> + \lambda_\eta<[d_r^*(a)]^2> \qquad (4)$$

The magnitudes assigned to the positive damping parameters λ_g, λ_d, and λ_η control the relative importance of departures from a tight fit, from surficial geostrophy, from smooth flow, and from frozen-flux.

For each sample of $(\lambda_g, \lambda_d, \lambda_\eta)$ studied, minimal Δ^2 was sought using a variant of Gauss's method to derive low degree coefficients in spherical harmonic expansions for the streamfunction $-T(c)$ and velocity potential $-U(c)$ giving $V_s^*(c) = \nabla_s T\times\hat{r} + \nabla_s U$ and for $d_r^*(c)$. An exact, closed form expression for the total derivative of Δ_r^2 with respect to the coefficients of T or U remains elusive, so numerical solution of the forward problem (3) on a 2^0 regular mesh with 10^{-2} year time steps was used to iteratively refine my approximation to such normal equations matrix elements–akin to the p_{ki} and q_{ki} of Voorhies [1986b]–as well as to accurately evaluate residuals. This procedure gave clearly convergent solutions–including stable global properties and flow patterns–after several iterations.

Appendix B offers a note on convergence and a response to a referee's query as to how I handle the triangle inequality [see, e.g., Bullard & Gellman, 1954]. Appendix C describes damped weighted least squares by contrast with other methods in order to answer a referee's query as to why I damp $<[d_r*(a)]^2>$ instead of $<[d_r*(c)]^2>$.

Differences between frozen-flux solutions ($\lambda_\eta = \infty$) such as Figure 3 and, say, Figure 7c of Voorhies [1986a] reflect changes in approach and method between 1983 and 1988: (i) a sequence of main field models at **a** is fitted instead of SV models at **c**; (ii) the flow is not constrained to generate minimal narrow scale SV at **c**; (iii) the secular change signal fitted is fully weighted; (iv) the non-linear inverse problem is solved; and (v) regularization is achieved primarily through damping instead of truncation. Previous studies have included some of these features: Voorhies [1986b] fitted fewer DGRFs without bias against narrow scale SV; Whaler [1986] introduced weights; Bloxham [1989] solved the non-linear inverse problem [Bloxham, 1987; Voorhies, 1987b]; and algorithms developed by Gire, LeMouel, & Madden [1986] appear indistinguishable from those of damped least squares.

Both flow damping and truncation of the expansions for $B_r(c,t_0)$, T, U, and d_r limit study to broad scale field and flow; yet characteristic values for flow speed, length scale, and η were not presumed. Indeed, each solution determines its own effective magnetic Reynolds number

$$R_m*(t) \equiv \{<[\nabla_s \cdot B_r(c,t)V_s*(c)]^2>/<[d_r*(c)]^2>\}^{1/2} \quad (5)$$

in accord with the values of (λ_g, λ_d, λ_η) assigned to it. R_m* can be identified as a special case of the specific magnetic Reynolds number defined by Ball, Kahle, & Vestine [1969, equation (23)]. As is well known, effective or specific magnetic Reynolds numbers can differ from the magnetic Reynolds number of dynamo theory [R_m, see, e.g., Gubbins & Roberts 1987, p25] due to differences between lateral and vertical length scales. Indeed, one might expect $R_m* \sim 1$ in the visco-magnetic core-mantle boundary layer despite $R_m \sim \mu_0 \sigma c < |V_s*|^2>^{1/2} \sim 700$.

6. RESULTS

Study of the trade-off volume (λ_g, λ_d, λ_η) focused on two surfaces. In both unconstrained ($\lambda_g = 0$) and surficially geostrophic (extremely large λ_g) cases, damping parameters λ_d and λ_η were varied to map a trade-off surface: misfit as a function of both flow roughness and diffusive contributions to SV. For steady frozen-flux flows, expansions for both T and U were truncated to degree and order 16. In the steady diffusive cases, both expansions for T and U were truncated to degree and order 13 and the expansion for d_r* was truncated to degree and order 10. Higher degrees for d_r* are not as easily estimable from the DGRFs. These truncation levels are high enough to ensure that damping, rather than truncation, dominates the regularization. On each iteration 576 parameters were derived for each frozen-flux flow ($\lambda_\eta = \infty$); only 510 were derived in the diffusive cases. These parameters are not all free due to damping.

Mapping Two Tradeoff Surfaces

Two trade-off surfaces were explored: one for ordinary steady flows and one for surficially geostrophic steady flows. Simply plotting suites of trade-off curves appears more revealing than other mappings of these surfaces. Figure 1 shows the trade-off

curves of "flow roughness" as a function of "normalized square weighted residual". The former measures curvature in both T and U and is Δ_d^2; the latter is not adjusted for the number of degrees of freedom, is just

$$NSWR \equiv \frac{\sum_{k=1}^{7} \sum_{i=1}^{120} \sum_{j=1}^{120} [g_i(t_k) - \gamma_i(t_k)]\Omega_{ij}(t_k)[g_j(t_k) - \gamma_j(t_k)]}{\sum_{k=1}^{7} \sum_{i=1}^{120} \sum_{j=1}^{120} [g_i(t_k) - g_i(t_0)]\Omega_{ij}(t_k)[g_j(t_k) - g_j(t_0)]} \quad (6)$$

and has a logarithmic scale. The 6 solid curves are from ordinary steady flows ($\lambda_g = 0$). The 6 dashed curves are from steady, surficially geostrophic flows. Each curve is at one value of diffusion damping parameter λ_η. As λ_η is reduced from ∞ to (in effect) 0, the curves march from right to left. In other words, as bias against steady radial flux diffusion is reduced, a tighter fit is obtained at fixed flow roughness; alternately, a smoother flow is obtained at fixed misfit. Yet the right-most solid curve in Figure 1 proves that steady core surface flow alone explicates over 99.5% of the weighted power in the definitive secular change signal.

Figure 2 plots "flow complexity" as a function of "significance of residual signal". The former is Δ_d; the latter is SRS $\equiv [\Delta_r^2/(7\times120)]^{1/2}$ in units of $\sigma*$ and, according to the weights, should be 1 for a tolerable fit (vertical lines in Figs. 1 and 2). Figure 2 shows that none of the solutions give a tight enough fit to the DGRFs—as was anticipated for such a simple earth model. The smallest residuals obtained are $4.78\sigma*$; more typical trajectories miss the error ellipsoids by about 7 times the typical radius of the ellipsoids. For each of the 5 pairs of curves derived with matched λ_η, both unconstrained (solid) and surficially geostrophic (dashed) curves naturally intercept the abscissa at the same point. The intercept of both the left-most solid and dashed curves with the abscissa reveals the modest increase in SRS implied by supposing steady flux diffusion and no flow ($\lambda_d \to \infty$)

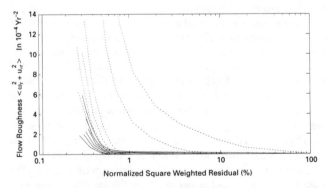

Fig. 1. Trade-off curves of flow roughness (mean square radial vorticity plus mean square downwelling, Δ_d^2) as a function of normalized square weighted residual (eqn. 6). The 6 solid curves are from steady flows; the 6 dashed curves are from steady, surficially geostrophic flows. Each curve marks a single diffusion damping parameter λ_η. As λ_η is reduced from ∞ to (in effect) 0, the curves march from right to left.

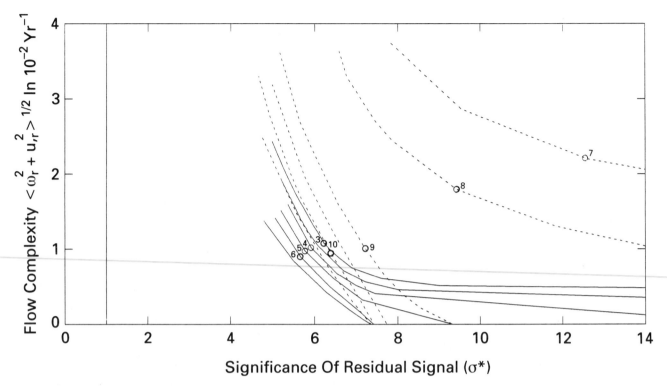

Fig. 2. Trade-off curves of flow complexity Δ_d as a function of the significance of the residual signal in σ^*. Numbered points map to flow Figs. 3-10.

instead of steady frozen-flux flow ($\lambda_\eta \to \infty$, right-most solid or dashed curves). The numbered points on Figure 2 correspond to the flow field Figures 3-6 and 7-10.

Steady frozen-flux flows near the knee of the rightmost solid curve yield normalized rms residuals (Nrms residuals $NSWR^{1/2}$) of about 7%; however, allowance for steady radial flux diffusion did not reduce such residuals to the anticipated level of about 3.6%. It did enable a much tighter fit for surficially geostrophic steady flows. Indeed, for the smoother, surficially geostrophic steady flows the allowance for diffusion reduced Nrms residuals by about 6% (from about 14% to 8%).

Flow Fields by the Core-Mantle Boundary

The first set of flows, Figures 3, 4, 5, and 6, show how reducing bias against steady radial diffusion, and thus decreasing the effective magnetic Reynolds number from infinity to less than unity, can alter the derived steady flow. These figures portray a slice through the $\lambda_g = 0$ trade-off surface with λ_η varying between figures and fixed λ_d giving fixed bias against swirl and confluence (Fig. 2, points 3-6). The flow does not change much until R_m^* drops from infinity to well below unity.

Table I lists some of the properties of these (and some of the other) steady flows derived, including: corresponding figure number (if shown); flow designation; R_m^* from (5) at $t = t_o = 1980$; rms speed $<|V_s^*(c)|^2>^{1/2}$ in km/yr; rms downwelling $<(\partial_r u)^2>^{1/2}$ in $(10^3 \text{ yr})^{-1}$; mean westward flow $-<w>$ in km/yr; bulk westward drift (from the first non-trivial coefficient of T) in

degrees per year; the unweighted rms residual in $B_r(a,1960)$, $<\{\Sigma_{i \leq 120} [g_i(1960) - \gamma_i(1960)]S_i\}^2>^{1/2}$, in nT (obtained via 20 years of steady diffusive motional induction on the mesh and upward continuation); the weighted variance reduction $(1 - NSWR)$; the Nrms residual $(NSWR)^{1/2}$; the apparent significance of the residual signal (SRS) in σ^*; the rms value of $d_r^*(a)$ in nT/yr (which can be compared with the 58.3 nT/yr characterizing the DGRFs); and, finally, the coefficient of correlation between diffusive and advective contributions to the predicted value of $\partial_t B_r(c,1980)$

$$C^* \equiv \frac{<[d_r^*(c)][-\nabla_s \cdot B_r(c,1980)V_s^*(c)]>}{\{<[d_r^*(c)]^2><[\nabla_s \cdot B_r(c,1980)V_s^*(c)]^2>\}^{1/2}} \quad (7)$$

Table I confirms that the fit is made but slightly tighter by allowing for steady flux diffusion. The bottom line of Table I shows that SV–the part of $\partial_t B_{rp}(c,t_o)$ –induced by steady flow is negatively correlated with that due to steady radial flux diffusion ($C^* < 0$). The anti-correlation increases as R_m^* decreases. The finding $C^* < 0$ was unexpected; however, it is easily explained by the core geodynamo hypothesis (see Appendix D). Solutions with $R_m^* < 1$ must be regarded with caution as other parameterizations seem more appropriate to the diffusively dominated regime. Solution SIVAO4 shows signs of having incompletely converged, but is included because it would have the rms value of $d_r^*(a)$ exceed the rms value of $\partial_t B_r(a,t)$ indicated by the DGRFs–a seemingly provocative notion.

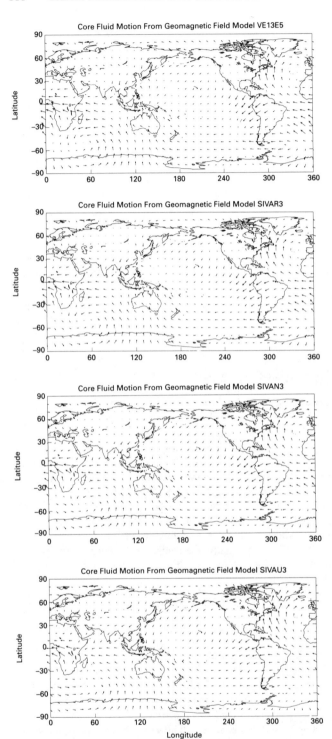

Figs. 3, 4, 5 & 6. Steady flows at the top of the core show how reducing bias against steady radial flux diffusion, and thereby reducing R_m^* from infinity to less than unity, can alter the derived flow. These unconstrained steady flows share the same flow damping parameter ($\lambda_g = 0$, fixed λ_d, Fig. 2 points 3-6). Values for R_m^* and rms speed in km/yr are, from top to bottom: Fig. 3 (∞, 8.78); Fig. 4 (2.01, 8.52); Fig. 5 (1.34, 8.48); Fig. 6 (0.96, 8.18). A vector length of 10° represents a speed of 22.24 km/yr.

The second set of flows, Figures 7, 8, 9, and 10, is analogous to the first, but the geostrophic radial vorticity balance is enforced (so downwelling implies poleward flow and vice-versa). Allowance for steady radial flux diffusion visibly decreases surficially geostrophic flow speed and complexity while substantially tightening the fit (see Tables II and III). At modest R_m^*, these surficially geostrophic steady flows can fit the DGRFs about as closely as an unconstrained steady frozen-flux flow of similar spatial complexity. Indeed, the last flow (Fig. 10) achieves this via reduced bias against flow complexity (Fig. 2, points 7-10). In the context of (statistically) steady core surface flow, the closer fit at fixed flow complexity enhances the plausibility of the surficially geostrophic flow hypothesis at the expense of the frozen-flux core approximation.

Some Further Comments

The surficially geostrophic steady frozen-flux flows derived from the DGRFs do not easily fit a part of that SV which is constant. Allowance for steady flux diffusion helped remedy this situation; yet the unconstrained steady flows show no appreciable tendency towards surficial geostrophy—even when steady radial flux diffusion is admitted. Indeed, my routine measure of departures from surficial geostrophy, $A_g \equiv \langle(\partial_r u\cos\theta + v\sin\theta/c)^2\rangle / \langle(\partial_r u\cos\theta)^2 + (v\sin\theta/c)^2\rangle$, stays within a few percent of unity unless damped by large λ_g. Either my simple suppositions conceal true core dynamics, other contributions to the radial vorticity balance counter the Coriolis term (perhaps the Lorentz term, but see Voorhies [1991]), or $A_g \approx 1$ results from both model simplicity and geodynamic complexity. The first view enjoys the support of SRS > $1\sigma^*$; yet the second view seems difficult to dismiss: A_g almost never exceeds unity—suggesting dynamical processes which barely foil, rather than wholly overwhelm, the Coriolis term.

Imposition of surficial geostrophy increases the residuals left by steady core surface flows [Voorhies, 1986c]. All else being equal, it basically halves the number of free flow parameters. The fact that NSWR for flow VE13E5 (0.472% from Table I) is slightly less than half that for surficially geostrophic flow GVC3E5 (1.102% from Table III) derived with the same λ_d seems to support suggestions [Bloxham 1988, 1989] that geostrophy increases relative misfit in the frozen-flux approximation. Yet diffusive solutions lead me to suggest this tendency dissappears, and might even reverse, when appreciable diffusion is admitted. Then surficially geostrophic steady diffusive flows might allow a tighter fit per degree of freedom than ordinary steady diffusive flows.

Such details of model efficiency are judged unimportant compared with the apparent significance of the residuals. Indeed, 840 pre-1980 DGRF coefficients were fitted with less than 576 effectively free parameters, so adjustments for the number of degrees of freedom would merely multiply SRS by a factor between 1 and 1.78; yet if the tens of thousands of observations underlying the DGRFs had been fitted directly, then the corresponding factor should be very close to unity. Residuals can be reduced by time dependent flux diffusion (the easy way), by a different initial condition, or perhaps by allowances for mantle

TABLE I. Some Properties of some Steady Diffusive Flows
(A cut through the trade-off surface)

Figure	3	–	4	5	6	–	–	–
Flow	VE13E5	SIVBB4	SIVAR3	SIVAN3	SIVAU4	SIVAX4	SIVAY4	SIVAO4
R_m*(1980)	∞	3.78	2.01	1.34	0.957	0.878	0.827	0.883
Rms speed (km/yr)	8.78	8.64	8.52	8.48	8.18	7.85	7.57	8.55
Rms downwelling (10^{-3} yr^{-1})	7.43	7.26	7.10	6.82	6.29	6.07	5.88	6.25
Mean westward flow (km/yr)	4.78	4.69	4.62	4.60	3.99	3.39	2.47	2.12
Bulk westward drift (°/yr)	0.101	0.100	0.100	0.100	0.089	0.077	0.056	0.049
Rms residual in $B_r(a,1960)$ (nT)	92	88	85	84	87	86	85	84
Weighted variance reduction (%)	99.528	99.550	99.574	99.595	99.609	99.612	99.619	99.633
Nrms residual (%)	6.87	6.70	6.52	6.37	6.25	6.23	6.17	6.06
Significance of residual signal (σ*)	6.23	6.08	5.92	5.78	5.67	5.65	5.60	5.50
Rms d_r*(a) (nT/yr)	0.0	1.1	2.4	4.7	12.0	18.3	45.9	95.9
Correlation between diffusive and advective contributions to $\partial_t B_{rp}(c,1980)$ (C*)	–	-0.29	-0.32	-0.38	-0.45	-0.48	-0.53	-0.63

conductivity, core asphericity, time dependent flow, and/or a laterally heterogeneous α-effect.

7. CONCLUSIONS: A FEW FINDINGS AT FINITE R_m*

It seems difficult to draw firm conclusions about large regions of Earth's interior from the experimental results of an analysis of sparsely distributed Earth observations or standard models thereof. Neither the magnetic field nor the fluid velocity near the top of Earth's core have ever been directly measured. If the true motion were known, and if acoustic, tidal, indeed all sub-annual fluctuations were filtered out, the resulting flow might well exhibit intense variations in time and across spatial scales far shorter than shown in figures resulting from the application of my rather primitive method of parameter estimation to my geophysically over-simplified interpretation of geomagnetic secular variation. Still, all the derived flows shown do explain over 98% of the square weighted signal in the secular change indicated by the DGRFs, and over 99.5% if surficial geostrophy is not enforced. So the geomagnetic evidence for such interesting variations contained in the weighted DGRFs seems rather less than might be hoped. Moreover, it is quite possible that the broad scale part of the fluid motion near the core surface looks like one of these Figures–perhaps 5 or 10 if not 3 or 7–when averaged over the past few decades. Furthermore, the rms speed and apparent scale height for vertical motion $\{<|V_s*|^2>/<[\partial_r u]^2>\}^{1/2}$ from Table I ($2.64\pm0.13\times10^{-4}$ m/s and 1258 ± 130 km) are about half those of Voorhies [1984, 1986a] and are thus in better accord with the predictions of Elsasser [1946], Frazer [1973], and whole outer core convection. The corresponding values for the more fully developed surficially geostrophic flows of Table III ($4.11\pm2.05\times10^{-4}$ m/s and 3138 ± 1293 km) may reflect the efficacy with which rotation supresses broad scale vertical motion.

The results in hand do establish some firm conclusions.

1. Steady frozen-flux core surface flows fitted the weighted DGRFs more closely than did steady radial flux diffusion. The steady, surficially geostrophic, frozen-flux flows did not.

2. Both steady flow and steady radial flux diffusion together fitted the DGRFs more closely than either separately.

3. For ordinary steady flows the inclusion of steady radial flux diffusion allowed a slightly tighter fit. Flow pattern and properties are robust for R_m* > 1; changes are confined to 3 areas (under the far western Pacific Ocean, the SW Indian ocean, and the NW coast of S. America). Larger change is found for R_m* < 1, but other parameterizations are better suited to a diffusively dominated regime.

4. The SV induced by the steady core surface flows is negatively correlated with that due to the steady radial diffusion (C* < 0). Stronger anti-correlation is found at smaller R_m*. This negative correlation is not expected in the dually damped least squares approach; it is to be expected for geodynamo action (see Appendix D). These facts corroborate core geodynamo theory.

5. For surficially geostrophic steady flows, the inclusion of steady radial flux diffusion allowed a dramatic tightening in fit. The broad flow pattern persists, but at fixed λ_d the flow speed and complexity can decay rapidly as R_m* is decreased from infinity to order unity. (This trend is expected as a viscous boundary layer is traversed towards the mantle–though viscosity could void surficial geostrophy). The fit can then be further tightened by decreasing λ_d so as to restore flow complexity. Figure 2 contours λ_η, not Δ_η^2, so weak knees in the dashed curves at weak λ_η are not necessarily troublesome.

6. The negative correlation between motional and diffusive contributions to $\partial_t B_{rp}(c,t)$ is also found in surficially geostrophic steady flows. It varies less with R_m* (for R_m* > 0.6), but nonetheless substantiates both finding 4 and what is arguably the central hypothesis of core geodynamo theory: motional induction opposes flux diffusion.

7. Several surficially geostrophic steady diffusive flows vaguely resemble unconstrained steady flows (e.g., compare Fig. 3 with Fig. 10 in the regions of faster flow). Allowance for steady diffusion did not make steady flows surficially geostrophic.

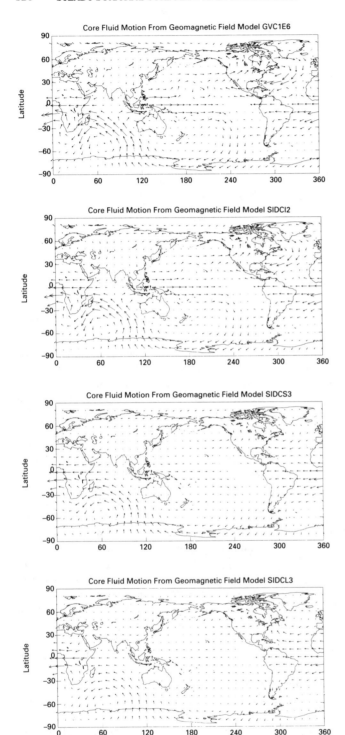

Figs. 7, 8, 9 & 10. Surficially geostrophic steady flows at the top of the core also show how reducing bias against steady flux diffusion can alter the derived flow. Values for R_m* and rms speed in km/yr are, from top to bottom: Fig. 7 (∞, 12.85); Fig. 8 (2.56, 11.49); Fig. 9 (0.89; 8.32); Fig. 10 (0.74, 7.51). Flow damping parameter for Fig. 10 is that used for Fig. 3-6 (30% that used for Figs. 7-9). Scale is as for Figs. 3-6.

8. The residual signal in the DGRFs apparently remains significant: the weighted residuals are 5 to 8 times larger than needed. Certain lighter weightings reduce this factor to about 2.5. Diagonally weighted residuals appear much more nearly tolerable, but the diagonalization destroys valuable information reflecting the distribution of geomagnetic data in space and time. The residual signal is also larger than, albeit less than twice, that anticipated on the basis of a few calculations concerning mantle conductivity and core asphericity.

9. As noted in the introduction, geodynamo theory would have motional induction sustain Earth's magnetic field against the long term decay implied by magnetic flux diffusion and Ohmic dissipation. Indeed, sometime and somewhere within a core geodynamo, motional induction must not only oppose, but overcome, diffusion. Moreover, diffusion should occasionally oppose motional induction somewhere near the core surface as described in Appendix D. Although my analysis of the DGRFs uses an over-simplified magnetic earth model, the experimental results are consistent with, and indeed show, geodynamo action occuring near the top of Earth's core. I believe this confirms the core geodynamo hypothesis.

10. Changes in the field at c must diffuse across the mantle before being observed. The derived radial flux diffusion is laterally heterogeneous and steady for but a few decades, so it may approximate diffusive contributions to SV from lateral variations in deep mantle conductivity rather than from the core. Indeed, the very weak form of Δ_η^2 is well suited to such reinterpretation. If and where deep mantle conductivity is very high, surficial geostrophy can fail; then the plain steady flows suggest that lateral variations in deep mantle conductivity need not appreciably alter the derived flow.

APPENDIX A

Length of day (LOD) data provided by B. Chao (pers. comm.) and by R. Hide & J. Dickey (pers. comm.) confirm changes of about ± 1.5 ms/d. For example, the annual mean LOD clearly tended to increase from 1961 to 1972 ($\Delta LOD/\Delta t = +1.83$ ms/d) and to then decrease from 1972 to 1986 ($\Delta LOD/\Delta t = -1.35$ ms/d).

Numerical values for the polar components of the moment of inertia tensors and angular velocity vectors considered here are taken from Stacey [1977]. Let Ω_m represent the bulk angular velocity of the mantle (7.292×10^{-5} rad/s) and Ω_c that of the core ($\Omega_c \sim \Omega_m$). Let I_m represent the moment of inertia of the mantle (7.1178×10^{37} kg m^2) and I_c that of the core (0.920×10^{37} kg m^2). The bulk eastward core speed at radius r and colatitude θ is $r\Omega_c \sin\theta$. Note $LOD \equiv 2\pi/|\Omega| \approx 2\pi/\Omega_m$, so the change in LOD is $\Delta LOD \approx -2\pi\Delta\Omega_m/\Omega_m^2$.

If angular momentum gained by the mantle is lost by the core, then

$$I_m \Delta\Omega_m = -I_c \Delta\Omega_c \qquad (A1)$$

where changes in I_m and I_c are omitted. For additional simplicity, suppose the bulk angular velocity lost by the core is homogeneously distributed throughout the core. Then the incremental westerly drift in the core at the equator is

$$\Delta w_0 \equiv r\Delta\Omega_c = -r\frac{I_m}{I_c}\Delta\Omega_m \approx r\frac{I_m\Omega_m^2}{2\pi I_c}\Delta LOD \qquad (A2)$$

With $\Delta LOD \approx -1.5$ ms and $r = c = 3480$ km near the top of the core, (A2) gives $\Delta w_0 \approx -3.42\times10^{-5}$ m/s $= -1.08$ km/yr. The average of $\Delta w_0 \sin\theta$ over the sphere $r = c$ is $\Delta w_0\pi/4 \approx -0.85$ km/yr; the rms is $(2\Delta w_0^2/3)^{1/2} \approx 0.88$ km/yr. The change in bulk core drift relative to the mantle is $\Delta\Omega_c - \Delta\Omega_m = (1 + I_c/I_m) = 1.13\Delta\Omega_c$, so the mean westward flow near the core surface increases by about 0.96 km/yr (0.99 km/yr rms).

The exchange of angular momentum between the liquid outer core and the imperfectly rigid mantle and inner core is more complicated than the foregoing simple example might suggest. Small changes in the bulk angular velocities of these bodies suggest very small changes in their ellipticities and inertia tensors; moreover, angular momentum shed (or gathered) by the core might be preferentially expelled (or absorbed) by its upper reaches, the depths of the outer core, or the inner core. The discovery by Jault, Gire, & LeMouël [1988] and further work of Jault & LeMouël [1989] clarify this process and show that the change in core angular momentum can be distributed amongst coaxial cylinders, each of which rotates as a rigid body (i.e., with its own homogeneous angular velocity). Yet from Table 1 of Jault et al. [1988] I calculate changes in the westward flow averaged over the core surface of at most 2.5 km/yr; moreover, there might be several competing dynamical processes operating. So a homogeneously distributed angular velocity change within the core still provides a fair, dynamically unbiased (and unenlightened) indication of the magnitude of core surface flow change accompanying decade fluctuations in LOD.

TABLE II. Properties of Some Surficially Geostrophic Steady Diffusive Flows
(A cut through the geostrophic trade-off surface)

Figure	7	8	9	–	–	–
Flow	GVC1E6	SIDCI2	SIDCS3	SIDCW3	SIDCK3	SIDCZ3
$R_m{}^*(1980)$	∞	2.56	0.89	0.68	0.55	0.22
Rms speed (km/yr)	12.85	11.49	8.32	6.79	5.28	2.33
Rms downwelling (10^{-3} yr^{-1})	4.12	3.29	1.82	1.26	0.92	0.62
Mean westward flow (km/yr)	2.55	2.53	2.93	2.97	2.73	0.38
Bulk westward drift (°/yr)	0.048	0.048	0.059	0.062	0.058	0.005
Rms residual in $B_r(a,1960)$ (nT)	209	149	105	105	102	96
Weighted variance reduction (%)	98.074	98.920	99.368	99.404	99.418	99.423
Nrms residual (%)	13.88	10.39	7.95	7.72	7.63	7.60
Significance of residual signal (σ^*)	12.58	9.42	7.21	7.00	6.92	6.89
Rms $d_r{}^*(a)$ (nT/yr)	0.0	7.0	21.0	28.5	36.1	55.1
Correlation between diffusive and advective contributions to $\partial_t B_{rp}(c,1980)$ (C*)	–	-0.45	-0.44	-0.34	-0.22	0.08

TABLE III. Properties of More Surficially Geostrophic Steady Diffusive Flows
(Two cuts through the geostrophic trade-off surface at weaker flow damping)

Figure	–	–	–	10	–	–	–	–
Flow	GVC3E5	SIDCD2	SIDCT3	SIDCL3	GVC1E5	SIDCC3	SIDBL3	SIDCA4
$R_m{}^*(1980)$	∞	3.24	1.13	0.74	∞	3.94	1.41	0.78
Rms speed (km/yr)	14.96	13.67	10.46	7.51	17.42	16.24	13.13	10.47
Rms downwelling (10^{-3} yr^{-1})	5.57	4.52	2.83	1.75	7.54	6.28	4.24	3.07
Mean westward flow (km/yr)	2.01	1.79	2.61	2.73	1.96	1.41	2.10	2.67
Bulk westward drift (°/yr)	0.034	0.031	0.050	0.057	0.032	0.017	0.036	0.054
Rms residual in $B_r(a,1960)$ (nT)	153	123	92	96	126	104	85	89
Weighted variance reduction (%)	98.898	99.246	99.476	99.505	99.270	99.444	99.580	99.606
Nrms residual (%)	10.50	8.68	7.24	7.04	8.55	7.46	6.48	6.28
Significance of residual signal (σ^*)	9.52	7.87	6.56	6.38	7.75	6.76	5.87	5.70
Rms $d_r{}^*(a)$ (nT/yr)	0.0	4.6	15.2	29.6	0.0	3.2	11.3	24.1
Correlation between diffusive and advective contributions to $\partial_t B_{rp}(c,1980)$ (C*)	–	-0.46	-0.48	-0.34	–	-0.47	-0.50	-0.35

APPENDIX B

Typical $V_s*(c)$ induce structure in $B_{rp}(c,t \neq t_o)$ at all lateral spatial scales via a bi-directional cascade; therefore, both broad and narrow scale structure in $B_{rp}(c,t \neq t_o)$ must be retained to accurately evaluate both residuals and elements of the normal equations matrix. In the spectral domain, the cascade towards higher wavenumbers follows from repeated application of the triangle inequality: a first application to (3), a second to the time derivative of (3), and so on ad infinitum. Logical induction proves that steady motion can induce all spherical harmonic degrees in $B_{rp}(c,t \neq t_o)$ despite truncation of $B_r(c,t_o)$. Some fine structure induced in B_{rp} is lost by truncation of the triples (the $X_{ijk} = -X_{jik}$ and Y_{ijk} of Voorhies [1986b]) whose relation to Gaunt-Elsasser factors is pointed out by Whaler [1986]. The present method, being an extension of previous numerical methods [Voorhies, 1984; 1986a,b], bypasses evalution of the triples; therefore, the triangle inequality never arises explicitly. Implicitly, some fine structure is lost due to coarseness of mesh. The azimuthal Nyquist wavenumber for a 2° mesh is spherical harmonic order 90.

The efficacy with which $V_s*(c)$ can cross-couple $\gamma_i(t)$ suggests that the inverse motional induction problem becomes highly non-linear and sensitive to initial conditions on time scales of order $<(\partial_r u)^2>^{-1/2}$. One might expect iterated linearized attempts to solve such a problem to exhibit convergence towards local, rather than global, minima or even fail to converge—particularly when the problem is over-parameterized and the normal equations matrix elements are but approximate. I found indications of such behavior only for the weaker flow dampings studied (smaller λ_d)—notably in the unconstrained frozen-flux case ($\lambda_g = 0$ and $\lambda_\eta = \infty$). Such "non-solutions" are not presented. I did use a learning algorithm to place an upper bound on Δ_r^2 as a function of Δ_d^2; such flows mark the perhaps too steep wall of the unconstrained frozen-flux trade-off curve (upper reaches of the right-most solid curve in Figs. 1 and 2), but need not be minima of Δ^2 and are thus neither mapped nor tabulated. Adequate regularization enabled the convergence characterizing the solutions presented.

The rough flow corresponding to the uppermost point on that curve offers an example of the cumulative effect of steady motional induction and the triangle inequality. The initial value gives $<B_r(c,1980)^2>^{1/2} = 3.02 \times 10^5$ nT; numerical advection by said flow yields $<B_{rp}(c,1945)^2>^{1/2} = 3.38 \times 10^5$ nT. The rms value of $B_{rp}(c,1945)$ when truncated to degree 10 is 2.99×10^5 nT; the rms value of the narrow scale component above degree 10 is about 1.57×10^5 nT. The latter may seem large for only 35 years of steady motional induction; yet it is not unreasonable for the core field [Langel & Estes, 1982; Voorhies, 1984, p154-5]. The rms value of such a narrow scale field at Earth's surface is very much less than 161 nT rms uncertainty in $B_r(a,t)$ assigned to the broad scale 1945 DGRF.

APPENDIX C

Selection of inversion strategies and damping reflect information received by the investigator. Such information precludes certain methods as follows. Calculations suggesting rms residuals from time-dependent flow, core asphericity and deep mantle conductivity seemed too primitive to sustain a full covariance analysis of these effects. Further research is also needed to find a theory of magnetohydrodynamic turbulence which predicts the amplitude of residuals when parameters describing narrow scale, statistically steady flow and flux diffusion near the top of Earth's core are not estimated. Covariances were therefore not adjusted for expectations regarding residuals attributable to unmodeled geophysical effects. I suspect this precludes a Gauss-Markov estimate: the weights employed reflect DGRF covariance, not deficiencies in the earth model.

Both Bayesian inference and stochastic inversion require prior information in the form of probability distribution functions (PDFs) for estimated parameters; the prior PDFs are more subjective in the former than in the latter [Backus, 1988; 1989]. In the case of core surface flow estimation, an approximation to stochastic inversion is possible by softening available hard bounds.

For example, geomagnetic core flow studies which omit displacement currents do not test the theory of relativity. The latter thus provides sufficiently reliable information for one to assume that the velocity of every particle in the core is sub-light; this also holds for the velocity of any fluid parcel, hence for the fluid velocity. Seismology suggests that the latter is at most Mach 1; the anelastic approximation would have Mach 1 as a hard upper bound on core flow speed everywhere. Either electromagnetically or acoustically quasi-steady approximations thus offer an infinite number of hard quadratic bounds on core flow estimates. These bounds require the surface kinetic energy density spectrum [Voorhies, 1984] to fall off faster than n^{-1} (n^{-2} by the continuum hypothesis)—perhaps after peaking at very high degree.

Softening these bounds into PDFs is objectionable because (i) neither the bounds nor the finitude of the surface kinetic energy norm imply the PDFs (which are thus personal) and (ii) normal PDFs assign non-zero probability to violation of the bounds [Backus, 1988, 1989]. Suitable non-normal PDFs remove objection (ii)—typically at the expense of the least squares algorithm. Objections to softening hard bounds (i) are fundamental; they may be muted by "pseudo-stochastic inversion": by choosing either uniform PDFs which vanish outside the hard bounds or, perhaps more normally, the broadest Gaussian PDFs needed to keep the estimate within bounds. The latter produces the roughest model compatible with the generalized data—a property enjoyed by the particle, if not the quantum, theory of matter.

The regularizing influence of such a slightly softened Mach 1 bound is insufficient for my algorithms to yield clearly convergent solutions to the non-linear inverse problem addressed. Lacking the PDFs to sustain stochastic inversion, and finding PDFs from slightly softened hard bounds offer insufficient regularization to entertain Bayesian inference, I used damped weighted least squares. In damped weighted least squares the physics of the problem addressed dictates the finitude of the quantities damped.

The balance between the continuum hypothesis and the effectively inviscid treatment of core surface flow allows lateral discontinuities in the flow—but only in a set of measure zero (e.g., a finite number of sheet vorticies abutting **c** or point singularities in downwelling); therefore, both the squared radial vorticity and the squared downwelling averaged over a non-zero area must be finite. Addition of a finite positive definite multiple of $<\omega_r^2 + (\partial_r u)^2>$ to Δ^2 (as in (4)) ensures a model of core surface flow which enjoys the latter property.

Because Δ_r^2 is normally interpreted as being proportional to the argument of the exponential in a Gaussian PDF, a Bayesian view of this action might, following Backus [1988, 1989], be that the multiple λ_d controls the variance of zero-mean normal PDFs assigned to coefficients of T and U. Zero-mean normal PDFs are biased against non-zero coefficients; smaller λ_d give broader PDFs; and the structure of the norm Δ_d^2 implies like PDFs for coefficients of like degree. This norm also damps jets, gyres, and plumes found previously; of course, jumps in ω_r or $\partial_r u$ across narrow fronts or plumes are not resolved when high degree coefficients of T and U are not estimated.

The finite electrical conductivity of the present core model offers finite current density. A similar argument applied to a finite number of lateral discontinuites in **J** implies finite $<[\hat{r}\bullet\nabla\times J]^2>$ at **c**. With non-zero σ this implies finite $<[d_r(c,t)]^2>$, hence finite $<[d_r^*(a)]^2>$ after vacuum upward continuation; indeed, if the set of lateral discontinuities in J_s at **c** is empty rather than merely non-Cantorian, then one can argue for finite $<|\nabla_s d_r(c)|^2>$. A Bayesian view might equate damping $<[d_r(c)]^2>$ with assigning like prior PDFs to all coefficients of $d_r(c)$: the prior spatial power spectra for all $d_r(c)$ sampled would be flat. Yet prior work with DGRF candidates gave no clearly significant sign of diffusion on very broad scales – certainly through degree 5 and perhaps 8 or more [Voorhies & Benton, 1982; Voorhies, 1984; Benton & Voorhies, 1987]. So I resisted an impulse to damp $<[d_r^*(c)]^2>$ and thus avoided uniform damping of the low degree part of $d_r^*(c)$ estimated. In contrast, damping $<[d_r^*(a)]^2>$ allows more rapid diffusion at higher degrees at **c**–in accord with many studies since Booker [1969] suggesting diffusion might be more important at narrower scales. It also eases access to the purely diffusive end-member of SV interpretation.

Despite its susceptibility to Bayesian reinterpretations, damped weighted least squares differs from Bayesian inference. For example, damping is introduced to investigate the weighted residuals induced by various estimates rather than impose a particular set of personal prior PDFs. Indeed, when both more objective PDFs and PDFs from slightly softened bounds offer insufficient regularization, all accessible damping parameters λ are judged too large; nevertheless, the range of accessible λ can be explored. Because a choice of λ within this range appears random, damped weighted least squares inversions have a stochastic quality–as in Bayesian inference with pseudo-random prior. One may adopt preferred solutions a posteriori. For example, solutions near the knee of the accessible trade-off curve may compromise preferences for a smooth model and a tight fit. Solutions portrayed in more detail here compromise preferences for a loose fit (due to simplifying suppositions) and a rough model (reflecting potential complexity in the real Earth).

Damped weighted least squares seems appropriate when appreciable misfit is anticipated due to restrictions on the model imposed by simplifying suppositions. Such problems often sugggest an infinite dimensional parameter space (e.g., an infinte number of spherical harmonic coefficients). Damped weighted least squares extracts finite samples from the parameter space which are more or less constrained to fit a finite amount of information. The number of parameters estimated, though finite, is large enough to mimic difficulties in inverting the underlying, infinte dimensional information matrix. The resulting sample information matrix is finite dimensional and is regularized by damping a necessarily finite quantity (instead of either further truncation or singular value decomposition). The damping parameter is varied to extract different samples; properties shared by many samples may be of interest. The utility of the covariance for any one set of estimated parameters is limited by: the significance of residuals; the basis of the regularization; the selection of estimated parameters; and the underlying physical suppositions. Such severe limitations dictate extreme caution before accepting such a covariance as a starting point for Bayesian inference.

APPENDIX D

Diffusive contributions to SV were explicitly damped–albeit homogeneously so at **a** rather than **c**. Motional contributions to SV were not explicitly damped, but it was clear both before and after the calculations that stronger damping of $<\omega_r^2 + (\partial_r u)^2>$ reduces both flow speed and motional contributions to SV. It follows that dual damping should give motional and diffusive contributions to SV which are not only small, but which cooperate so as to fit the weighted definitive secular change. Such anticipation of $C^* \approx 1$ need not be fully realized. At **c** the diffusion damping is very much weaker at narrower scales (higher degrees up to 10), so diffusion tends to dominate narrower, and motional induction broader, scales of the predicted SV at **c**. This drives C^* towards 0. The combined effects indicate bias toward $C^* > 0$; yet this expected sign was found only in the case of very slow flow at suspiciously low R_m^* (SIDCZ3, Table II).

Fluid motion within the core which maintains a non-trivial, and non-monotonically changing, global absolute (or unsigned) flux linkage is an important feature of a geodynamo. Such changes in the signed flux linking areas bounded by contours of zero normal field component are considered an essential feature of the geodynamo. Compelling evidence for such changes is not easily established: both global absolute and patchwise flux linkages are model dependent. Though recent field models show robust, monotonic decline in Earth's pole strength, the absolute flux linking the core shows little, and perhaps no significant, variation [Voorhies & Benton, 1982, Benton & Voorhies, 1987]. Hard scientific proof of patchwise flux non-conservation near **c** remains elusive: although widespread, all evidence for it since Booker [1969] has been challenged–if not dismissed.

Consider a small but non-zero area δA fixed at **c** with non-zero flux linkage $B_r\delta A$. As is clear from (3) with $d_r = 0$, downwelling $\partial_r u > 0$ tends to increase $|B_r\delta A|$ by sucking in field line footpoints, upwelling tends to decrease $|B_r\delta A|$ by blowing away field line footpoints, and advection $(V_s\bullet\nabla_s B_r)$ changes $|B_r\delta A|$ in accord with the local structure of B_r. Such lateral transport of normal flux density changes neither patchwise nor global absolute flux linkage [Backus, 1968]. As should be well known, entrainment of tangentially magnetized fluid into the resistive boundary layer can alter the absolute flux linking the underlying, free-streaming core—specifically via vertical motion under contours of null radial field [e.g., Voorhies, 1984, p53]. Diffusion is needed to actually separate field lines from the fluid in the boundary layer—enabling the patchwise flux linking the core to change. Such emerging flux regions (EFRs) are widely held to be an essential feature of geodynamo action and have been simulated by Bloxham [1986].

Anelastic upwelling and downwelling need not always imply purely compressive effects at depth; moreover, my solutions remain suggestive of whole outer core flow. So it can be supposed that upwelling occasionally indicates upflow at depth (and vice-versus).

Suppose upflow under an EFR accompanies upwelling at the surface. The upwelling decreases $B_r \delta A$ (and $|B_r \delta A|$) on the side of the null-flux contour where $B_r > 0$ and increases $B_r \delta A$ (but decreases $|B_r \delta A|$) on the side where $B_r < 0$. Diffusion must be doing precisely the opposite in order to have established the sign of B_r in the EFR. So one should expect the contributions from motional induction and flux diffusion to be negatively correlated in an EFR—most notably when both flow and flux diffusion have been steady for some time. Of course, downflow under a null-flux curve suggests downwelling, radial field intensificiation on either side of the curve, a steepening lateral gradient in B_r, currents, a braking of runaway steepening via diffusive reconnection, and negative correlation. Downwelling does not tend to form low flux regions as does upwelling, so coincidence of null-flux contours and downwellings might seem rare; yet the exponential growth of $|B_r|$ implied by downwelling at extrema of B_r and at stagnation points must eventually be offset by negatively correlated diffusion. Clearly, one can expect the contributions to $\partial_t B_r(c,t)$ from motional induction and magnetic flux diffusion to be negatively correlated in the case of core geodynamo action.

Because the dual damping biases steady solutions towards $C^* > 0$, the only factor driving $C^* < 0$ is the fit to the weighted DGRFs at Earth's surface. So the finding of $C^* < 0$ does in fact corroborate the core geodynamo hypothesis.

Curiously, patterns of flux diffusion which are either perfectly correlated or perfectly anti-correlated with the motional contribution to $\partial_t B_{rp}$ could be reproduced by motional induction alone and would thus not confirm geodynamo action. This is impossible when both flow and flux diffusion are steady and the cryptic set has an empty interior.

Acknowledgements. My thanks to J. Dickey [Eubanks et al., 1983] for motivating a reckoning of the magnitude of core surface flow changes. Special thanks to M. Nishihama for helping to run my code; to M. Ford for helping to prepare the figures; to R. Langel & T. Sabaka for DGRF covariances and advice on the ms.; to the groups contributing candidates to the DGRFs; to the many great geomagnetists who helped measure the field; and to the people of the United States of America who supported this work through their National Aeronautics and Space Administration. This work is dedicated to the memory of Professor Edward R. Benton.

REFERENCES

Backus, G. E., Kinematics of the geomagnetic secular variation in a perfectly conducting core, *Phil. Trans. Roy. Soc. Lond., A263*, 239-266, 1968.

Backus, G. E., Bayesian inference in geomagnetism, *Geophys. J., 92*, 125-142, 1988.

Backus, G. E., Confidence set inference with a prior quadratic bound, *Geophys. J., 97*, 119-150, 1989.

Backus, G. E., and J.-L. LeMouel, The region of the core-mantle boundary where a geostrophic velocity field can be determined from frozen-flux magnetic data, *Geophys. J. Roy. astr. Soc. 85*, 617-628, 1986.

Ball, R. H., A. B. Kahle, and E. H. Vestine, Determination of surface motions of the earth's core, *J. Geophys. Res., 74*, 3659-3679, 1969.

Benton, E. R., On the coupling of fluid dynamics and electromagnetism at the top of Earth's core, *Geophys. Astrophys. Fluid Dyn., 33*, 315-330, 1985.

Benton, E. R., and C. V. Voorhies, Testing recent geomagnetic field models via magnetic flux conservation at the core-mantle boundary, *Phys. Earth Planet. Inter., 48*, 350-357, 1987.

Bloxham, J., The expulsion of flux from the earth's core, *Geophys. J. R. astr. Soc., 87*, 669-678, 1986.

Bloxham, J., Steady core surface motions derived from models of the magnetic field at the core-mantle boundary, (abstract), IUGG XIX General Assembly Abstracts, V.2, 437, 1987.

Bloxham, J., The dynamical regime of fluid flow at the core surface, *Geophys. Res. Lett., 15*, 585-588, 1988.

Bloxham, J., Simple models of fluid flow at the core surface derived from geomagnetic field models, *Geophys. J. Int., 99*, 173-182, 1989.

Bloxham, J., and D. Gubbins, Thermal core-mantle interactions, *Nature, 325*, 511-513, 1987.

Bloxham, J., and A. Jackson, Fluid flow near the surface of Earth's outer core, *Rev. Geophys., 29*, 97-120, 1991.

Booker, J. R., Geomagnetic data and core motions, *Proc. Roy. Soc. Lond., A309*, 27-40, 1969.

Bullard, E., and H. Gellman, Homogeneous dynamos and terrestrial magnetism, *Phil. Trans. Roy. Soc. Lond., A247*, 213-278, 1954.

Elsasser, W. M., Induction effects in terrestrial magnetism. Part I: Theory, *Phys. Rev., 69*, 106-116, 1946.

Eubanks, T. M., J. O. Dickey, and J. A. Steppe, A spectral analysis of the Earth's angular momentum budget (abstract), *Eos Trans. AGU, 64*, 205, 1983.

Frazer, M. C., Temperature gradients and convective velocity in the earth's core, *Geophys. J. R. astr. Soc., 34*, 193-201, 1973.

Gire, C., J.-L. LeMouel, and T. Madden, Motions at the core surface derived from SV data, *Geophys. J. R. astr. Soc., 84*, 1-29, 1986.

Gubbins, D., and P. H. Roberts, Magnetohydrodynamics of Earth's core, in *Geomagnetism*, J. A. Jacobs (editor), Academic Press, New York, V.2, 1-183, 1987.

Hide, R., and S. R. C. Malin, On the determination of the size of Earth's core from observations of the geomagnetic secular variation, *Proc. Roy. Soc. Lond., A374*, 15-33, 1981.

Hide, R., Fluctuations in the earth's rotation and the topography of the core-mantle interface, *Phil. Trans. R. Soc. Lond., A328*, 351-363, 1989.

Hide, R., and J. O. Dickey, Earth's variable rotation, *Science, 253*, 629-637, 1991.

IAGA Division I Working Group 1, International Geomagnetic Reference Field Revision 1987, *Phys. Earth Planet. Inter., 50*, 209-213, 1988.

Jacobs, J. A. (editor), *Geomagnetism*, Academic Press, New York, Vol. 1, 627 pp. and Vol. 2, 579 pp., 1987.

Jault, D., C. Gire, and J.-L. LeMouel, Westward drift, core motions and exchanges of angular momentum between core and mantle, *Nature, 333*, 353-356, 1988.

Jault, D., and J.-L. LeMouel, The topographic torque associated with a tangentially geostrophic motion at the core surface and inferences on the flow inside the core, *Geophys. Astrophys. Fluid Dyn., 48*, 273-296, 1989.

Jault, D., and J.-L. LeMouel, Core-mantle boundary shape: constraints inferred from the pressure torque acting between the core and the mantle, *Geophys. J. Int.,, 101*, 233-241, 1990.

Jault, D., and J.-L. LeMouel, Exchange of angular momentum between the core and the mantle, *J. Geomag. Geoelectr., 43*, 111-129, 1991.

Krause. F., and K. H. Radler, *Mean Field Magnetohydrodynamics and Dynamo Theory*, Pergamon Press, Oxford, 271 pp., 1980.

Langel, R. A., The main field, in *Geomagnetism*, J. A. Jacobs (editor), Academic Press, New York, V.2, 249-512, 1987.

Langel, R. A., D. R. Barraclough, D. J. Kerridge, V. P. Golovkov, T. J. Sabaka, and R. H. Estes, Definitive IGRF Models for 1945, 1950, 1955, and 1960, *J. Geomag. Geoelectr., 40*, 645-702, 1988.

LeMouel, J.-L., Outer core geostrophic flow and secular variation of Earth's magnetic field, *Nature, 311*, 734-735, 1984.

McLeod, M. G., Stochastic processes on a sphere, *Phys. Earth Planet. Inter., 59*, 89-97, 1986.

Moffat, H. K., *Magnetic Field Generation in Electrically Conducting Fluids*, Cambridge Univ. Press, 343 pp., 1978.

Stacey, F. D., *Physics of the Earth*, John Wiley & Sons, New York, 414 pp., 1977.

Stephenson, F. R., and L. V. Morrison, Long term changes in the rotation of the Earth: 700 B.C. to A.D. 1980, *Phil Trans. Roy. Soc. Lond., A313*, 47-70, 1984.

Voorhies, C. V., *Magnetic Location of Earth's Core-Mantle Boundary and Estimates of the Adjacent Fluid Motion*, Ph.D. thesis, 348 pp., University of Colorado, Boulder, 1984.

Voorhies, C. V., Steady flows at the top of Earth's core derived from geomagnetic field models, *J. Geophys. Res., 91*, 12,444-12,466, 1986a.

Voorhies, C. V., Steady surficial core motions: an alternate method, *Geophys. Res. Lett., 13*, 1537-1540, 1986b.

Voorhies, C. V., Steady, pseudo-geostrophic flows at the top of Earth's core, (abstract), *Eos Trans. AGU, 67*, 263, 1986c.

Voorhies, C. V., Separating steady flux diffusion from steady motional induction: a kinematic corollary to the steady motions theorem (abstract), IUGG XIX General Assembly Abstracts, V.2, 436, 1987a.

Voorhies, C. V., Steady surficial core motions: a non-linear inverse problem, (abstract), IUGG XIX General Assembly Abstracts, V.2, 437, 1987b.

Voorhies, C. V., Magnetic probing of planetary core motions? Steady induction theorems for giant planets (abstract), *Eos Trans. AGU, 71*, 490, 1990.

Voorhies, C. V., Coupling an inviscid core to an electrically insulating mantle, *J. Geomag. Geoelectr., 43*, 131-156, 1991.

Voorhies, C. V., On the information about secular variation (SV) contained in the Definitive Geomagnetic Reference Field models (DGRFs 1980-1945) (abstract), XX General Assembly IUGG/IAGA Program and Abstracts, p6, 1991b.

Voorhies, C. V., and E. R. Benton, Pole strength of the earth from MAGSAT and magnetic determination of the core radius, *Geophys. Res. Lett., 9*, 258-261, 1982.

Voorhies, C. V. and G. E. Backus, Steady flows at the top of Earth's core from geomagnetic field models: the steady motions theorem; *Geophys. Astrophys. Fluid Dyn., 32*, 163-173, 1985.

Whaler, K. A., Geomagnetic evidence for fluid upwelling at the core-mantle boundary. *Geophys. J. R. astr. Soc., 86*, 563-588, 1986.

C. V. Voorhies, Geodynamics Branch, Code 921/NASA, Goddard Space Flight Center, Greenbelt, MD, 20771 USA.

Degree One Heterogeneity at the Top of the Earth's Core, Revealed by SmKS Travel Times

SATORU TANAKA and HIROYUKI HAMAGUCHI

Faculty of Science, Tohoku University, Sendai 980, Japan

Lateral heterogeneity at the top of the Earth's core was investigated using travel times of SmKS (m=1,2,3) seismic phases. Residuals of the differential travel times $\Delta T_{S2KS-SKS}$ and $\Delta T_{S3KS-S2KS}$ in the Pacific hemisphere were found to be positive near the equatorial region and negative near the polar regions. Those in the other hemisphere including the Atlantic Ocean, Africa and the Indian Ocean indicated the reverse pattern. Spherical harmonic analysis indicated that the power of degree 1 component of the residual was about 1.5 times as much as that of degree 2 and was the most predominant feature. Average residuals were +0.6 sec in the Pacific hemisphere and −0.6 sec in the other one, which suggests the existence of ±0.3% P velocity heterogeneity in the outermost 200 km of the core. This hemispherical pattern coincides with the lateral temperature variation immediately beneath the CMB derived from a geomagnetic study.

INTRODUCTION

A large number of studies has been made on the outermost core called the E' region in order not only to understand the nature of the core-mantle boundary (CMB) [Lay, 1989] but also to perceive the meaning of the source of the geodynamo [Busse, 1975; Braginski, 1977; Fearn and Loper, 1981]. Interpretations of the geomagnetic secular variation have been controversial for the existence of a stably stratified layer and the strong Lorenz force in the E' region [e.g. Bloxham, 1990]. A well established model of P velocity structure in this region has not yet appeared. There is no definitive model precluding 5% variations of the P wave velocity immediately beneath the CMB [Jeffreys, 1939; Randall, 1970; Hales and Roberts, 1971]. Although it was pointed out that an existence of some degree of heterogeneity immediately beneath the CMB [Stevenson, 1987; Lay, 1989], it has been believed for a long time that lateral heterogeneity of seismic structure in the outer core is too small to detect owing to fluidity of the outer core [Stevenson, 1987]. Therefore Schweitzer and Müller [1986] discussed mantle heterogeneity for preferred probable cause of anomalies of amplitude ratios and differential travel time residuals between SKS and SKKS waves found in the Pacific region.

Recently, revised seismic velocity models of the E' region were proposed [Lay and Young, 1990; Souriau and Poupinet,

1991] and a latitudinal variation of the structure in this region was discussed [Souriau and Poupinet, 1990]. Their results were insufficient to discuss a global feature in the E' region because of a bias of the analyzed region. Thus, the analyses using global extended data are required to obtain a unique model of the outermost core. We therefore analyzed not only SKKS-SKS travel times of the epicentral distance ranges being from 100° to 135° but also SKKKS-SKKS travel times from 135° to 180°. From now on, we call SKKS as S2KS and SKKKS as S3KS. After correction for lower mantle heterogeneity [Dziewonski, 1984] as well as ellipticity at the CMB [Bullen and Haddon, 1973], we obtained the residuals of the differential travel times of S2KS–SKS and S3KS–S2KS and discussed relationship between the travel time residuals and other geophysical data.

DATA AND METHOD

We selected about 300 recording of long-period waveforms for 71 deep earthquakes ($h > 200$ km), which were mainly located on the region circling the Pacific Ocean and Hindu Kush. Long period seismograms were retrieved from Global Digital Seismograph Network (GDSN) and Worldwide Standardized Seismograph Network (WWSSN). We selected 13 earthquakes and about 100 recordings from the WWSSN data. The long-period WWSSN recordings were digitized and sampled at 1 second interval by the Image Processing System of the Computer Center of Tohoku University [Morita, 1985]. The GDSN data were available on optical disks (CD-ROM) provided by the United States Geological Survey (USGS). We selected about 200 recordings with good signal to noise ratio from the data set. The SRO, ASRO and DWWSSN data were mainly

Dynamics of Earth's Deep Interior and Earth Rotation
Geophysical Monograph 72, IUGG Volume 12

analyzed because their stations were widely distributed. The difference of a phase lag among the instruments was several seconds for DWWSSN, SRO and ASRO and more than 10 seconds between the WWSSN and SRO. In order to pick up the onsets of SmKS phases on an equal condition, we removed the instrumental responses in the frequency domain. After removing these responses of the seismographs, a Butterworth band-pass filter with cut-off periods of 10 and 32 seconds was digitally applied to the seismograms in both forward and backward directions in the time domain to avoid distortion of the waveform due to phase shift. After appropriate rotation of horizontal components to synthesize radial components, we obtained 245 good records of SmKS phases.

The differential travel times were widely used to remove the contribution of the mantle heterogeneities and event mislocations [e.g. Hales and Roberts, 1971; Schweitzer and Müller, 1986; Lay and Young, 1990; Souriau and Poupinet, 1990, 1991]. For measurement of lag times between two phases, we corrected $\pi/2$ phase shift resulting from Hilbert transform [Choy and Richards, 1975]. The $+\pi/2$ phase shift was operated on an SKS phase when we subtracted an arrival time of an SKS phase from that of an S2KS phase ($T_{S2KS-SKS}$), using the cross correlation method. The $-\pi/2$ phase shift was operated on an S2KS phase when we subtracted an arrival time of an S2KS phase from that of a SKS phase ($T_{SKS-S2KS}$). The cases for measurement between S2KS and S3KS ($T_{S3KS-S2KS}$ and $T_{S2KS-S3KS}$) were the same as those for SKS and S2KS. Figure 1 shows examples of measurements for S2KS–SKS and S3KS–S2KS. The differential travel times are defined as a mean value of the two subtracted arrival times in each measurement. The uncertainty of the differential travel times is typically ±1 sec. All observation data and hypocenter and station parameters will be presented elsewhere [Tanaka and Hamaguchi, in preparation]

We obtained 174 data S2KS–SKS and 71 times for S3KS–S2KS. Since it is difficult to remove the contribution of heterogeneity in the lowermost mantle by only taking the differential travel times, travel times in the mantle were corrected for a lower mantle heterogeneity assuming $\delta v_S = 2\delta v_P$ following Revenough and Jordan [1989], where δv_P is calculated using the model L02.56 [Dziewonski, 1984]. The ellipticity at the CMB, based on the hydrostatic assumption [Bullen and Haddon, 1973], was also corrected. Gwinn et al. [1986] showed dynamical ellipticity of fluid core was 500 m larger than that in hydrostatic equilibrium. However, the excess ellipticity results only travel time residuals of about 0.05 sec. Therefore, we considered only hydrostatic ellipticity. The epicentral distances used in this study were shifted for hypocenters to correspond to the common focal depth at 500 km because that more than half of events used in this study occurred in the depths greater than 500 km. As references of differential travel times for S2KS–SKS and S3KS–S2KS, the 2nd order polynomials of epicentral distance Δ were fitted to the corrected travel times T by the least square method,

$$T_{S2KS-SKS} = 71.74 + 2.933\Delta' + 0.02438\Delta'^2 \qquad (1)$$

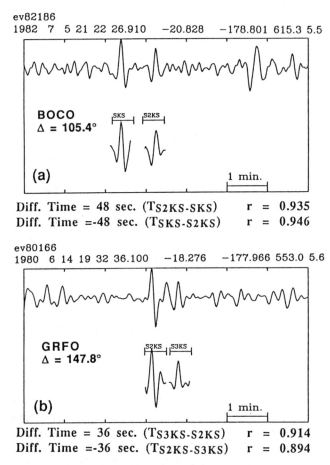

ev82186
1982 7 5 21 22 26.910   −20.828   −178.801 615.3 5.5

BOCO
$\Delta = 105.4°$

SKS   S2KS

(a)   1 min.

Diff. Time = 48 sec. ($T_{S2KS-SKS}$)   r = 0.935
Diff. Time =−48 sec. ($T_{SKS-S2KS}$)   r = 0.946

ev80166
1980 6 14 19 32 36.100   −18.276   −177.966 553.0 5.6

GRFO
$\Delta = 147.8°$

S2KS S3KS

(b)   1 min.

Diff. Time = 36 sec. ($T_{S3KS-S2KS}$)   r = 0.914
Diff. Time =−36 sec. ($T_{S2KS-S3KS}$)   r = 0.894

Figure 1. Examples for measurement of differential travel times. (a) SKS and S2KS. The upper trace is a long period seismogram recorded at GDSN station BOCO in the south America with epicentral distance 105.4° from South Fiji event of which hypocenter is taken from NEIC and indicated above the box. A right lower trace is an inverse Hilbert transformed SKS phase after a relevant time shift determined by cross correlation method and a left one is a Hilbert transformed S2KS phase. Numerals indicated below the box are differential travel times measured by the cross correlation method. The upper one is time of S2KS-SKS, the lower one is that of SKS-S2KS. Correlation coefficients between observed and transformed waveforms are also shown. (b) Same as (a) except for the case of S2KS and S3KS observed at station GRFO in Germany with distance 147.8°.

$$T_{S3KS-S2KS} = 44.89 + 1.139\Delta' + 0.00369\Delta'^2 \qquad (2)$$

where $\Delta' = \Delta-115°$ for S2KS–SKS and $\Delta' = \Delta-155°$ for S3KS–S2KS. Both references of differential travel times for S2KS–SKS and S3KS–S2KS were 1~2 sec larger than those calculated from PREM [Dziewonski and Anderson, 1981]. Residuals of the differential travel times, $\Delta T_{S2KS-SKS}$ and $\Delta T_{S3KS-S2KS}$, were defined as deviations from Eqs. (1) and (2) for $T_{S2KS-SKS}$ and $T_{S3KS-S2KS}$, respectively. Scattering of the residuals was significantly larger than that expected from the observational uncertainty of 1 sec. About 90% of residuals were within ±3 and ±2 sec for $\Delta T_{S2KS-SKS}$ and $\Delta T_{S3KS-S2KS}$, respectively. Because the distances of the ray paths between SKS and S2KS are larger than those for S2KS and S3KS, $\Delta T_{S2KS-SKS}$ can include the effects of the structural ambiguity in the D" region more than $\Delta T_{S3KS-S2KS}$. In the following, we will discuss a lateral variation in the outermost core using residuals ΔT of the differential travel times from the references.

RESULTS

The residuals $\Delta T_{S2KS-SKS}$ and $\Delta T_{S3KS-S2KS}$ along ray paths through the outer core were projected on the world map in Figure 2, which revealed remarkable regional variations of the residuals on a global scale. The bottoms of the ray paths of both S2KS for distances 100°~135° and S3KS for 135°~175° are within the outermost 100~350 km of the core. Therefore, the regional variations of the residuals $\Delta T_{S2KS-SKS}$ and $\Delta T_{S3KS-S2KS}$ commonly resulted from structural anomalies in the outermost 100~350 km of the core.

The geographical distribution of $\Delta T_{S2KS-SKS}$ and $\Delta T_{S3KS-S2KS}$ complementarily covered the surface of the core. $\Delta T_{S2KS-SKS}$ densely covered the regions beneath the Pacific Ocean, central Eurasia, India, the Indian Ocean and the north Atlantic Ocean. $\Delta T_{S3KS-S2KS}$ densely covered north-east Eurasia and the south

Indian Ocean. East Asia was covered by the both residuals where we can find similar values of $\Delta T_{S2KS-SKS}$ and $\Delta T_{S3KS-S2KS}$. Large positive residuals were found beneath the south east Pacific Ocean and slight positive ones beneath the central and western Pacific Ocean and east Asia. On the other hand, strong negative residuals were found beneath the Indian Ocean and slight negative ones beneath the Atlantic Ocean, Africa and the north Pacific Ocean. This pattern of a regional variations suggested that a existence of a longitudinal dependence which was exhibited by a difference between the residuals beneath the Pacific Ocean and those beneath the other regions.

To check a longitudinal difference, we divided a sphere into two hemispheres by the 120°E meridian line. One hemisphere including the Pacific Ocean ranged eastward from 120°E to 60°W (hereafter, we call the P-hemisphere). The other hemisphere included the Atlantic Ocean, Africa and the Indian Ocean (the A-hemisphere). According to the same classification as Souriau and Poupinet [1990], the latitudes of the S2KS reflection points at the CMB were used for sorting the residuals into 9 latitudinal ranges. Figures 3 (a) and (b) show latitudinal dependences of the residuals in the P-hemisphere and the A-hemisphere, respectively. The residuals in the P-hemisphere varied from 1~2 sec near the equator to

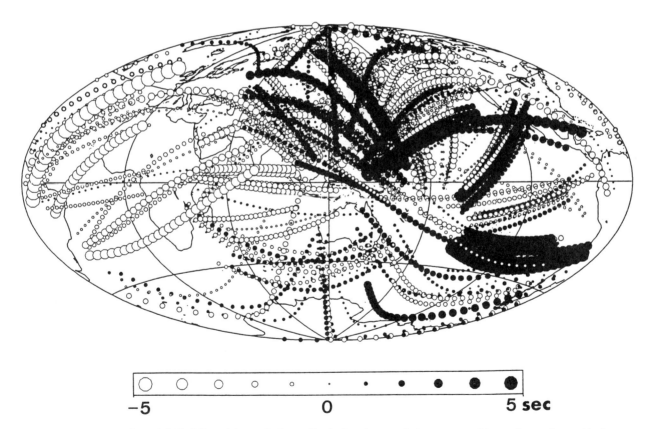

−5   0   5 sec

Figure 2. Distribution of 245 differential travel times. Hatched and open circles mean positive and negative residuals, respectively. Radii of circles are proportional to residuals. The rays through the core are projected of the surface. This map is Aitof projection centered on (0°, 120°E).

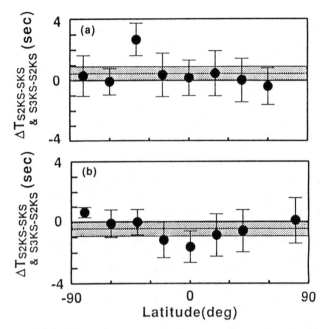

Figure 3. Latitudinal variations of residuals in two hemispheres. Solid circles mean average values in each 20° latitude ranges. Error bars are 1 standard deviation. (a) For the case of the P-hemisphere of which latitude ranges from 120°E eastward to 60 °W. Average values of solid circles is 0.5 sec and hatched region indicates the range of 1 standard deviation. (b) For the case of the A-hemisphere. Average residual is -0.5 sec.

0~–0.5 sec near the high latitude regions. This pattern was very similar to the results using SKS and S2KS from GEOSCOPE data by Souriau and Poupinet [1990]. On the other hand, the residuals in the A-hemisphere varied from –1~–2 sec near the equator to 0~+0.5 sec near the high latitude regions. The largest difference between the patterns found in the two hemispheres appeared near the equatorial regions, suggesting the presence of large scale P wave heterogeneity in the outermost core.

Table 1. Spherical harmonic coefficients of residuals of differential travel times S2KS-SKS and S3KS-SKS up to degree and order 3. σ is standard deviation.

l	m	A_{lm}	σ	B_{lm}	σ
0	0	−0.137	0.27		
1	0	−0.307	0.14		
1	1	−0.635	0.12	−0.118	0.19
2	0	0.221	0.09		
2	1	0.368	0.15	0.179	0.12
2	2	−0.061	0.01	−0.349	0.15
3	0	0.031	0.04		
3	1	−0.255	0.09	0.124	0.10
3	2	0.019	0.08	0.150	0.13
3	3	0.113	0.06	0.236	0.10

Spherical harmonic expansion has been used to express many geophysical observations such as geomagnetic field, geoid and seismic heterogeneity. In this study, this technique was applied to the geographical distribution of the differential travel time residuals of $\Delta T_{S2KS-SKS}$ and $\Delta T_{S3KS-S2KS}$,

$$\Delta T(\theta, \phi) = \sum_{l=0}^{L} \sum_{m=0}^{l} \{A_{lm}\cos m\phi + B_{lm}\sin m\phi\}p_l^m(\cos\theta), \quad (3)$$

where θ is co-latitude and ϕ is longitude and

$$p_l^m(\cos\theta) = [(2-d_{m,0})(2l+1)\frac{(l-m)!}{(l+m)!}]^{1/2} P_l^m(\cos\theta),$$

P_l^m is the associated Legendre function of degree l and order m [Abramowitz and Stegun, 1972]. In order to avoid overweighting regions and to reduce the contribution of small scale heterogeneity in the lowermost mantle, the surface of the CMB was partitioned into 104 rectangles of approximately equal area (20° X 20° at the equator) the same as Creager and Jordan [1986]. The locations of S2KS reflection points at the CMB were used to sort the residuals into the rectangles. The 57 block data, composed of average residuals with standard deviations, were used to determinate the coefficients up to degree and order 3 by the least squares method. The obtained coefficients and their standard deviations are listed in Table 1. The maps of the observed differential travel time residuals are shown in Figure 4 for the pattern composed of all degrees and those for each degree.

The observed residuals expressed by spherical harmonics can be compared with those predicted by models of CMB topographies [Morelli and Dziewonski, 1987; Doornbos and Hilton, 1989] and of the lower mantle heterogeneity assuming $\delta v_S = 2\delta v_P$ [Dziewonski, 1984]. The spherical harmonic coefficients of the predicted residuals are listed in Table 2 and their patterns shown in Figure 5.

The power W_l of each degree was defined as,

$$W_l = \sum_{m=0}^{l} (A_{lm}^2 + B_{lm}^2), \quad (4)$$

and we find that the power of degree 1 component of the observed residual was about 1.5 times as much as that of degree 2 as shown in Figure 6 (a). The powers of each degree of the predicted residuals (from the CMB topographies and the lower mantle heterogeneity) were much smaller than the observed degree 1 residual as shown in Figure 6 (a). Figure 6 (b) shows the degree by degree cross-correlation coefficient r_l between the observed residuals and the predicted ones defined as,

$$r_l = \frac{\sum_{m=0}^{l} (A_{lm}C_{lm} + B_{lm}D_{lm})}{\sqrt{\sum_{m=0}^{l} (A_{lm}^2 + B_{lm}^2)} \sqrt{\sum_{m=0}^{l} (C_{lm}^2 + D_{lm}^2)}}, \quad (5)$$

where C_{lm} and D_{lm} are spherical harmonic coefficients of predicted residuals. The correlation coefficients for degree 1

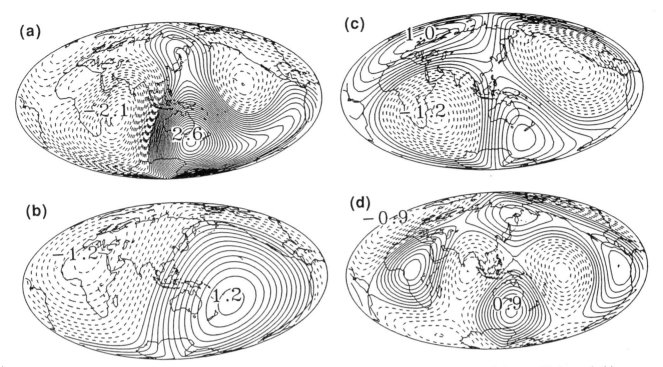

Figure 4. Maps of the differential travel time residuals represented by spherical harmonics for (a) all degrees, (b) degree 1, (c) degree 2 and (d) degree 3 components. Solid and broken lines mean positive and negative values. Contour interval is 0.1 sec.

Table 2 (a). Spherical harmonic coefficients of differential travel times residuals calculated from the model of the topography of the core-mantle boundary by Morelli and Dziewonski (1987).

l	m	A_{lm}	B_{lm}
0	0	−0.03	
1	0	−0.06	
1	1	0.10	−0.18
2	0	−0.04	
2	1	−0.01	−0.01
2	2	0.01	−0.04
3	0	0.02	
3	1	0.02	0.02
3	2	0.03	0.05
3	3	−0.11	0.00

Table 2 (b). Same as (a) except for the model by Doornbos and Hilton (1989).

l	m	A_{lm}	B_{lm}
0	0	0.02	
1	0	−0.06	
1	1	0.05	−0.09
2	0	0.03	
2	1	−0.01	0.00
2	2	0.00	−0.03
3	0	0.03	
3	1	−0.01	0.02
3	2	0.01	0.03
3	3	−0.02	−0.03

Table 2 (c). Same as (a) except for the model of lower mantle heterogeneity by Dziewonski (1984).

l	m	A_{lm}	B_{lm}
0	0	0.12	
1	0	−0.11	
1	1	0.15	−0.27
2	0	−0.06	
2	1	−0.25	0.05
2	2	0.11	−0.05
3	0	0.20	
3	1	0.13	0.01
3	2	−0.04	−0.08
3	3	0.16	−0.25

were −0.16, −0.04 and −0.12 for three models, indicating no correlation in degree 1 component between the observed and predicted residuals. On the other hand, degree 2 and 3 components showed some correlation with the residuals from the CMB topography by Doornbos and Hilton [1989] (r_2= 0.54) and with those from the lower mantle heterogeneity by Dziewonski [1984] (r_2= −0.43, r_3 = −0.48).

The degree 1 component had peaks of +1.2 sec in the south Pacific Ocean and −1.2 sec in the north Africa as shown Figure 4 (b). Average residuals were 0.6 sec in the P-hemisphere and − 0.6 sec in the A-hemisphere. The boundary of the polarity was located along the rim of the Pacific Ocean. Tanaka and Hamaguchi [1992] showed that radial P velocity structure in the outermost 200 km in the core was slightly inhomogeneous

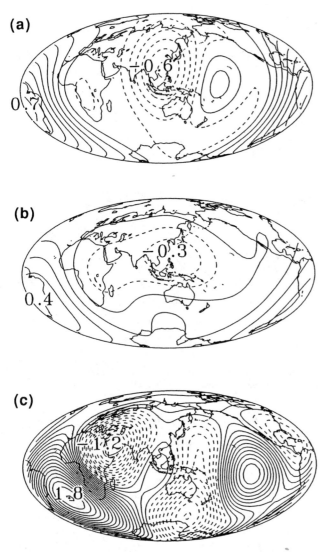

Figure 5. Maps of the predicted differential travel time residuals calculated using the models of the core-mantle topographies of (a) Morelli and Dziewonski [1987] and (b) Doornbos and Hilton [1989], using the model of (c) lower mantle heterogeneity [Dziewonski, 1984].

where index η [Bullen and Bolt, 1985] was larger than unity. Therefore, if the P wave velocity heterogeneity concentrates in the outermost 200 km in the core, these residuals corresponded to a manifestation of ±0.3% variation of the P wave velocity.

DISCUSSION AND CONCLUSIONS

We revealed the existence of a degree 1 residual with ±0.3% P wave velocity heterogeneity in the outermost 200 km in the core. This small heterogeneity in the outermost core hardly effects the PKP travel times. When PKP rays cross the CMB at either the P- or A-hemisphere, the heterogeneity yields the travel time residual of, at most, 0.2 sec. Therefore, the residual of PKP travel times from the outermost core is swamped by other effects above the CMB.

Possibility of heterogeneity in the outermost core has been qualitatively suggested by Stevenson [1987] and Wahr and de Vries [1989]. However, the pattern of heterogeneity in the outermost core has been controversial. Many patterns of fluid

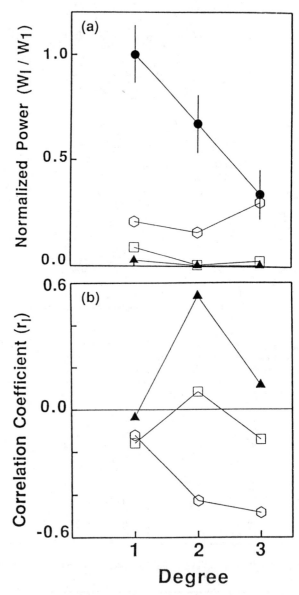

Figure 6. (a) Power in each harmonic degree as defined by Eq. (4). Each plot is normalized by the value of degree 1 of the observed residual. Closed circle means the observed residual. An error bar indicates a range of a standard deviation. Closed triangle, open square and hexagon mean power of the residuals predicted from the model of CMB topography by Doornbos and Hilton [1989], by Morelli and Dziewonski [1987] and of lower mantle heterogeneity by Dziewonski [1984], respectively. (b) Degree by degree cross-correlation coefficient between spherical harmonic coefficients of observed and calculated residuals. Closed triangle, open square and hexagon mean correlation between the observed residuals and those predicted from the model of CMB topography by Doornbos and Hilton [1989], by Morelli and Dziewonski [1987] and of lower mantle heterogeneity by Dziewonski [1984], respectively.

(a)

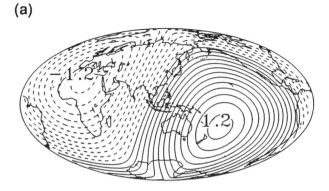

(b)

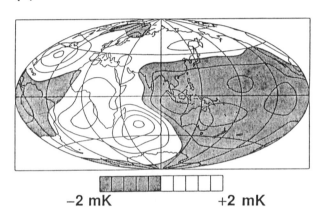

−2 mK   +2 mK

Figure 7. (a) The geographical distribution of the degree 1 residual. (b) Temperature distribution at the CMB derived from geomagnetic study [Bloxham Jackson, 1991]. The both features are quite similar each other in the predominance of the pattern of degree 1.

flow at the core surface were proposed because of the inherent non-uniqueness in using the geomagnetic secular variations (SV), although there are some large scale features widely recognized such as weak flow beneath the Pacific Ocean and westward flow beneath Africa [for a review, Bloxham and Jackson, 1991]. On the other hand, Yukutake [1992] suggested that the Pacific Ocean was not any particular place where the non-dipole was weak. Temperature and density heterogeneities just beneath the CMB were estimated using the tangential components of the induction equation [Bloxham and Jackson, 1990; Gubbins, 1991]. Jault and Le Mouël [1991], however, proposed that these heterogeneities could not be estimated from SV data alone. Souriau and Poupinet [1991] supported, from seismic observations, a differential rotation in the core proposed by Jault et al. [1988].

The pattern of the degree 1 residual presented in this study coincides with the pattern of temperature distribution at the surface of the core, which was derived from the geomagnetic studies [Bloxham and Jackson, 1990, 1991] (Figure 7). Therefore, the seismic heterogeneities in the outermost core are discussed in reference to adiabatic hemispherical distributions of temperature. The temperature distribution from

the geomagnetic study showed that a degree 1 component was dominant and that the maximum temperature difference from the adiabatic temperature were ±2 mK in the A- and P-hemispheres. The differences were dependent on the thermal expansion coefficient in the core, which was one of unknown parameters in the outer core. We could compare this pattern with that translated from the heterogeneity of P wave velocity in the outermost core, because seismic velocity is related to the temperature gradient.

Adiabatic gradient of temperature in terms of seismic velocity given by

$$(dT/dr)_S = -g\gamma T/\phi \tag{6}$$

[Bullen and Bolt, 1985] where T is adiabatic temperature, r is a distance from the Earth's center, g is the gravity acceleration, γ is the Grüneisen parameter, ϕ is the seismic parameter ($= v_P{}^2 - 4v_S{}^2/3$) and S is entropy. The integration of Eq. (6) is given by

$$\ln\frac{T_{CMB}}{T_{ref}} = -\int_{r_{ref}}^{r_{CMB}} \frac{g\gamma}{\phi}\,dr, \tag{7}$$

where T_{CMB} is the adiabatic temperature at r_{CMB}, T_{ref} is that at a reference depth r_{ref}. Assuming $\gamma = 1.5$ [Melchior, 1986] through the outer core, adopting g values of PREM, we considered only ±0.3% perturbations of P velocity in the outermost core for the P- and A-hemispheres to calculate T_{CMB}/T_{ref} for each hemisphere. When $r_{CMB} - r_{ref} = 200$ km was assumed

$$T_{CMB}/T_{ref} = 0.953280 \text{ for the P-hemisphere,}$$
$$T_{CMB}/T_{ref} = 0.953642 \text{ for the A-hemisphere.}$$

The temperature differences between the two hemispheres at the CMB were

$$\Delta T_{CMB} = T_{CMB}(A) - T_{CMB}(P) = 3.62 \times 10^{-4}\,T_{ref}.$$

Therefore, T_{CMB} in the A-hemisphere was higher than that in the P-hemisphere. If T_{ref} is assumed to be 3000 K, then $\Delta T_{CMB} = 1.1$ K.

It is unlikely that the variation of g and γ exceed more than one order. If the variation of P wave velocity is less than order of 0.1%, we can not detect the seismic heterogeneity in the outermost core. Hence, the absolute value of ΔT_{CMB} is entirely dependent on the assumed skin depth. Although the skin depth in the outermost core was an unknown parameter and the definition of the temperature difference in the present study was different from that in the geomagnetic study, it is likely that the temperature differences deduced from the seismic heterogeneity and the SV data are not far from each other. Therefore, it is suggested that low temperature in the P-hemisphere relates to very little flow in the Pacific region, and that high temperature in the A-hemisphere relates to vigorous flows such as strong westward motion beneath Africa and two large gyres under the north Atlantic Ocean and the Indian Ocean.

In summary, we investigated lateral variation in the outermost core by using travel times of SmKS (m=1,2,3) seismic phases. Spherical harmonic analysis indicated that the degree 1 component of the residuals was most dominant and could not be explained by known lower mantle heterogeneity and CMB topography. We found that the average residuals of differential travel times of S2KS–SKS and S3KS–S2KS were 0.6 sec and –0.6 sec in the P- and the A-hemisphere, respectively. Thus we concluded that there was a degree 1 heterogeneity in the outermost 200 km of the outer core, of which amplitude was ±0.3% of the P velocity. We suggested that adiabatic temperatures at the CMB were relatively low in the P-hemisphere and high in the A-hemisphere.

Acknowledgements. We thank Dr. I. S. Sacks for critical reading of the manuscript, Dr. Y. Morita for discussions and two anonymous reviewers for comments. This work was partly supported by Grant-in-Aid for Scientific Research (B) No. 03452055 and Grant under the Monbushou International Scientific Research Program No. 02041008.

REFERENCES

Abramowitz, M. and I. A. Stegun, *Handbook of Mathematical Functions, with Formulas, Graphs, and Mathematical Tables*, 9th ed., 1046 pp, Dover Pub., New York, 1972.

Bloxham, J., On the consequences of strong stable stratification at the top of Earth's core, *Geophys. Res. Lett*, 17, 2081-2084, 1990.

Bloxham, J. and A. Jackson, Lateral temperature variations at the core-mantle boundary deduced from the magnetic field, *Geophys. Res. Lett.*, 17, 1997-2000. 1990.

Bloxham, J. and A. Jackson, Fluid flow near the surface of Earth's outer core, *Rev. Geophys.*, 29, 97-120, 1991.

Braginski, S. I., Nearly axisymmetric model of the hydromagnetic dynamo of the Earth I, *Geomagn. Aeron.*, 15, 225-231, 1975.

Bullen, K. E. and B. A. Bolt, *An Introduction to the Theory of Seismology*, 4th ed., 499 pp., Cambridge Univ. Press, Cambridge, 1985.

Bullen, K. E., and R. A. W. Haddon, The ellipticities of surfaces of equal density inside the Earth, *Phys. Earth Planet. Inter.*, 7, 199-202, 1973.

Busse, F. H., A model of the geodynamo, *Geophys. J. R. astr. Soc.*, 42, 437-459, 1975.

Choy, G. L. and P. G. Richards, Pulse distortion and Hilbert transformation in multiply reflected and refracted body waves, *Bull. Seismol. Soc. Am.*, 65, 55-70, 1975.

Creager, K. C. and T. H. Jordan, Aspherical structure of the core-mantle boundary from PKP travel times, *Geophys. Res. Lett.*, 13, 1497-1500, 1986.

Doornbos, D. J. and T. Hilton, Models of the core-mantle boundary and the travel times of internally reflected core phases, *J. Geophys. Res.*, 94, 15741-15751, 1989.

Dziewonski, A. M., Mapping the lower mantle: determination of lateral heterogeneity in P velocity up to degree and order 6, *J. Geophys. Res.*, 89, 5929-5952, 1984.

Dziewonski, A. M. and D. L. Anderson, Preliminary reference Earth model, *Phys. Earth Planet. Inter.*, 25, 297-356, 1981.

Fearn, D. R. and D. E. Loper, Compositional convection and stratification of the Earth's core, *Nature*, 289, 393-394, 1981.

Gubbins, D., Dynamics of the secular variations, *Phys. Earth Planet. Inter.*, 68, 170-182, 1991.

Gwinn, C. R., T. A. Herring, and I. I. Shapiro, Geodesy by radio interferometry: Studies of the forced nutations of the Earth 2. Interpretation, *J. Geophys. Res.*, 91, 4755-4765, 1986.

Jault, D., C. Gire, and J. L. Le Mouël, Westward drift, core motions and exchanges of angular momentum between core and mantle, *Nature*, 333, 353-356, 1988.

Jault, D. and J. L. Le Mouël, Physical properties at the top of the core and core surface motions, *Phys. Earth Planet. Inter.*, 68, 76-84, 1991.

Jeffreys, H., The times of the core waves, *Mon. Not. R. astr. Soc., Geophys. Suppl.*, 4, 548-561, 1939.

Hales, A. L. and J. L. Roberts, The velocities in the outer core, *Bull. Seism. Soc. Am.*, 61, 1051-1059, 1971.

Lay, T., Structure of the core-mantle transition zone: A chemical and thermal boundary layer, *EOS, Trans. Am. Geophys. Un.*, 70, 49-59, 1989.

Lay, T. and C. J. Young, The stable-stratified outer most core revisited, *Geophys. Res. Lett.*, 17, 2001-2004, 1990.

Melchior, P., *The Physics of the Earth's Core: An Introduction*, 256 pp., Pergamon Press, Oxford, 1986.

Morelli, A. and A. M. Dziewonski, Topography of the core-mantle boundary and lateral homogeneity of the liquid core, *Nature*, 325, 678-683, 1987.

Morita, Y., Seismological study on the core-mantle boundary beneath the Hawaiian hotspot, *Thesis of Sc. Doctor*, 159 pp, Tohoku Univ., 1985.

Randall, M. J., SKS and seismic velocities in the outer core, *Geophys. J. R. astr. Soc.*, 21, 441-445, 1970.

Revenough, J. and T. H. Jordan, A study of mantle layering beneath the western Pacific, *J. Geophys. Res.*, 94, 5787-5813, 1989.

Schweitzer, J. and G. Müller, Anomalous difference traveltimes and amplitude ratios of SKS and SKKS from Tonga-Fiji events, *Geophys. Res. Lett.*, 13, 1529-1532, 1986.

Souriau, A. and G. Poupinet, A latitudinal pattern in the structure of the outermost liquid core, revealed by the travel times of SKKS-SKS seismic phases, *Geophys. Res. Lett.*, 17, 2005-2007, 1990.

Souriau, A. and G. Poupinet, A study of the outermost liquid core using differential travel times of SKS, SKKS and S3KS phases, *Phys. Earth Planet. Inter.*, 68, 183-199, 1991.

Stevenson, D. J., Limits on lateral density and velocity variations in the Earth's outer core, *Geophys. J. R. astr. Soc.*, 311-319, 1987.

Tanaka, S. and H. Hamaguchi, Velocities and stable stratification in the outermost core (abstract), The Third SEDI Symposium, Mizusawa, Japan, July 6-10, 1992.

Yukutake, T., The geomagnetic non-dipole field in the Pacific (abstract), The Third SEDI Symposium, Mizusawa, Japan, July 6-10, 1992.

S.Tanaka and H.Hamagichi, Observation Center for Prediction of Earthquakes and Volcanic Eruptions, Faculty of Science, Tohoku University, Sendai 980, Japan

Aspherical Structure of the Mantle, Tectonic Plate Motions, Nonhydrostatic Geoid, and Topography of the Core-Mantle Boundary

ALESSANDRO M. FORTE, ADAM M. DZIEWONSKI,
and ROBERT L. WOODWARD

*Department of Earth and Planetary Sciences, Harvard University,
Cambridge, MA 02138*

Seismic imaging of global lateral heterogeneity in the mantle has made significant progress with the recent joint inversions of large data sets combining differential travel times and waveforms of long-period body waves and mantle waves. We describe here a recently obtained model of 3-D shear velocity heterogeneity, called $SH8/U4L8$, which shows striking examples of plume-like features characteristic of thermal convective flows. We have employed the lateral density variations derived from model $SH8/U4L8$ to predict the main convection-related observables: the nonhydrostatic geoid, the tectonic plate motions, and the CMB topography. These predictions, obtained using viscous flow models of a compressible mantle, are in good agreement with the corresponding observed fields. The variance reductions which are obtained with the predicted geoid and predicted plate motions are generally in the range 65% to 80%. We find that such good fits may only be obtained with viscous flow models which assume a-priori a whole-mantle style of flow. Flow models which explicitly include a barrier to radial flow at the 670 km seismic discontinuity can provide fair fits to the geoid data only by invoking anomalous $\partial ln\rho/\partial lnv_S$ proportionality factors and such models provide very poor fits to the plate-motion data.

In the course of fitting the geoid and plate motion data we have inferred a new radial viscosity profile of the mantle and we have also inferred new depth-dependent $\partial ln\rho/\partial lnv_S$ proportionality factors. The nonhydrostatic geoid data were used to constrain the relative variation of viscosity with depth in the mantle and we thus inferred a thin layer of strongly reduced viscosity at the base of the upper mantle and a region of significantly increased viscosity within the lower mantle. A significant viscosity increase within the lower mantle is needed to provide acceptable fits to the Y_2^0-component of the satellite-observed nonhydrostatic geoid. The absolute value of mantle viscosity was determined using the plate-motion data and we thus infer a viscosity of about 1×10^{21} Pa s for most of the upper mantle (with the exception of the low-viscosity zone at the base of the upper mantle). On the basis of the nonhydrostatic geoid data we infer $\partial ln\rho/\partial lnv_S \approx 0.35$ in the depth range 400-670 km, $\partial ln\rho/\partial lnv_S \approx 0.20 - 0.30$ in the depth range 670-1000 km, and $\partial ln\rho/\partial lnv_S \approx 0.10$ in the remainder of the lower mantle. The $\partial ln\rho/\partial lnv_S$ values inferred for the top 400 km of the mantle are not well constrained by the geoid data and they are generally rather sensitive to the particular choice of lateral heterogeneity model and also to the details of the assumed radial viscosity profile.

The flow-induced CMB topography calculated using model $SH8/U4L8$ agrees well with the pattern displayed by previous models of CMB topography inferred from PcP and PKP travel time residuals [Morelli and Dziewonski, 1987] and from the observed splitting of normal-mode multiplets [Li et al., 1991]. Our predicted CMB topography is dominated by a Y_2^2 pattern manifested by the characteristic ring of depressed topography below the circum-Pacific margin. The root-mean-square amplitude of the predicted CMB topography (in the degree range $l = 2 - 8$) is approximately 1 km and individual 'bumps' on the CMB may reach peak amplitudes of about 3 km.

1. INTRODUCTION

Over the past decade seismic tomographic studies of global aspherical Earth structure have been based upon the separate inversion of rather different classes of data such as P-wave travel-time residuals [e.g. Dziewon-

Dynamics of Earth's Deep Interior and Earth Rotation
Geophysical Monograph 72, IUGG Volume 12

ski, 1975,1982,1984; Dziewonski et al., 1977; Clayton and Comer, 1983; Morelli and Dziewonski, 1986, 1987, 1991; Gudmundsson, 1989; Inoue et al., 1990], apparent central-frequency shifts of unresolvably-split normal mode multiplets [Masters et al., 1982], surface wave dispersion measurements [e.g. Nakanishi and Anderson, 1982,1983,1984; Nataf et al., 1984,1986; Wong, 1989; Montagner and Tanimoto, 1991], normal mode splitting functions [e.g. Giardini et al., 1987,1988; Ritzwoller et al., 1986,1988; Li et al., 1991], and complete waveforms of long-period surface waves and body waves [e.g. Woodhouse and Dziewonski, 1984,1986; Tanimoto, 1987,1990; Su and Dziewonski, 1991]. Increasingly useful and important information on aspherical mantle structure is now being provided by studies of lateral variations of intrinsic attenuation in the mantle [e.g. Durek et al., 1989; Romanowicz, 1990]. A useful review of the development and recent progress in seismic imaging techniques may be found in Woodhouse and Dziewonski [1989] and Romanowicz [1991].

It is noteworthy that although the frequency range of the data employed in the previous seismic tomographic studies spans several orders of magnitude (1 Hz to 1 mHz) the inferred models of global lateral heterogeneity can be quite similar to each other [e.g. Giardini et al., 1987]. This consistency provides the justification for recent studies in which, for example, normal mode multiplet splitting data is jointly inverted with seismic travel-time residuals to obtain models of three-dimensional mantle heterogeneity [e.g. Woodward and Masters, 1992]. The joint inversion of very different classes of seismic data has made substantial progress in the recent studies by Dziewonski and Woodward [1992] and Woodward et al. [1992], in which waveforms of long-period surface waves and body waves are combined with $SS - S$ and $ScS - S$ differential travel-time residuals to obtain models of global shear-velocity heterogeneity in the Earth's mantle. These recent models provide a much greater horizontal and vertical resolution of three-dimensional mantle structure than has hitherto been possible.

The earliest models of mantle heterogeneity [e.g. Dziewonski et al., 1977; Dziewonski, 1984; Clayton and Comer, 1983 (reported in Hager and Clayton [1989]); Woodhouse and Dziewonski, 1984] exhibited a large-scale three-dimensional structure which was reasonably well correlated to the major surface manifestations of mantle convection, namely the large-scale nonhydrostatic geoid [Hager et al., 1985] and the large-scale tectonic plate motions [Forte and Peltier, 1987a,b]. These early models of mantle heterogeneity thus helped to show that seismic tomography could indeed resolve the field of lateral temperature variations which drives the mantle flow responsible for the "drift" of the continents [Peltier and Forte, 1984]. The normal stresses produced by the mantle flow at the core-mantle boundary (CMB) will also deflect this major chemical discontinuity away from its hydrostatically defined reference position. The derivation of the Morelli and Dziewonski [1987] model of CMB topography, obtained by inverting PcP and PKP_{BC} travel-time residuals, made it possible to subject both the Morelli and Dziewonski model and the early mantle-heterogeneity models to an important consistency test: If the CMB topography model of Morelli and Dziewonski (henceforth referred to as the MD model for convenience) represents the actual dynamic topography resulting from mantle flow then it should agree, in principle, with the CMB topography predicted on the basis of the seismic models of internal mantle heterogeneity.

The earliest predictions of flow-induced CMB topography by Hager et al. [1985], employing the P-velocity heterogeneity model of Clayton and Comer [1983], did not appreciably resemble the MD model (see also Hager and Clayton [1989]). This difficulty led to an extensive study of CMB topography by Forte and Peltier [1989,1991a] in which it was argued that the MD model was indeed a plausible representation of the CMB topography induced by a mantle-wide convective flow. In this latter study Forte and Peltier employed the P-velocity heterogeneity model of Dziewonski [1984] and they pointed out that the discrepancy between their CMB topography predictions and those of Hager et al. must be due to the differing models of lower-mantle heterogeneity employed in the two studies. Such discrepancies indicate the significant uncertainties in the early models of global seismic velocity heterogeneity, particularly those models [e.g. Clayton and Comer, 1983; Dziewonski, 1984] derived from seismic travel times recorded in the Bulletins of the International Seismological Centre (ISC). The MD model was also obtained from an inversion of ISC travel times and, in spite of the mantle-flow plausibility arguments given by Forte and Peltier [1989,1991a], there remained the perception [e.g. Gudmundsson, 1989] that the ISC Bulletins could not be relied upon to provide sufficiently accurate images of either the CMB topography or the heterogeneity in the mantle. Such doubts concerning the validity of the MD model were accentuated when subsequent studies of CMB topography using the ISC Bulletins yielded rather different images of CMB topography [e.g. Gudmundsson, 1989; Doornbos and Hilton, 1989]. In this regard one must appreciate that the seismic determination of CMB topography to an accuracy of 1 or 2 km is an order of magnitude more difficult than mapping three-dimensional velocity anomalies in the lower mantle. Even so, such discordant inferences of the CMB topography are unfortunate for, as shown in Forte and Peltier [1989,1991a] this geophysical "observable" provides important and useful constraints on the geometry and dynamic state of the convective circulation in the deepest portions of the mantle.

The topography on the CMB has also been identified as an important ingredient in the description of the perturbations of the rotational dynamics of the solid Earth. Fluid motions in the Earth's outer core associated with the magneto-hydrodynamic dynamo generate pressure gradients which act on the topographic undulations of the CMB thus producing torques which transfer angular momentum between the core and mantle. Hide [1969] proposed that this exchange of angular momentum may account for length-of-day (l.o.d.) changes over decade periods (see Hide [1989] for a recent and extensive review). The action of dynamic core pressures on the CMB topography may also contribute to polar wander [e.g. Hide, 1989; Hinderer et al., 1990] and, in addition to the dynamic effects of the atmosphere and oceans [e.g. Munk and MacDonald, 1960; Wahr, 1982], such core pressures may also be important in maintaining the Chandler wobble [Hinderer et al., 1990].

The initial calculations of l.o.d. changes due to topographic torques on the CMB by Speith et al. [1986] indicated that the MD model would yield excessively large changes in l.o.d. over the past 20 years. This difficulty led to critical reanalyses of the core-mantle coupling problem [e.g. Roberts, 1988] and it also added once again to the doubts concerning the validity of the large-amplitude topographic undulations in the MD model [e.g. Hide, 1989]. The latter difficulty has led some to advocate unusual mantle viscosity profiles with a rather thick (300 km) layer of very low viscosity at the base of the mantle [Hager and Richards, 1989] or to postulate the existence of a low-viscosity layer of liquid silicate slag separating the mantle from the outer core [Hide, 1989]. The necessity for adopting such scenarios has been put into serious question by the recent work of Jault and Le Mouël [1990] who show that, with only small alterations, the MD model gives rise to core-mantle torques of sufficiently small amplitude to explain the observed decade-scale l.o.d. variations. In a similar vein Bloxham [1991] finds that, with small alteration, the core flows inferred from magnetic secular variation data produce acceptable core-mantle torques. A particularly important component of the studies by Jault and Le Mouël [1990] and Bloxham [1991] is the recognition that, in order to maintain acceptable levels of angular momentum transfer between mantle and core, the core flows must somehow vary in time in a manner dictated by the CMB topography itself. It is perhaps important that Jault and Le Mouël [1990] find that the MD model is quite "near" to the space of topographies producing zero torque. It now seems, as argued initially by Forte and Peltier [1989,1991a], that the MD model is indeed a plausible representation of the dynamic CMB topography after all. It is because of the important influence of the CMB topography on the rotational dynamics of the Earth and on the dynamics of core flow that we have conducted the following study of mantle flow and corresponding dynamic CMB topography.

The plan of our discussion is as follows. In section 2 we describe the very recent model $SH8/U4L8$ which describes the global S-velocity heterogeneity $\delta v_S/v_S$ in the mantle obtained from a joint inversion of long-period waveforms of mantle waves and body waves and $SS-S$ and $ScS-S$ differential travel times [Dziewonski and Woodward, 1992]. In this section we also analyze the direct correlation between the $\delta v_S/v_S$ heterogeneity described by $SH8/U4L8$ and the nonhydrostatic geoid and plate motions. Such an analysis will help identify which depth ranges in the mantle are important sources for the observed geoid and plate motions and it will thus help to constrain the viscous flow models of the mantle which describe these surface observables. In section 3 we employ the field of lateral density variations inferred from $SH8/U4L8$ to predict the

large-scale plate motions and nonhydrostatic geoid using the gravitationally consistent theory of viscous flow in a compressible mantle developed by Forte and Peltier [1991b]. As we shall see, the agreement between the predictions and observations is very good. Having thus demonstrated the plausibility of the mantle-flow field predicted with $SH8/U4L8$ we then calculate in section 4 the flow-induced CMB topography. This predicted CMB topography agrees well with the MD model and it also agrees very well with the CMB topography predicted on the basis of the P-velocity heterogeneity model $L02.56$ of Dziewonski [1984]. When we consider that models $SH8/U4L8$ and $L02.56$ are obtained from two very different sets of data (which are sensitive to two rather different aspects of the internal elastic structure: shear velocities and compressional velocities) the agreement between the CMB predictions is important in establishing their credibility. Finally in section 5 we will summarize our principal conclusions. We wish to point out that the geodynamic modelling results presented below were obtained very recently and have not appeared in print elsewhere. The following discussion therefore provides us with the first opportunity to provide a detailed account of these new results.

2.   A Model of Three-Dimensional Shear-Velocity Variations in the Mantle

2.1   Definition of the model, data and inversion

There are, basically, two approaches to representing a 3-D perturbation to a spherically symmetric (one-dimensional) Earth model. One is to divide the medium into a 3-D array of cells in each of which the perturbation is constant. For example, Inoue et al. [1990] divided the Earth into 16 spherical shells of varying thickness, and then subdivided each shell into 2,048 spherical rectangles (32×64) each $5.6° \times 5.6°$ in size. A similar parameterization, but involving many fewer elements, was used by Dziewonski [1975] and, with even more cells, by Clayton and Comer [1983].

The other approach is to use basis functions. Dziewonski [1982] proposed to represent 3-D variations in compressional velocity using spherical harmonics to describe horizontal variations and Legendre polynomials for variations with radius:

$$\delta v(r,\vartheta,\varphi)/v = \sum_{k=0}^{K}\sum_{\ell=0}^{L}\sum_{m=0}^{\ell} f_k(r)\, p_\ell^m(\vartheta)$$
$$({}_kA_\ell^m \cos m\varphi + {}_kB_\ell^m \sin m\varphi); \quad (1)$$

where p_l^m is the normalized associated Legendre polynomial:

$$p_l^m(\vartheta) = \left[\frac{(2l+1)(2-\delta_{m,0})(l-m)!}{(l+m)!}\right]^{1/2} P_{lm}(\vartheta). \quad (2)$$

This parameterization has been used, among others, by Dziewonski [1984], Woodhouse and Dziewonski [1984] and Giardini et al. [1987] even though these studies used entirely different data: travel times of body waves, waveforms and splitting of normal modes, respectively. One of the advantages of the representation above is that it provides a natural means for filtration of effects of different wavelength or symmetry. Tanimoto [1990] used a hybrid representation in which he applied spherical harmonics to express horizontal variations but divided the mantle into 11 spherical shells to describe variations with depth; there was no variation with depth within individual shells.

Several types of data have been used to derive 3-D Earth models. Bulletins of International Seismological Centre (ISC) are the most important source of travel times used in mapping the P-wave speed [Dziewonski, 1975, 1982, 1984; Dziewonski et al., 1977; Clayton and Comer, 1983; Morelli and Dziewonski, 1986, 1991]. After 25 years of compilation by the ISC there are now many millions of arrival times of P-waves (including the core phases) and considerably fewer of S-waves. Experience shows that the data for P-waves are not only the most numerous but are also the most reliable. Interpretation of secondary arrivals is subject to uncertainties and attempts to derive S-wave speed models using ISC Bulletins have not been successful.

Introduction of digital recording in the mid-1970's in two global networks: International Deployment of Accelerometers (IDA) [Agnew et al., 1976] and Global Digital Seismographic Network (GDSN) [Peterson et al., 1976] prepared the ground for use of waveform data in seismic tomography. In some cases these waveforms are pre-processed to derive parameters such as the shifts in the spectral peaks of normal modes [Masters et al., 1982] or phase velocities of mantle waves [Nakanishi and Anderson, 1983, 1984]. Dziewonski and Steim [1982] and Woodhouse and Dziewonski [1984] introduced waveform inversion techniques which allow derivation of perturbations to a model through direct operation on a waveform. The method of Woodhouse and Dziewonski [1984] was particularly powerful, since it allows simultaneous interpretation of all the modes contained in a particular time window as well as derivation of both even and odd terms in the spherical harmonic expansion of lateral heterogeneity.

The following text describes the results presented by Dziewonski and Woodward [1992] in their inversion of a combined data set. Figure 1a compares an observed mantle wave with a synthetic seismogram computed by a spherically symmetric Earth model (top) and a 3-D model (bottom). Figure 1b shows a similar comparison for a wavetrain containing several separate arrivals of body wave phases and a dramatic improvement of the fit that can be accomplished by 3-D modeling. The theory is presented in detail in Woodhouse and Dziewonski [1984]. The basic idea is one of an 'average structure' along the minor arc (shorter path between the epicenter and the station along the great circle connecting the two), $\delta\tilde{m}$, and the entire great circle, $\delta\hat{m}$. Such structures can be represented by superposition of K parameters as in eq. (1). A small perturbation in the k-th parameter is associated with a 'differential seismogram'. The average structures $\delta\tilde{m}$ and

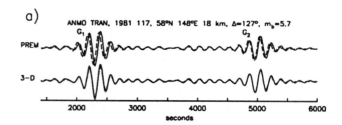

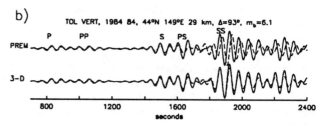

Fig. 1. Comparison of observed waveforms (solid line) with the synthetic seismograms (broken line) for a spherically symmetric PREM (top) and a 3-D model (bottom). (a) Mantle waves (Love waves) with periods in excess of 135s; (b) body waves, identified by the appropriate code, with periods longer than 45 s.

$\delta\hat{m}$ can be derived by finding the linear combination of the differential seismograms that minimizes the difference between the observed seismogram and that predicted for the spherical Earth, such as shown in the top parts of Figure 1a and 1b. Multiple iterations are necessary since the problem is nonlinear. It is also possible to reformulate the problem in such a way (see Section 4 in Woodhouse and Dziewonski [1984]) that the data for all the paths are considered at once and solution is obtained directly for the coefficients $_kA_l^m$ and $_kB_l^m$ in eq. (1). A set of some 15,000 waveforms, used as described above, is one of the two subsets of data used in this study.

The other subset of data consists of differential travel times gathered by Woodward and Masters [1991a, b]. Figure 2 shows the ray paths of S and ScS and identifies these two phases in a sample waveform . The differential time $t_{ScS} - t_S$ is obtained by cross-correlation of the two wavelets. Because the S and ScS ray paths in the upper mantle are very similar, their differential travel times are primarily sensitive to the structure in the lower mantle. $ScS - S$ residuals of Woodward and Masters [1991b], averaged in 5° spherical caps, are shown in the upper part of Figure 3 and the predictions of the model derived in this study are shown below. Visual inspection of these maps indicates the presence of a very large-scale structure. In addition to $ScS - S$ we also use $SS - S$, which is most sensitive to the structure at the mid-path reflection of the SS ray.

The frequency content of the data used here roughly span the range 5 mHz to 50 mHz. Comparison of our results to those obtained from data in different frequency ranges, such as from normal modes, provides an important cross-check. Woodhouse and Giardini [1985] developed a systematic approach to interpretation of splitting of normal

a) ScS–S Ray Path Geometry

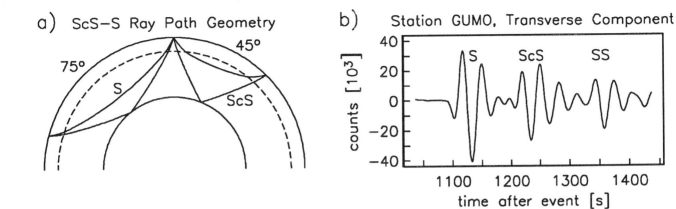

b) Station GUMO, Transverse Component

Fig. 2. The ray paths for the phases ScS and S and a sample seismogram showing both phases, as well as phase SS.

modes by lateral heterogeneity through introduction of the concept of a splitting function. For details, see Giardini et al. [1987, 1988] and Li et al., [1991]. On the subject of anomalous splitting of normal modes see also Masters and Gilbert [1981] and Ritzwoller et al. [1986, 1988]. To the first order, the splitting depends only on the even-l part of the lateral heterogeneity, and is therefore of limited use in describing a complete 3-D model. However, it provides useful auxiliary information, such as on the possible dependence on frequency. Normal modes used in sources quoted above span a range of frequencies from 0.3 mHz to 5 mHz; the body wave waveforms have peak energy at about 20 mHz; the waveforms used to derive differential travel times—40 mHz; the P-wave arrival times from the ISC Bulletins are typically 1 Hz data. It has been demonstrated [Dziewonski and Woodhouse, 1987, Figure 9; Woodhouse and Dziewonski, 1989, Plate 2 b, e, h and c, f, i] that very similar patterns of heterogeneity are obtained, for even l, separately from all these data. This also confirms the fact that spatial aliasing is not an important problem: one of the advantages of the normal mode data is that they represent excellent averages over the entire volume of the mantle, while 1 Hz P-wave paths have a ray tube that is only some tens of kilometers in diameter. The normal mode data are not used in deriving the model presented here.

We shall illustrate our formulation of the inverse problem using the travel time data. Application to the waveform inversion is not substantially different, except that for a given earthquake-station/component pair instead of considering a single residual, we sum the differences between the observed and computed seismogram at all the discrete values within the time window.

A deviation of the observed travel time from that predicted by a spherically symmetric reference Earth model, δt, can be expressed as:

$$\delta t = \int_{\mathbf{x}_e}^{\mathbf{x}_r} \delta v(\mathbf{x})\, G(s)\, ds \qquad (3)$$

where $\mathbf{x}_e$ and $\mathbf{x}_r$ are the coordinates of the source and receiver, $\delta v(\mathbf{x})$ is perturbation in wave speed, $G(s)$ is the differential kernel and ds is the element of the ray path. Equation (3) contains two important approximations. One

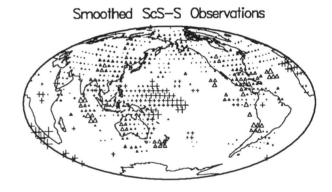

Smoothed ScS–S Observations

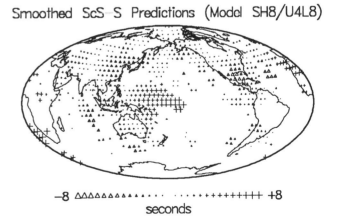

Smoothed ScS–S Predictions (Model SH8/U4L8)

Fig. 3. Observed differential travel times $ScS - S$ averaged in 5° spherical caps versus the predictions of model $SH8/U4L8$ derived in this study. Predictions of the model remove, roughly, 80% of the variance. Notice that a very long-wavelength pattern predominates both observations and model predictions.

is that the perturbation in the travel time is linear with respect to perturbation in the model parameters. The other is that the kernel G, computed for the reference model, does not depend on lateral heterogeneity and that the ray path is that predicted by the reference model. This is justified by Fermat's principle, but the assumption could be relaxed in subsequent iterations, where the ray path could be traced through the previously estimated model of lateral heterogeneity. The first assumption leads to errors of the second order in δv, which is acceptable for perturbations of the order of several percent; its effect would also be diminished in subsequent iterations.

For the i-th source and the j-th station substitution of eq. (1) into eq. (3) and numerical integration along the ray path to obtain constants α and β (see Section 4 in Dziewonski [1984]) leads to:

$$\delta t_{ij} = \sum_k \sum_l \sum_m \left({}_k\alpha_{lm}^{(ij)} \, {}_kA_l^m + {}_k\beta_{lm}^{(ij)} \, {}_kB_l^m \right). \quad (4)$$

In the n-th iteration ($n = 0, 1, \ldots$) of this non-linear inverse inverse problem the a-posteriori distribution is minimized:

$$L = \sum_{i,j} w_{ij} \left[\delta t_{ij}^{(n-1)} - \sum_{k,l,m} \left({}_k\alpha_{lm}^{(ij)} \, \delta_n({}_kA_l^m) \right. \right.$$
$$\left. + {}_k\beta_{lm}^{(ij)} \, \delta_n({}_kB_l^m) \right) \Big]^2$$
$$+ \sum_{k,l,m} \eta(k,l) \left[\left(\delta_n({}_kA_l^m) + {}_k^{(n-1)}A_l^m \right)^2 \right.$$
$$\left. + \left(\delta_n({}_kB_l^m) + {}_k^{(n-1)}B_l^m \right)^2 \right]; \quad (5)$$

where w_{ij} are weights, $\delta t_{ij}^{(n-1)}$ are the observed residuals after $n-1$ iterations, $\delta_n(A)$ and $\delta_n(B)$ are the unknown contributions to the coefficients A and B, ${}^{(n-1)}A$ and ${}^{(n-1)}B$ are coefficients accumulated in the previous $n-1$ iterations and η is the penalty function assumed to have a form:

$$\eta = \gamma_0 + \gamma_l l(l+1) + \gamma_k k^2; \quad (6)$$

which for small γ_0, effectively, seeks to minimize the squared gradient of δv integrated over the mantle volume. The values of γ_0, γ_l and γ_k are selected empirically.

Because it is assumed that the structure may be discontinuous at a depth of 670 km (the boundary between upper and lower mantle), separate parameterization for these two shells is adopted. Chebyshev polynomials are chosen for the radial basis functions :

$$f_k(r) = \bar{T}_k(x); \quad (7)$$

where $T_k(x)$ is defined in the interval (-1, 1) through the recurrence relationship: $T_{k+1} = 2xT_k - T_{k-1}$ with $T_0 = 1$ and $T_1 = x$. $\bar{T}$ is normalized,

$$\bar{T}_k = \left[\frac{(2k)^2 - 1}{(2k)^2 - 2} \right]^{1/2} T_k \quad (8)$$

so that

$$\int_{-1}^{1} [\bar{T}_k(x)]^2 \, dx = 1. \quad (9)$$

The mapping of the variable r into x is:

$$x = (2r - r_{670} - r_{moho})/(r_{moho} - r_{670}); \quad (10)$$

for the upper mantle and :

$$x = (2r - r_{CMB} - r_{670})/(r_{670} - r_{CMB}); \quad (11)$$

for the lower mantle, with r_{CMB}, r_{670} and r_{moho} being the radii of core-mantle boundary (CMB), 670 km discontinuity and Mohorovičić discontinuity, respectively.

Dziewonski and Woodward [1992] have combined, roughly, 15,000 waveform data with 5,000 measurements of differential travel times of $SS - S$ and 2,500 measurements of $ScS - S$. For the upper mantle they chose $K = 4$ and for the lower mantle $K = 8$. For both shells $L = 8$ has been chosen, which is in accordance with the inference of Su and Dziewonski [1991] that the power drops abruptly for orders greater than 6–8. Thus, there were 1,134 unknown coefficients. The weights w_{ij} (eq. 5) applied to the different subsets of data have been selected through experimentation. Because the problem is nonlinear, the solution was obtained in three iterations. In calculation of synthetic seismograms through summation of normal modes, eq. (16) of Woodhouse and Dziewonski [1984] was used to avoid inaccuracies introduced by linearization of the oscillatory terms.

2.2 Model SH8/U4L8

Table 1, after Dziewonski and Woodward [1992], lists the spherical harmonic coefficients of the perturbation in crustal thickness and the scaled coefficients of our new three-dimensional model of relative deviations of shear velocities (SH), $10^3 \times \delta v_S/v_S$, in the mantle. The model's name, $SH8/U4L8$, identifies the parametrization adopted: $L = 8$, $K = 4$ in the upper (U) mantle and $K = 8$ in the lower (L) mantle. Corrections for crustal thickness are described in Woodhouse and Dziewonski [1984, eq. 37], and the coefficients listed here are identical to those given in Table 3 of that paper. In order to obtain relative perturbation in shear velocity at coordinates (r, ϑ, φ) one should use eq. (1) with p_l^m defined in eq. (2) and $f_k(r)$ in eq. (7) and (10) or (11) depending on whether $r_{670} < r < r_{moho}$ or $r_{CMB} < r < r_{670}$, respectively; the numerical result must be multiplied by 10^{-3} to remove the scaling adopted in Table 1.

Figure 4 are the maps of $\delta v_S/v_S$, in percent, at eight depths in the mantle. The dark shading is used to show speeds faster than average, light shading for slower. The plate boundaries are shown on the maps. At a depth of

Table 1. Coefficients of a three-dimensional model SH8/U4L8, of shear velocities in the mantle defined in eq. (1); for details see text. Units are $10^3 \times \delta v/v$.

ℓ	m	CRUST A	B	UM k=0 A	B	k=1 A	B	k=2 A	B	k=3 A	B	k=4 A	B	LM k=0 A	B	k=1 A	B	k=2 A	B	k=3 A	B	k=4 A	B	k=5 A	B	k=6 A	B	k=7 A	B	k=8 A	B
0	0	0.00		-0.57		1.80		-0.06		0.10		0.36		-0.49		0.50		1.16		0.66		0.60		0.56		0.42		0.26		0.17	
1	0	3.92		3.26		2.64		0.31		0.00		-0.06		0.39		-0.46		-0.08		-0.43		0.01		0.03		-0.03		0.04		0.08	
1	1	3.04	1.83	4.86	2.39	0.62	0.76	-0.50	0.32	-0.24	-0.02	-0.04	-0.12	-0.97	0.37	0.12	0.06	-0.22	0.68	-0.57	0.05	0.22	0.26	-0.09	0.08	-0.28	-0.01	-0.12	0.04	-0.07	-0.09
2	0	3.16		1.14		2.47		-0.23		-0.36		0.00		0.26		-1.18		0.08		0.09		-0.03		-0.22		-0.21		-0.04		0.02	
2	1	1.35	1.70	-1.00	-0.59	0.62	-1.88	-0.16	-2.19	0.01	-0.70	0.08	0.25	0.22	-0.14	0.04	-0.20	0.28	0.10	-0.03	-0.30	0.06	-0.14	-0.04	-0.03	-0.03	0.04	0.09	0.02	0.04	0.04
2	2	-1.66	-1.00	2.12	-4.27	1.38	0.16	1.40	0.28	0.35	0.15	-0.04	0.05	-2.23	-1.63	2.00	0.59	0.19	0.15	0.05	0.11	0.16	0.02	0.27	0.03	0.23	0.06	-0.02	0.08	-0.15	0.02
3	0	0.70		1.35		0.18		0.17		0.05		0.01		-0.84		0.63		0.09		-0.04		0.11		0.01		-0.03		0.00		-0.02	
3	1	-1.08	0.18	-0.35	0.61	0.36	-0.10	0.50	0.50	0.19	-0.16	-0.12	0.02	-0.75	-0.56	0.90	-0.17	0.35	0.10	0.27	-0.37	0.06	-0.20	0.10	0.05	0.07	0.02	-0.03	0.15	-0.02	-0.11
3	2	-2.00	2.79	-0.06	1.68	-0.81	1.73	0.47	0.04	0.09	-0.11	-0.01	0.00	-0.36	1.45	0.04	-0.33	0.35	0.22	0.60	-0.37	0.10	0.10	0.07	0.07	0.02	0.06	0.08	0.15	0.05	-0.11
3	3	0.54	2.71	0.19	1.38	-1.59	-0.19	-0.48	-0.10	0.09	0.16	0.05	0.02	-0.56	0.08	0.07	0.51	0.15	0.22	-0.10	-0.13	-0.03	0.18	0.19	-0.12	0.06	-0.01	0.10	-0.08	0.14	-0.08
4	0	2.12		1.76		2.05		0.70		-0.05		-0.17		-0.34		0.42		-0.09		-0.19		-0.06		0.01		-0.12		-0.02		0.00	
4	1	-1.03	-0.44	-0.52	0.16	0.34	0.38	0.53	0.34	0.10	0.10	-0.05	-0.03	0.03	-0.07	-0.08	-0.37	-0.10	-0.06	0.31	0.34	0.06	0.06	0.26	0.13	0.13	0.08	0.03	0.03	-0.01	-0.02
4	2	-2.13	-0.53	-2.41	0.44	-1.26	0.41	0.04	-0.56	0.22	-0.36	0.08	0.01	0.45	-0.52	-0.45	-0.45	0.24	-0.25	-0.21	-0.34	-0.23	-0.23	0.03	0.18	0.06	-0.03	0.12	0.07	0.13	0.08
4	3	-1.55	-0.07	-0.06	0.20	-1.38	0.14	-0.49	0.20	0.03	0.07	0.07	0.03	0.18	0.72	-0.19	0.04	0.55	0.37	0.07	0.04	-0.07	-0.07	0.19	0.06	0.05	-0.07	-0.03	-0.02	0.00	-0.08
4	4	-0.61	3.36	-0.21	3.39	0.59	2.92	-0.04	0.37	0.00	-0.28	0.00	-0.20	0.02	0.78	0.33	0.12	-0.56	0.52	-0.10	-0.01	-0.08	-0.08	-0.12	-0.06	0.22	-0.04	-0.01	0.12	-0.05	0.07
5	0	-2.73		-2.14		-1.02		0.34		0.15		-0.01		0.04		-0.04		0.05		0.12		0.16		0.09		-0.12		0.11		0.05	
5	1	-0.10	-0.42	-1.69	0.28	1.13	-0.02	0.02	0.27	-0.08	0.00	0.01	-0.09	0.80	-0.11	-0.34	0.18	0.47	0.26	0.43	-0.35	-0.42	-0.11	0.02	-0.08	0.05	0.21	-0.02	-0.02	-0.05	0.03
5	2	-0.58	-0.86	-1.11	-0.05	-1.66	-0.30	-0.13	-0.13	-0.08	-0.04	0.01	0.01	-0.03	-0.30	0.11	-0.18	0.29	0.28	-0.06	-0.06	-0.11	-0.24	-0.08	0.17	0.08	-0.13	-0.02	0.20	0.03	0.04
5	3	-0.35	0.72	-1.11	1.56	-0.17	1.52	0.11	0.14	0.09	-0.05	0.01	0.03	-0.44	0.81	0.06	-0.02	-0.23	0.28	-0.08	-0.06	-0.11	-0.28	-0.03	0.06	0.15	0.06	-0.08	0.15	-0.02	0.02
5	4	3.21	-0.90	3.32	-2.15	2.26	-1.77	-0.05	-0.11	-0.21	0.14	-0.04	0.03	0.83	0.82	0.12	-0.16	-0.02	-0.02	0.00	0.24	-0.23	-0.28	-0.05	0.00	0.00	0.19	-0.06	-0.14	-0.13	0.03
5	5	0.07	1.11	1.24	0.97	0.66	0.26	0.03	0.24	0.02	0.01	-0.01	-0.04	-0.14	0.15	0.07	0.26	-0.15	-0.32	-0.14	0.10	-0.01	0.12	-0.03	0.05	-0.06	0.19	-0.09	-0.07	-0.01	-0.18
6	0	1.24		0.65		0.95		0.25		0.02		-0.05		0.00		-0.12		-0.10		-0.02		0.01		0.10		0.07		0.04		-0.11	
6	1	-0.08	-0.32	-0.78	-0.83	-0.50	-0.84	-0.05	0.11	0.36	0.19	0.05	0.04	-0.08	-0.28	0.02	0.25	0.23	0.30	0.20	0.03	-0.09	-0.14	-0.16	-0.07	0.14	0.05	0.13	0.00	-0.07	-0.01
6	2	-0.38	-0.11	-1.33	0.42	-1.35	1.13	-0.55	0.47	0.01	-0.04	0.05	-0.08	-0.15	0.33	-0.21	-0.18	0.30	0.12	0.04	0.03	-0.02	-0.12	-0.21	-0.07	-0.06	0.05	0.10	0.00	0.03	-0.02
6	3	0.20	1.16	0.07	2.19	0.20	1.00	0.11	-0.34	0.09	-0.18	0.03	0.00	0.19	-0.17	-0.04	-0.24	0.06	-0.29	0.04	0.08	0.02	0.14	0.03	0.11	-0.03	0.05	-0.01	-0.15	-0.09	-0.03
6	4	0.72	-0.54	0.15	0.92	0.90	0.34	0.48	0.11	-0.12	0.07	-0.12	0.00	0.34	-0.17	0.09	-0.34	0.07	-0.08	-0.02	-0.18	-0.07	-0.04	0.03	-0.06	-0.05	0.05	-0.02	0.03	0.00	-0.01
6	5	-0.29	-0.79	1.40	0.29	1.62	0.84	0.43	0.11	-0.13	-0.07	-0.11	0.03	-0.22	-0.24	0.42	-0.33	0.22	-0.10	0.00	0.10	0.01	0.21	0.07	0.08	0.00	0.14	-0.06	0.12	-0.06	-0.05
6	6	0.23	0.43	0.25	0.72	-0.40	0.43	-0.27	-0.10	-0.11	-0.05	0.00	-0.01	0.16	-0.16	0.08	0.23	-0.02	0.06	0.08	-0.10	-0.01	-0.01	-0.09	-0.02	0.04	0.00	0.07	-0.03	-0.01	0.10
7	0	-1.15		-1.03		-0.44		0.08		0.06		0.00		-0.18		-0.08		0.09		0.09		0.04		-0.08		-0.02		0.00		-0.04	
7	1	-0.11	0.90	-0.32	1.17	-0.22	0.53	-0.04	-0.10	-0.03	-0.03	-0.01	0.00	-0.35	0.35	0.11	0.14	0.09	-0.11	0.00	-0.04	0.08	0.08	-0.10	-0.10	0.11	0.10	0.08	0.02	-0.04	-0.02
7	2	0.62	0.36	-0.30	1.16	0.05	0.76	-0.15	-0.01	0.03	-0.09	0.02	0.05	-0.25	0.14	0.16	0.13	0.13	-0.09	0.23	0.08	0.08	-0.20	-0.19	-0.11	-0.09	0.04	-0.07	-0.03	0.04	0.07
7	3	-0.52	0.42	-0.30	0.78	-0.63	0.40	-0.06	-0.02	-0.02	-0.02	0.02	0.02	0.00	-0.10	0.18	0.14	0.07	-0.21	-0.22	-0.09	-0.16	0.04	0.04	-0.11	0.03	-0.02	-0.03	-0.02	-0.06	0.07
7	4	-1.09	0.11	0.36	-0.13	-0.40	0.63	-0.08	0.17	0.09	0.05	-0.01	0.02	0.18	0.52	0.00	-0.28	0.07	-0.12	0.22	-0.09	0.10	0.20	0.03	0.04	-0.13	0.14	-0.19	0.12	-0.06	0.07
7	5	0.43	-0.35	0.47	-0.98	-0.36	-0.58	0.07	-0.27	-0.07	-0.07	0.03	-0.01	-0.42	0.15	0.23	-0.42	0.22	0.16	-0.04	-0.18	0.32	0.21	0.07	0.08	0.06	0.14	-0.02	0.07	-0.11	0.02
7	6	-0.21	-1.11	-0.23	-1.09	-0.18	-0.68	-0.04	0.09	-0.01	0.13	0.03	0.03	-0.14	-0.17	-0.17	-0.47	-0.36	-0.22	-0.13	0.10	0.02	0.26	-0.03	-0.03	-0.02	0.14	0.06	0.12	0.09	0.15
7	7	-0.31	-0.74	-0.68	0.02	-0.32	-0.03	0.06	-0.03	0.04	0.00	0.02	-0.01	-0.86	-0.17	-0.24	0.01	0.00	-0.01	-0.05	0.06	0.20	0.11	-0.10	-0.08	-0.07	-0.08	-0.02	-0.03	-0.01	0.10
8	0	-0.72		-1.11		-0.91		-0.03		0.08		0.05		-0.04		0.16		-0.15		-0.32		0.03		0.09		0.10		0.08		0.02	
8	1	-0.07	-0.30	-0.40	0.16	-0.53	-0.51	-0.14	-0.11	-0.03	0.00	0.06	0.00	0.14	0.43	0.30	0.10	-0.05	-0.18	0.00	0.14	0.18	-0.03	-0.22	0.04	-0.16	0.06	0.01	-0.04	0.01	-0.12
8	2	0.17	0.11	-0.75	0.62	-0.73	0.11	-0.14	-0.04	0.06	0.00	0.07	0.04	-0.06	0.24	-0.19	-0.01	-0.05	0.13	0.23	0.14	0.03	-0.03	0.18	0.02	-0.01	-0.04	0.01	-0.09	0.01	-0.00
8	3	0.17	0.32	0.36	0.78	0.70	0.57	0.31	0.05	0.06	-0.01	-0.05	-0.04	0.21	0.45	-0.06	0.29	-0.27	-0.21	0.13	0.19	0.07	-0.04	0.04	-0.02	-0.10	-0.01	-0.13	-0.08	-0.05	-0.05
8	4	-0.41	0.84	-0.24	1.03	0.40	0.24	0.43	-0.16	0.04	-0.01	-0.04	-0.01	0.04	0.01	-0.26	-0.17	0.22	0.13	0.16	0.05	-0.01	-0.04	0.05	-0.02	0.04	0.02	0.04	-0.02	-0.05	-0.05
8	5	0.57	-0.35	-0.76	-0.20	-0.41	-0.34	0.12	-0.02	-0.04	0.00	0.00	-0.01	0.14	0.13	-0.22	0.04	-0.36	0.13	0.03	0.05	-0.11	-0.16	-0.05	-0.05	-0.10	-0.02	-0.13	0.09	-0.11	-0.01
8	6	0.99	0.38	0.61	-0.37	-0.09	0.04	-0.04	0.03	-0.02	0.01	0.14	0.03	0.06	0.20	-0.17	-0.26	0.15	0.36	0.15	0.19	0.31	0.03	-0.05	-0.01	0.09	-0.07	0.04	-0.02	0.09	0.01
8	7	-0.22	-0.17	0.77	-0.12	-0.32	0.04	-0.66	-0.01	-0.02	0.01	-0.17	0.03	-0.14	0.16	-0.17	-0.26	0.15	0.00	0.15	0.19	-0.09	0.03	0.02	-0.17	0.01	-0.09	0.02	0.06	0.00	-0.04
8	8	-0.31	-0.39	0.23	-0.12	-0.09	-0.34	-0.12	-0.22	0.03	0.04	0.14	0.02	0.13	0.32	-0.24	-0.21	-0.17	-0.02	-0.02	-0.08	0.28	-0.16	0.14	-0.05	-0.08	0.04	0.10	-0.01	-0.02	0.02

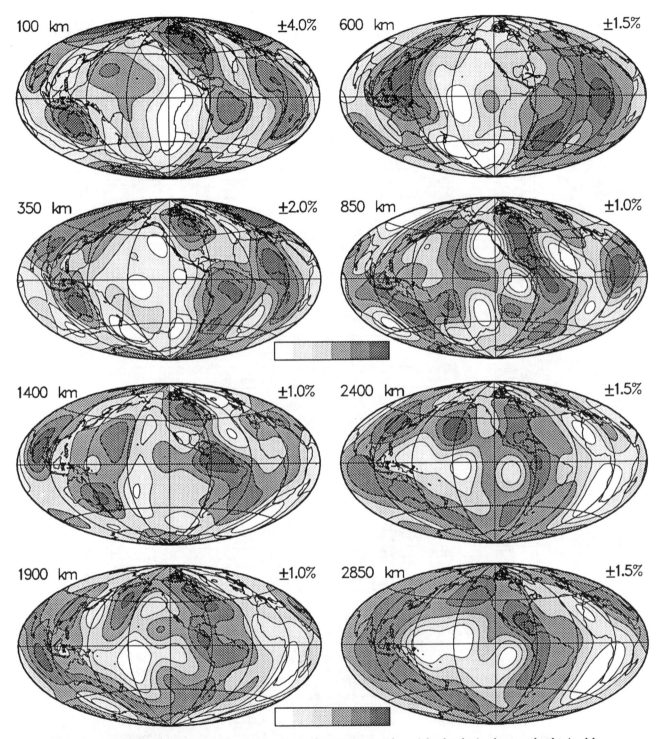

Fig. 4. Maps of relative deviations from the average shear wave speed at eight depths in the mantle obtained by the synthesis of the coefficients of model $SH8/U4L8$ listed in Table 1. The depths and minimum and maximum scale values are shown with each map. The scale bar is shown at the bottom and the minimum and maximum values are represented by the lightest and darkest shades, respectively.

100 km the velocity anomalies follow the tectonic signature at the surface. Mid-oceanic ridges are slow; this is most clear for the fast spreading East Pacific Rise, but also the mid-Atlantic ridge and mid-Indian Rise are clearly visible. Slowness of back arc basins is also obvious at this depth. On the other hand, stable continental areas are several percent faster than normal. Such results were expected from regional studies as well as from the current

understanding of the physical processes involved. The fact that the tomographic imaging was able to resolve these features without any prior information on the tectonic regime and age of the ocean crust, is very reassuring.

The principal expression of lateral heterogeneity at a depth of 350 km is the difference between continents and oceans. Here, correlation with age of the oceanic lithosphere is much less obvious than at 100 km. Instead, entire ocean basins (the Pacific and Indian Ocean in particular) are slow, while the continents are fast. This ocean–continent asymmetry disappears in the transition zone. Instead, the power is shifted to lower degrees, particularly degree 2 [Masters et al., 1982], although harmonics of degree 1 are also present. High velocities at this depth under the western Pacific and South America could be related to the accumulation of subducted material. Thus there are marked changes with depth in the pattern of lateral heterogeneity in the upper mantle.

Another such change takes place once we cross the 670 km discontinuity. The pattern at 850 km depth is very different both in terms of location of specific features (high velocities in the western Pacific disappear, for example) as well as the spectral content: the dominance of shorter-wavelength features is evident. Certain features, such as high velocities under North and South America or Australia, continue at 1400 km depth, but others (e.g. the

strong high under Africa) have disappeared. There is only a small decrease in amplitude with depth in the middle mantle.

A feature well known from earlier tomographic studies: a ring of high velocities surrounding the Pacific basin, begins to emerge at 1900 km. There is also a slight increase in the amplitude. This is much clearer in the maps at 2400 km and at the core-mantle boundary. Most of the increase in total amplitude is due to the growth in amplitude of the long-wavelength anomalies. Note the correspondence between this ring and the travel time residuals of $ScS - S$ shown in Figure 3.

Figures 5a and 5b show great-circle cross-sections through model $SH8/U4L8$. Each cross-section is made along the particular great-circle, shown as the heavy black line in the inset map, and passes through the center of the Earth. In Figure 5a we observe striking 'mega-plumes' rising from the core-mantle boundary below southern Africa and below the central Pacific. The almost antipodal location of these two plumes is a manifestation of the dominant degree-2 symmetry of lateral heterogeneity in much of the lower mantle. The African plume is apparently impeded near the 670 km seismic discontinuity by the seismically fast (perhaps more viscous) 'root' below the African continent. In contrast the Pacific plume does cross the 670 km seismic discontinuity, above which it appears to undergo

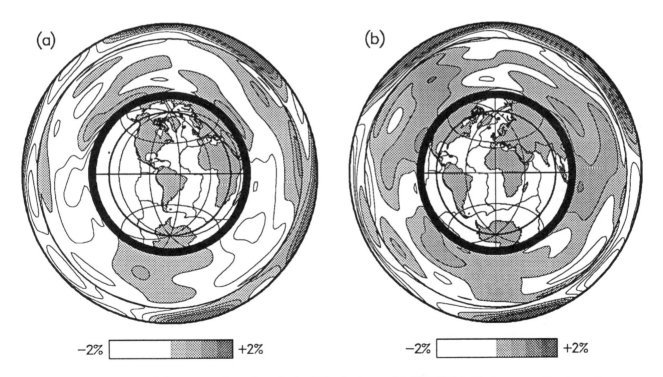

Fig. 5: Two great-circle cross-sections through the 3-D velocity model $SH8/U4L8$. Each cross-section is made along a particular great circle (the heavy black line in the inset map) and passes through the center of the Earth. The outermost ring corresponds to the top of the lithosphere, the innermost corresponds to the CMB. The 670 km seismic discontinuity is shown by a solid black ring concentric with the inner and outer rings. The scale bar corresponds to ±2.0%, with significant saturation of the scale possible in the upper mantle.

significant horizontal shearing. In Figure 5b we observe that the major continents, such as North America, Antarctica, and Eurasia (with the notable exception of Africa–see Figure 5a), all overlie portions of the lower mantle which are characterized by higher-than-average seismic velocities. Such high-velocity 'roots' are much deeper than the chemically-distinct, high-velocity, continental tectosphere which has been envisaged by Jordan [1975, 1979]. In this cross-section we also observe the 'pooling' and horizontal shearing of seismically slower-than-average features in the vicinity of the 670 km seismic discontinuity. This behaviour is found in many of the cross-sections we have observed and its potential dynamical significance is considered below in section 3.1.

In Figure 6 we plot the so-called degree-variance function $g_l(r)$ (i.e. the root-mean-square amplitude of $\delta v_S/v_S$ calculated at each degree l for a given radius r in the mantle) which is defined by the expression :

$$g_l(r) = \left[\sum_{k,m} \left(f_k(r) \; _kA_l^m \right)^2 + \left(f_k(r) \; _kB_l^m \right)^2 \right]^{1/2} . \quad (12)$$

Figure 6 summarizes well the general nature of the variation of lateral heterogeneity with depth:
- the rapid decrease in amplitude of the high-l harmonics within the top 400 km;
- relatively large amplitudes of degrees 1 and 2 in the transition zone;
- the overall decrease in amplitude, but relatively even contributions across the range of l from 2 to 8, in the depth range from 800 to 1800 km;
- large increase in amplitude of degree 2, in particular, but also degrees 3 and 4, in the last 1000 km of the mantle.

Such results are of great importance for modeling convection in the mantle. Our results indicate a different style of flow in the upper and lower mantle; there is an indication that the flux across the 670 km discontinuity may be impeded. The dominating presence of very large-wavelength heterogeneity in the lowermost mantle may be indicative of the relative importance of heating from within (radioactivity), as pointed out by Jarvis and Peltier [1986,1988].

2.3 Correlations Between Mantle Heterogeneity, Plate Motions, and Nonhydrostatic Geoid

Prior to attempting any detailed mantle flow calculations it will be instructive to consider how the $\delta v_S/v_S$ heterogeneity in model $SH8/U4L8$ is correlated, at various depths in the mantle, to the principal signatures of mantle convection: the tectonic plate motions and the nonhydrostatic geoid. This analysis will be carried out in the spherical-harmonic spectral domain and it will take the form of the degree-correlation versus depth plots employed by Forte and Peltier [1987b] (who will be referred to as FP for convenience). These degree-correlation diagrams are helpful (barring any strong nonlinearities in the actual mantle-flow mechanism which would entirely obscure any degree-by-degree relationships) in identifying which depth ranges in the mantle are likely to be important sources for the observed plate motions and nonhydrostatic geoid. This information is useful in constraining the mantle-flow models (especially the radial viscosity profile of the mantle) which are employed in sections 3 and 4.

The degree-correlations between the $\delta v_S/v_S$ heterogeneity in model $SH8/U4L8$ and the so-called $TRENCH$ and $RIDGE$ scalars, defined in FP, are shown in Figure 7 (the $TRENCH$ field is simply the rate of plate convergence at the principal subduction zones while the $RIDGE$ field is the rate of plate divergence at the mid-oceanic ridges–see FP for further details). In Figure 7 we observe that the $RIDGE$ field is strongly correlated to $\delta v_S/v_S$, across most spherical harmonic degrees, in the top 200 km of the upper mantle. The negative sign of these degree correlations shows that the regions of plate spreading at mid-oceanic

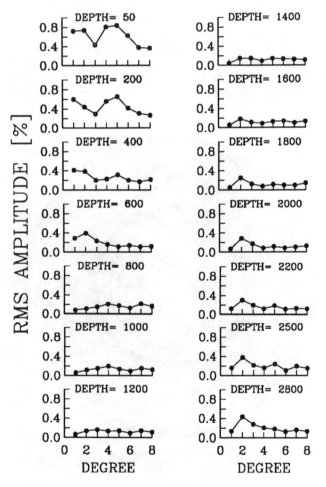

Fig. 6. Degree variances, calculated according to expression (12) in the text, of the $\delta v_S/v_S$ heterogeneity in model $SH8/U4L8$ at various depths in the mantle. The units of variance (rms amplitude) are per cent.

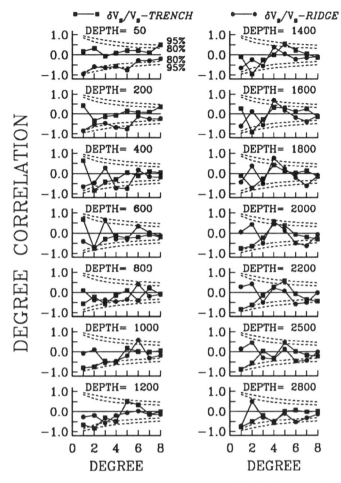

Fig. 7. Degree correlations, as a function of depth, between the $\delta v_S/v_S$ heterogeneity in model $SH8/U4L8$ and the $TRENCH$ and $RIDGE$ components of the plate-velocities [Forte and Peltier, 1987b].

ridges (which overlie hotter-than-average mantle) directly overlie regions of (seismically) slower-than-average mantle. This correlation was of course evident in the maps depicted in Figure 4 and points to an obvious thermal origin for the $\delta v_S/v_S$ heterogeneity in the top 200 km of the mantle. The $\delta v_S/v_S$-$RIDGE$ correlations at degrees 4 and 5 (and to a lesser extent at degree 2) persist to the base of the upper mantle and continue into the lower mantle. This observation is important since, as shown in FP, the dominant power in the plate-divergence field is found at harmonic degrees 4 and 5. It thus appears that the dominant scale of plate divergence is linked to deeply-rooted hot upwellings which may originate in the lower mantle. The most striking correlation between $\delta v_S/v_S$ and the $TRENCH$ field is at degree 2, appearing at a depth of 300 km and extending, almost without interruption, to the very bottom of the lower mantle. The negative sign of this degree-2 correlation shows that the degree-2 component of the rate of

plate convergence at subduction zones (which presumably overlie colder-than-average mantle) overlies a corresponding pattern of (seismically) faster-than-average mantle. The dominant power in the rate of plate-convergence field is found at degree 2 and thus appears to be driven by an extremely broad scale cold downwelling extending throughout the mantle. In the bottom half of the lower mantle there are prominent $\delta v_S/v_S$-$RIDGE$ correlations at degrees 3,6, and 7. The negative sign of these correlations is reasonable (i.e. plate-spreading overlying [seismically] slower-than-average mantle which is presumably hotter-than-average) and is again in accord with a thermal origin for the $\delta v_S/v_S$ heterogeneity in the lower mantle.

In Figure 8 we show the degree-correlations between the $\delta v_S/v_S$ heterogeneity in model $SH8/U4L8$ and the GEM-T2 geoid [Marsh et al., 1990] which has removed from it the hydrostatic rotational flattening [e.g. Jeffreys, 1963; Nakiboglu, 1982]. The most prominent $\delta v_S/v_S$-geoid

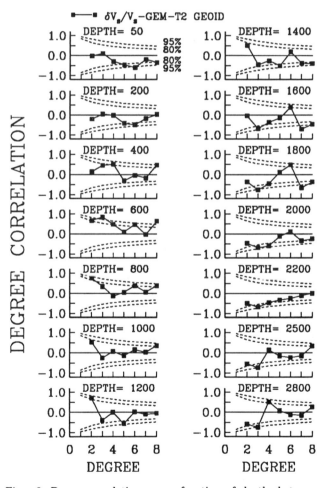

Fig. 8. Degree correlations, as a function of depth, between the $\delta v_S/v_S$ heterogeneity in model $SH8/U4L8$ and the GEM-T2 geoid [Marsh et al., 1990] filtered by removal of the hydrostatic flattening [e.g. Jeffreys, 1963; Nakiboglu, 1982].

correlations in Figure 8 appear in the transition-zone region (from 400 to 670 km depth). As we explain below, it is may be quite significant that the magnitude of these correlations peaks at the bottom of the upper mantle. The positive sign of these correlations indicates that geoid highs overlie (seismically) faster-than-average regions at the base of the upper mantle. If the assumption of a thermal origin for seismic velocity perturbations is generally valid, as argued above, then it also follows that geoid highs overlie denser-than-average regions at the base of the upper mantle.

Figure 8 shows that in the lower mantle the strongest correlation between the degree-2 heterogeneity and the geoid is found in the vicinity of the 670 km seismic discontinuity. At greater depths (in the range 1400-1800 km) we observe a fundamental overall change in the sign of the $\delta v_S/v_S$-geoid correlations which is now mostly negative whereas at the bottom of the upper mantle they were generally positive. Such changes in the sign may readily be interpreted in terms of flow in a mantle with a viscosity increasing with depth [e.g. Richards and Hager, 1984]. At the very bottom of the lower-mantle the most prominent $\delta v_S/v_S$-geoid correlations are at degrees 2 and 3. The mantle heterogeneity at these great depths is largely dominated by degree-2 variations. This correlation between lower-mantle degree-2 structure and the degree-2 geoid was already known on the basis of the earliest lower-mantle models [e.g. Dziewonski et al., 1977] and its negative sign may again be interpreted in terms of a convecting mantle [e.g. Pekeris, 1935; Dziewonski et al., 1977; Richards and Hager, 1984].

The observation of strong $\delta v_S/v_S$-geoid correlations (especially at degree 2) in a depth interval encompassing the 670 km seismic discontinuity is an important result. Such correlations were first suggested by Masters et al. [1982] on the basis of their inferences of upper-mantle degree-2 heterogeneity in the transition zone. These correlations suggest that seismically inferred heterogeneity in the upper-mantle transition zone is at least as important as seismically inferred lower-mantle heterogeneity in explaining the long wavelength geoid. If we assume a chemically stratified mantle in which the 670 km seismic discontinuity is a barrier to radial flow, then any density loads placed at this boundary are entirely compensated by the deformation of this boundary and produce no geoid signal. The strong correlation that we instead observe, in the vicinity of the 670 km discontinuity, between the geoid and $\delta v_S/v_S$ provides compelling evidence for a flow which, on the largest scales, is whole-mantle in style. It is important to recognize, however, that there are localized regions (e.g. below the western Pacific margin) where the seismically inferred heterogeneity appears to "pile up" at the bottom of the upper mantle. This evidence for an apparent localized layering of the mantle flow has recently been interpreted by Woodward et al. [Woodward, R.L., A.M. Dziewonski, and W.R. Peltier, "Comparison of seismic heterogeneity models

and convective flow simulations", submitted to *Nature*, 1992] in terms of the dynamic effects due to the spinel to post-spinel phase change at 670 km depth.

3. VISCOUS FLOW CALCULATIONS OF THE NONHYDROSTATIC GEOID AND TECTONIC PLATE MOTIONS

The well-known ability of the mantle to creep indefinitely over geological time scales is understood in terms of the existence of atomic-scale defects in the lattice of crystal grains [e.g. Nicolas and Poirier, 1976]. If the ambient mantle temperature is sufficiently high, the imposition of nonhydrostatic stresses causes the lattice defects to propagate and thus allows the mantle rocks to creep or 'flow' slowly. This process may be characterized in terms of a single parameter, namely an effective viscosity [e.g. Stocker and Ashby, 1973; Weertman and Weertman, 1975]. In general the effective viscosity of the mantle will depend on the effective grain size of mantle rocks (in the case of diffusion-controlled creep), the prevailing stress or strain rate (in the case of dislocation-controlled creep), the pressure, and (most importantly) the temperature. For a more detailed review of mantle rheology the interested reader will find the discussion of Weertman [1978] especially helpful. Over the past decade there have been numerous studies which have also indicated the potential importance of the chemical environment on the effective viscosity of the mantle [e.g. Kohlstedt and Hornack, 1981; Ricoult and Kohlstedt, 1985; Karato et al., 1986; Borch and Green, 1987; Ryerson et al., 1989]. We should therefore expect that the effective viscosity of the mantle will vary (perhaps strongly) in three dimensions as a result of the lateral- and depth-variation of the thermodynamic state variables mentioned above.

There have been numerous studies investigating the effects of lateral variations of viscosity on the style and dynamics of numerically simulated thermal convective flows. A proper review of these numerical studies would unfortunately lead us well beyond the scope of the present paper; we therefore limit ourselves to providing only a bare outline of previous relevant studies. The importance of lateral viscosity variations for understanding the plate-tectonic nature of the lithospheric flow has been discussed in Hager and O'Connell [1981]. A thorough investigation of the effects of stress-, temperature-, and pressure-dependent rheology on thermal convective flows in 2-D Cartesian geometry has been carried out by Christensen [1984]. The coupling between the various Fourier-harmonic components of the flow field due to lateral viscosity variations, in 2-D Cartesian geometry, was considered by Richards and Hager [1989]. It is important to recognize, however, that such studies confined to 2-D geometries are fundamentally limited because they cannot describe the excitation of toroidal flows and their coupling to poloidal flows. Such an investigation of poloidal-toroidal coupling induced by lateral viscosity variations requires a fully three-dimensional geometry. The important effects of this poloidal-toroidal coupling on the flow-induced surface topography (and hence the nonhydrostatic geoid) were pointed out in Forte and Peltier [1987b]. Studies of the effects of lateral viscosity variations on buoyancy-induced flows in a realistic spherical geometry has so far been quite limited in their scope [e.g. Stewart, 1992]. We expect that significant progress will eventually be made on the basis of recent formulations which allow us to explicitly investigate the effects of lateral viscosity variations in the lithosphere [e.g. Ricard et al., 1988; Forte, 1992; Ribe, 1992] and, in general, throughout the mantle [e.g. Forte, 1992].

In this study we will instead employ a simplified treatment of mantle flow in which we assume that the mantle viscosity varies only in the radial direction. It should therefore be understood that the radial viscosity profile we shall propose, in the course of fitting the geoid and plate motion data, is likely to reflect the horizontally-averaged effects of lateral variations in temperature, stress, and chemistry.

The characterization of the creep properties of the mantle in terms of an effective viscosity allows us to model the slow flow of the mantle with the conventional hydrodynamic field equations which express the principles of mass and momentum conservation. The previous treatments of mantle-flow in spherical shells with radially-varying viscosity [e.g. Ricard et al., 1984; Richards and Hager, 1984; Forte and Peltier, 1987b] have assumed that the mantle is incompressible. In recent studies of mantle flow in spherical geometry by Forte [1989] and Forte and Peltier [1991b] it was shown that the effects of the finite compressibility of the mantle are not negligible. In particular, it has been demonstrated that the very-long wavelength nonhydrostatic geoid is quite sensitive to the effects of mantle compressibility [Forte and Peltier, 1991b]. In the viscous flow predictions of the geoid, plate motions, and CMB topography presented below we shall therefore employ the compressible-flow theory described in Forte and Peltier [1991b].

The principal assumption we shall make in modelling the mantle flow is that the 670 km seismic discontinuity is NOT a barrier to radial flow. This assumption is motivated by the previous degree-correlation analysis in section 2.3 which provided clear evidence in favour of a whole-mantle style of flow. The cross-section views of the $\delta v_S/v_S$-heterogeneity in model $SH8/U4L8$, shown in Figure 5, also provided examples of plume-like features extending across the 670 km seismic discontinuity. In addition we have found (results not presented here) that all our attempts to model the geoid and plate motions with a single layered-flow model, which explicitly includes a barrier at 670 km depth, have resulted in rather unsatisfactory fits to the observations. We find, in particular, that fair fits to the observed nonhydrostatic geoid can be achieved only by invoking anomalous $\partial ln\rho/\partial lnv_S$ values in the transition

zone and the top of the lower mantle (e.g. $\partial ln\rho/\partial lnv_S \approx 3$ in the transition zone and $\partial ln\rho/\partial lnv_S \approx -1$ for the top 300 km of the lower mantle). The transition-zone value is an order of magnitude greater than the expected values (see below) and is substantially greater than the already large values which have been advocated by Anderson [1987b]. We also find that layered-flow models which provide fair fits to the nonhydrostatic geoid provide poor fits to the plate motions and vice-versa. In contrast, as we show below, the assumption of whole-mantle flow allows us to derive a single model which provides very good fits to both the observed nonhydrostatic geoid and the plate motions.

3.1 The Predicted Nonhydrostatic Geoid

The solution of the hydrodynamic field equations in spherical geometry [e.g. Richards and Hager, 1984; Ricard et al., 1984; Forte and Peltier, 1987b,1991b] expresses the three-dimensional flow velocity $\mathbf{u}$ excited by the internal buoyancy forces due to the density perturbations $\delta\rho(r,\theta,\varphi)$. On the basis of these theoretical flow predictions it is possible to obtain the so-called 'kernels' which relate, in the spherical-harmonic spectral domain, the various flow-related geophysical surface observables (e.g. the nonhydrostatic geoid) to density perturbations in the mantle. The nonhydrostatic geoid kernels $G_l(r)$ relate the spherical harmonic coefficients of the geoid δN_l^m to the radially-varying spherical harmonic coefficients of the perturbed density $\delta\rho_l^m(r)$ as follows :

$$\delta N_l^m = \frac{3}{(2l+1)\bar{\rho}} \int_b^a G_l(r)\delta\rho_l^m(r)dr, \qquad (13)$$

in which $\bar{\rho} = 5.515 Mg/m^3$ is the average density of the Earth, $b = 3480$ km is the radius of the CMB, and $a = 6368$ km is the radius of the solid surface. The geoid kernels $G_l(r)$ have been shown [e.g. Richards and Hager, 1984] to be rather sensitive to the radial variation of mantle viscosity and to the assumed surface boundary condition (i.e. free-slip or no-slip). The geoid calculations we present below will assume that on the Earth's solid surface and core-mantle boundary free-slip boundary conditions obtain. It is important to remember that the geoid kernels are only sensitive to relative viscosity variations $\eta(r)/\eta_0$ (where $\eta(r)$ is the actual viscosity profile and η_0 is a reference viscosity) and therefore the nonhydrostatic geoid data alone cannot constrain the absolute value of mantle viscosity [Forte and Peltier, 1987b].

The viscosity-sensitivity of the predicted nonhydrostatic geoid is explicitly indicated by its Fréchet kernels. As described in Forte et al. [1991a], these Fréchet kernels describe the relationship between perturbations in the harmonic coefficients of the geoid and perturbations in the radial viscosity profile as follows:

$$\delta\left(\delta N_l^m\right) = \int_b^a \delta L(r)G_l^m(r)dr, \qquad (14)$$

in which $\delta L(r) = \delta\eta(r)/\eta(r)$, where $\delta\eta(r)$ is a perturbation to the original (i.e. starting) viscosity profile $\eta(r)$, and $G_l^m(r)$ are the Fréchet kernels associated with the geoid perturbations $\delta(\delta N_l^m)$. In Figure 9 we show the geoid Fréchet kernels calculated for a starting model with a factor of 5 jump in viscosity at 670 km depth (this jump is intermediate, in a logarithmic sense, between a model with no jump [i.e. an isoviscous mantle] and a model with a factor of 25–30 jump [e.g. Forte et al., 1991b]). The Fréchet kernels were calculated by solving the perturbed propagator equations for viscous flow in a gravitationally consistent, compressible mantle:

$$\frac{d\delta\mathbf{v}(r)}{dr} = \mathbf{M}(r)\delta\mathbf{v}(r) + \delta\mathbf{M}(r)\mathbf{v}(r), \qquad (15)$$

in which the vector $\mathbf{v}(r)$ and the matrix $\mathbf{M}(r)$ (defined, respectively, in equations (21a) and (20) in Forte and Peltier [1991b]) define the viscous flow solution for the starting viscosity model, $\delta\mathbf{M}(r)$ is obtained from $\mathbf{M}(r)$ by perturbing the starting viscosity profile, and the vector $\delta\mathbf{v}(r)$ is the perturbation in the viscous-flow solution arising from this viscosity perturbation. The geoid Fréchet kernels

shown in Figure 9 were calculated using the $\delta v_S/v_S$-heterogeneity in model $SH8/U4L8$ and also by assuming $\partial ln\rho/\partial ln v_S = 0.4$ [e.g. Anderson et al., 1968] throughout the mantle.

The geoid Fréchet kernels in Figures 9a and 9b show that the very long wavelength geoid is most sensitive to viscosity changes in the bottom of the upper mantle and in the bottom half of the lower mantle. By adjusting the starting viscosity profile in these regions of sensitivity we can (by repeated forward modelling) improve the fit between the predicted and observed nonhydrostatic geoid. In this manner we obtain, for example, the two viscosity profiles shown in Figures 10a and 10c which provide nearly the same good match to the nonhydrostatic geoid data. It is clear from the large radial averaging widths of these Fréchet kernels that the geoid data will not be sensitive to the details of the viscosity adjustments needed to provide acceptable matches to this data. This explains why the somewhat different viscosity models in Figure 10 provide nearly the same fit to the geoid data. Unless stated otherwise, the viscosity profile in Figure 10a will be employed in the viscous flow modelling presented below.

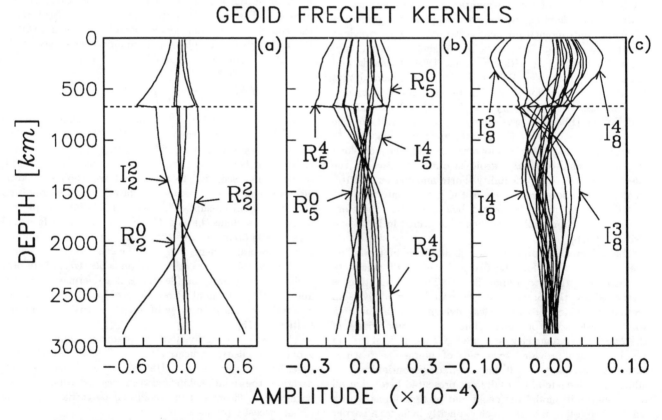

Fig. 9. The viscosity Fréchet kernels, defined in equation (14) of the text, for: (a) The degree-2 geoid, (b) The degree-5 geoid, and (c) The degree-8 geoid. The kernels are calculated for a compressible mantle with a factor of 5 jump in viscosity at 670 km depth. All the kernels shown here have been multiplied by a factor of 10^4. The kernels are identified by the labels R_l^m and I_l^m which correspond, respectively, to the real and imaginary parts of $G_l^m(r)$, defined in equation (14). The dashed horizontal line identifies the location of the 670 km seismic jump.

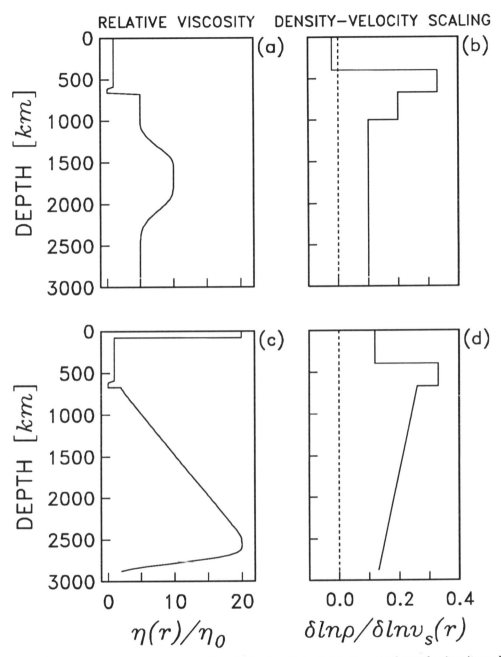

RELATIVE VISCOSITY   DENSITY–VELOCITY SCALING

(a) (b) (c) (d)

DEPTH [km]

$\eta(r)/\eta_0$

$\delta ln\rho/\delta ln v_s(r)$

Fig. 10. (a) and (c) The relative viscosity profile $\eta(r)/\eta_0$ in which $\eta(r)$ is the actual mantle viscosity and η_0 is a reference mantle viscosity. (b) and (d) The depth-varying proportionality between density perturbations and shear velocity perturbations in the mantle. The viscosity profiles and density-velocity scaling, $\partial ln\rho/\partial ln v_S$, were inferred on the basis of the nonhydrostatic geoid data (see text for details).

The most notable feature of the two viscosity profiles in Figure 10 is the zone of strongly reduced viscosity in a thin (70 km thick) layer at the base of the upper mantle. The a-priori choice for such a thin layer is based on the peak sensitivity (in the degree range $l = 2 - 5$) of the geoid data to viscosity changes at this depth and also by the noticeable reduction of the seismic velocity gradients in the depth interval 600 to 670 km in the PREM model [Dziewonski and Anderson, 1981]. Such reduced seismic velocity gradients suggest the possible existence of a hot thermal boundary layer at the base of the upper mantle and hence a corresponding layer of reduced viscosity. This thermal boundary layer is reminiscent of the hypothesized hot thermal boundary layer at the base of the lower mantle [e.g. Yuen and Peltier, 1980; Stacey and Loper, 1983]. As shown by Stacey and Loper [1983] the variation of viscosity across such a hot thermal boundary layer is likely to be of the form:

$$\eta(r)/\eta_0 = (\eta_{670}/\eta_0) \exp\left[\frac{r - r_{670}}{h}\right], \quad (16)$$

in which r_{670} is the radius of the 670 km seismic discontinuity, η_{670}/η_0 is the normalized viscosity at the bottom of the upper mantle, and h is the scale height for the viscosity variation. The quantities η_{670}/η_0 and h are selected so that the viscosity at 600 km depth is 100 times greater than the viscosity at the bottom of the upper mantle. The nonhydrostatic geoid data seem to strongly favour such steep decreases of viscosity at the base of the upper mantle. In fact, if we increase the scale-height h so that $\eta_{670}/\eta_{600} = 1/10$ (rather than 1/100) we find that our fit to the longest wavelength (i.e. $l=2$ and 3) geoid is completely lost.

The existence of a thermal boundary layer at the base of the upper mantle, which would be expected if there were separate upper- and lower-mantle circulations, is somewhat puzzling given that we assume a WHOLE-MANTLE flow to model the geoid. A possible explanation may be the local distortion of the geotherm arising from the latent-heat release accompanying the continual flux of hot upwelling mantle (e.g. see Figure 5a) across the spinel - post-spinel phase-change boundary (this is the so-called "Verhoogen effect" described by Verhoogen [1965] and Turcotte and Schubert [1971]). An alternative explanation may lie with the long-term thermal effect of the extensive "pooling" of hotter-than-average mantle in the vicinity of the 670 km seismic discontinuity which is quite apparent in the cross-section views of the $\delta v_S/v_S$-heterogeneity in model $SH8/U4L8$ (see Figure 5). The horizontally-averaged effect of these hotter-than-average "pools" on the strongly temperature-sensitive mantle viscosity might lead to an effective low-viscosity layer at these depths. Such hypotheses are of course only speculative but it is worth recalling that in the PREM radial Earth model [Dziewonski and Anderson, 1981] the region from 600 to 670 km depth is characterized by a noticeable decrease in the radial gradients of density, P-velocity, and S-velocity. Such reduced gradients would be expected if there were an increased (superadiabatic) temperature gradient at the base of the upper mantle. In fact the ratio $\delta ln\rho/\delta lnv_S = 0.41$ calculated directly from the radial gradients in PREM in the depth-range 600-670 km is similar to the laboratory measurements of $(\partial ln\rho/\partial lnv_S)_P$ on mantle minerals. From the high-temperature measurements on forsterite by Isaak et al. [1989] we find $(\partial ln\rho/\partial lnv_S)_P = 0.47$ and from the data for ($MgAl_2O_4$) spinel and (MgO) periclase gathered by Anderson et al. [1968] we find $(\partial ln\rho/\partial lnv_S)_P = 0.42$ and $(\partial ln\rho/\partial lnv_S)_P = 0.39$, respectively. Such agreement between the PREM gradients and the laboratory data suggests that the assumption of a superadiabatic temperature increase in the narrow depth-range 600-670 km may not be unreasonable.

In Figure 11 we now show a selection of geoid kernels, defined in (13), calculated for the viscosity profile $\eta(r)/\eta_0$ in Figure 10a. Here we show geoid kernels for a mantle which is assumed to be incompressible and for a mantle which is (PREM) compressible. The effects on the geoid arising from the finite compressibility of the mantle are evidently important and a more detailed discussion will be found in Forte and Peltier [1991b]. The most important feature of the geoid kernels shown in Figure 11 is their peak amplitude found at 670 km depth. This behaviour is a direct consequence of the low viscosity zone at the base of the upper mantle and is in accord with the peak correlations between the geoid and $\delta v_S/v_S$-heterogeneity near 670 km depth (see Figure 8). The depth-variation of the sign of the geoid kernels also matches the overall depth-variation of the $\delta v_S/v_S$-geoid correlations discussed in section 2.2 and this explains why the viscosity profile in Figure 10a provides a good match between the predicted and observed geoid, as we show below.

An issue of great importance in the modelling of mantle flow with seismically-inferred heterogeneity concerns

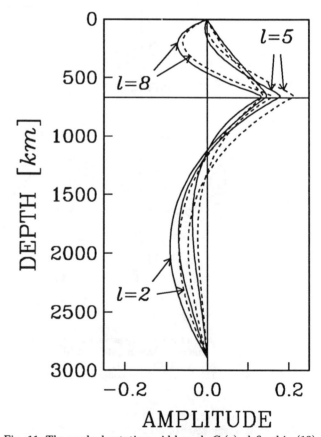

Fig. 11. The nonhydrostatic geoid kernels $G_l(r)$, defined in (13), calculated for the viscosity profile shown in Figure 10a. The solid lines represent the geoid kernels for a compressible mantle with a PREM [Dziewonski and Anderson, 1981] radial density profile and the dashed lines represent the geoid kernels for an incompressible mantle. The location of the 670 km seismic discontinuity is indicated by the heavy horizontal line.

the choice of appropriate density-velocity proportionality factors $\partial ln\rho/\partial lnv_S$ which are required to translate the $\delta v_S/v_S$-heterogeneity in model $SH8/U4L8$ into an equivalent density heterogeneity field:

$$\delta\rho_l^m(r) = \rho_o(r)\left(\frac{\partial ln\rho}{\partial lnv_S}\right)(r)\left(\frac{\delta v_S}{v_S}\right)_l^m(r), \qquad (17)$$

in which $\rho_o(r)$ is the PREM radial density profile. We have assumed here that the $\partial ln\rho/\partial lnv_S$ is only a function of radius and thus we are explicitly ignoring the possibility of substantial lateral variations in this quantity. Such lateral variations, if they are indeed important [e.g. Forte et al., 1991b], may be greatest in the top 400 km of the mantle in which the chemically distinct "roots" of continents hypothesized by Jordan [1975] may possess $\partial ln\rho/\partial lnv_S$ values which are quite different from the ambient mantle (see Jordan [1979] for a detailed discussion). We will follow the example of Hager and Clayton [1989] and infer $(\partial ln\rho/\partial lnv_S)(r)$ directly from the nonhydrostatic geoid data itself. The procedure is quite straightforward and involves substituting (17) into (13), using the $\delta v_S/v_S$-heterogeneity model $SH8/U4L8$, and using the geoid kernels shown in Figure 11. If we further assume for simplicity that $\partial ln\rho/\partial lnv_S$ is constant in the depth range 0-400 km, 400-670 km, 670-1000 km, and 1000-2891 km, the problem reduces to a simple linear least-squares inversion of the nonhydrostatic geoid data. The $\partial ln\rho/\partial lnv_S$ values that are thus inferred using the viscosity profile in Figure 10a are shown in Figure 10b and they are:

$$\partial ln\rho/\partial lnv_S = \begin{cases} -0.02, & 0\text{--}400 \text{ km}, \\ +0.33, & 400\text{--}670 \text{ km}, \\ +0.20, & 670\text{--}1000 \text{ km}, \\ +0.10, & 1000\text{--}2891 \text{ km} \end{cases} \qquad (18)$$

In Figure 10d we also show the $\partial ln\rho/\partial lnv_S$ values inferred by least-squares fitting to the geoid data using the viscosity profile in Figure 10c. In this particular inversion the depth-variation of $\partial ln\rho/\partial lnv_S$ was constrained to be linear in the lower mantle. The inferred $\partial ln\rho/\partial lnv_S$ values in Figure 10d are very similar (except in the top 400 km) to those in Figure 10b: $\partial ln\rho/\partial lnv_S = 0.12$ from 0-400 km depth, 0.33 from 400-670 km depth, and varies linearly from a value of 0.26 at the top of the lower mantle to a value of 0.13 at the core-mantle boundary. The geoid (not shown here) predicted on the basis of the viscosity profile in Figure 10c and the density-velocity scaling in Figure 10d accounts for 67% of the variance in the observed nonhydrostatic geoid. This fit to the geoid data is nearly identical to that obtained using the viscosity profile in Figure 10a (see below).

In (18) the $\partial ln\rho/\partial lnv_S$ value in the transition zone (400-670 km depth) is somewhat smaller than the value $\partial ln\rho/\partial lnv_S \approx 0.4$ expected on the basis of laboratory measurements [e.g. Anderson et al., 1968; Isaak et al. 1989] and, considering the uncertainties in model $SH8/U4L8$, this difference is probably not significant. An important feature of the inferences in (18) is the marked decrease in the value of $\partial ln\rho/\partial lnv_S$ from the upper mantle to the bottom of the lower mantle. This decrease may reflect the important effect of increasing pressure with depth. The effect of the high-pressure environment of the deep mantle on the temperature derivatives of the seismic velocities has been discussed in detail by Anderson [1987] who concludes, among other things, that the in-situ derivatives will be significantly smaller in magnitude than those obtained in standard-pressure laboratory measurements. This conclusion has received further support from recent high-pressure laboratory studies [e.g. Chopelas, 1988, 1990; Duffy and Ahrens, 1992] and also from theoretical lattice-dynamic models [e.g. Agnon and Bukowinski, 1990]. Since the temperature derivatives of the seismic wave speeds are expected to decrease with increasing depth, it is clear from (18) that the coefficient of thermal expansion must decrease even more rapidly. The measurements by Chopelas and Boehler [1989] show that the coefficient of thermal expansion will decrease significantly (perhaps by as much as an order of magnitude) across the depth-range of the mantle. This important result has also been discussed in Chopelas [1990], Anderson et al. [1990], and Reynard and Price [1990].

The $\partial ln\rho/\partial lnv_S$ value inferred for the top 400 km of the mantle is anomalous both in its sign and in its amplitude. One possible explanation for this anomalous value is that we are inferring an effective horizontal average of a laterally-varying $\partial ln\rho/\partial lnv_S$ which may reflect the effects of partial-melting below mid-ocean ridges and/or the effects of the continent-ocean differences envisaged by Jordan [1975]. Another, rather different, explanation for the small magnitude (relative to laboratory data) of the $\partial ln\rho/\partial lnv_S$ inferred for the top 400 km may lie in the frequency-dependent correction to the seismic heterogeneity which might be expected on the basis of the lateral variations in upper-mantle Q-structure [Romanowicz, 1990]. When considering such explanations it is important to remember that the $\delta v_S/v_S$ heterogeneity is poorly correlated to the geoid in the top 400 km of the mantle (see Figure 8) and therefore our least-squares inversion of the geoid data has penalized this portion of the mantle by assigning it a very small $\partial ln\rho/\partial lnv_S$ value. It also appears that the $\partial ln\rho/\partial lnv_S$ value inferred for the top 400 km of the mantle is sensitive to the details of the assumed radial viscosity profile (compare Figures 10b and 10d for example) and is therefore rather uncertain.

On the basis of the $\partial ln\rho/\partial lnv_S$ values in (18), and the geoid kernels shown in Figure 11, we can now calculate the nonhydrostatic geoid expected on the basis of the $\delta v_S/v_S$ heterogeneity in model $SH8/U4L8$. The results of this calculation are shown in Figure 12 in which we compare the predicted geoid, in Figure 12b, with the GEM-T2 nonhydrostatic geoid in Figure 12a. The predicted and observed geoids evidently agree rather well and this agreement may be quantified in terms of the 65% variance reduction obtained with our geoid prediction.

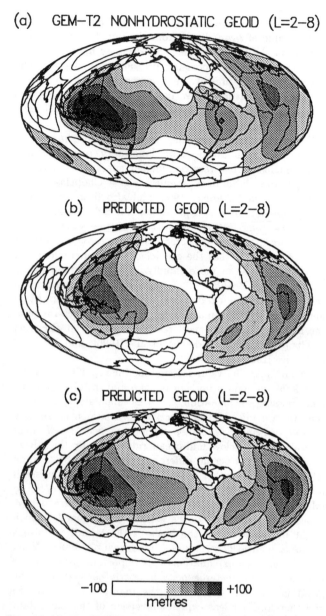

(a) GEM–T2 NONHYDROSTATIC GEOID (L=2–8)

(b) PREDICTED GEOID (L=2–8)

(c) PREDICTED GEOID (L=2–8)

−100 [   ] +100
metres

Fig. 12. (a) The GEM-T2 geoid [Marsh et al., 1990], filtered by removal of the hydrostatic flattening [e.g. Jeffreys, 1963; Naki-boglu, 1982], in the degree range $l = 2 - 8$. (b) The nonhy-drostatic geoid, in the degree range $l = 2 - 8$, predicted with model $SH8/U4L8$ using the viscosity profile in Figure 10a and the $\partial ln\rho/\partial lnv_S$ values in (18). (c) The nonhydrostatic geoid of (b) which has been adjusted by setting its Y_2^0 coefficient equal to the value of the observed Y_2^0 coefficient in (a). The contour interval in all cases is 25 metres.

The greatest source of misfit between the predicted and observed geoids is the poorly predicted N_2^0 geoid coefficient (predicted N_2^0 = -10.8 metres) which is much smaller than the corresponding GEM-T2 value (nonhydrostatic GEM-T2 N_2^0 = -27.6 metres). If we adjust the predicted

N_2^0 coefficient to be equal to the GEM-T2 nonhydrostatic value the resulting geoid, shown in Figure 12c, now agrees more closely with the GEM-T2 geoid and, in fact, the variance reduction we obtain with this adjustment is 82%. The importance of the mantle heterogeneity in the vicinity of the 670 km seismic discontinuity is confirmed by the observation that the seismically-inferred density contrasts (derived from model $SH8/U4L8$) in the depth-range 400-1000 km account for 60% of the variance in the geoid prediction shown in Figure 12b.

The difficulty in matching the excess ellipticity of the observed nonhydrostatic geoid has been reported previously on the basis of earlier seismic heterogeneity models [e.g. Forte, 1989; Ricard and Vigny, 1989; Forte and Peltier, 1991b]. This difficulty does not, however, seem to arise in the geoid modelling studies of Hager and Clayton [1989] and Hager and Richards [1989] in which the Clayton and Comer [1983] model of lower-mantle P-velocity heterogeneity is employed. As discussed in detail by Forte and Peltier [1991b], it appears that the Clayton and Comer model is characterized by an anomalous Y_2^0-heterogeneity which is substantially larger than the Y_2^0-heterogeneity in Dziewonski's [1984] P-velocity model $L02.56$. It is apparently because of this anomaly in the Clayton and Comer model that Hager and Richards [1989] are able to obtain such excellent ($\approx$ 80-90%) variance reductions with their geoid predictions. Indeed, if we employ the mantle viscosity model 'WL' of Hager and Richards [1989] to predict the geoid, using the $\delta v_S/v_S$ heterogeneity in model $SH8/U4L8$ (and also using the optimal $\partial ln\rho/\partial lnv_S$ values inferred from the geoid using the model 'WL' geoid kernels), we obtain a poor fit (variance reduction = 38%) to the GEM-T2 nonhydrostatic geoid. In this latter case we find that the misfit is again mostly due to the poorly predicted N_2^0 geoid coefficient.

The lower-mantle Y_2^2-heterogeneity in model $SH8/U4L8$ and in Dziewonski's [1984] model $L02.56$ is comparable in size to the lower-mantle Y_2^0-heterogeneity. It is important to appreciate that this result is not unique to models $L02.56$ and $SH8/U4L8$. It is also evident from the maps and/or the corresponding harmonic coefficients of many other published lateral heterogeneity models [e.g. Giardini et al., 1987; Woodhouse and Dziewonski, 1989; Tanimoto, 1990; Su and Dziewonski, 1991; Li et al., 1991; Morelli and Dziewonski, 1991] that the amplitudes of the lower-mantle Y_2^2- and Y_2^0-heterogeneity are similar. This general agreement between numerous lateral heterogeneity models, obtained from rather different data sets and inversion techniques, suggests that the dominance of the Y_2^0-component over the Y_2^2-component in the Clayton and Comer [1983] model is indeed anomalous.

It is important to consider the possible reason(s) for the discrepancy in the relative importance of the Y_2^0-heterogeneity in the Clayton and Comer [1983] model compared to other models of lower-mantle lateral het-erogeneity. As indicated above, such a discrepancy is important from our perspective because the nonhydrostatic Y_2^0-geoid coefficient is by far larger than all other geoid coefficients and accounts for a substantial fraction of the 41.0 metre root-mean-square amplitude of the total (in the degree-range $l = 2 - 32$) nonhydrostatic geoid. The amplitude of the Y_2^0 mantle heterogeneity that is obtained from global travel-time inversions will of course depend strongly on the manner by which we account for the hydrostatic ellipticity of Earth. In particular, as pointed out by Dziewonski and Gilbert [1976], the traditional ellipticity correction of Bullen [1937] yields significantly different values for the $\tau_1(\Delta)$ factor which appears in the general expression for the effects of ellipticity on the seismic travel times (see Figure 2 and eq. (26) in Dziewonski and Gilbert [1976]). In their travel-time inversion Clayton and Comer [1983] (reported in Hager and Clayton [1989]) use the ISC travel-time residuals with respect to the Jeffreys-Bullen tables (which include the approximate Bullen [1937] ellipticity correction), but later they add to their final solution the inverted difference between the Dziewonski-Gilbert and Bullen correction. The $\tau_1(\Delta)$ factor in the Dziewonski-Gilbert ellipticity correction is associated with a term which varies as $\cos \xi$ (where ξ is the azimuth from the epicentre to a station) and the effect of a North-South shift in the epicentral coordinates is also proportional to $\cos \xi$. Therefore, the difference between the Dziewonski-Gilbert and Bullen ellipticity correction could to a large extent be accommodated by the relocation of earthquakes. Clayton and Comer do not relocate their earthquakes and so the difference in the ellipticity corrections, which would translate itself into a change in epicentral coordinates, is instead absorbed into the final structure they obtain. This may well explain the anomalous Y_2^0 mantle heterogeneity in their final model. The importance of relocating earthquakes is also illustrated by the contrasting inferences of the size of the Y_2^0-heterogeneity in the study by Dziewonski et al. [1977], who committed effectively the same mistake in using the ISC locations without applying relocations, and the study by Dziewonski [1984], in which earthquakes are relocated. In the study by Dziewonski et al. [1977; see their Figure 5] it is apparent that the Y_2^0-heterogeneity dominates the Y_2^2-heterogeneity while in the study by Dziewonski [1984; see his Figure 11a] the Y_2^0 and Y_2^2 heterogeneity are comparable in size.

The ultimate resolution of the Y_2^0-geoid misfit problem is of significant geodynamic interest. In the early 1960's, soon after the first indications of substantial excess ellipticity in the satellite inferences of the geoid, it was postulated that this excess ellipticity was in fact a 'fossil' rotational bulge associated with the significantly greater Earth-rotation rates in the geologic past [e.g. Munk and MacDonald, 1960]. Such a 'fossil' bulge was later interpreted by MacDonald [1965] and McKenzie [1966] as evidence for very large mantle viscosities. These large viscosities provide the mantle with sufficient long-term strength to resist the relaxation that would otherwise occur in response to the

deceleration of rotation rate arising from tidal friction. It was later pointed out [e.g. Goldreich and Toomre, 1969; O'Connell, 1971] that the analysis of MacDonald and McKenzie was marred by the bias arising from their choice of reference axes. If, however, our present attempts to account for the observed ellipticity of the nonhydrostatic geoid (in terms of the mantle-flow models) should fail, then we might be obliged to reconsider the possibility of additional long-term nonhydrostatic stress in the mantle arising from other sources (e.g. the incomplete relaxation envisaged by MacDonald and McKenzie).

We emphasize that the principal difficulty we have in modelling the observed Y_2^0 nonhydrostatic geoid is in finding a single radial viscosity model that provides equally good fits to ALL the long wavelength spherical harmonic components of the geoid. This difficulty is readily appreciated by examining the geoid Fréchet kernels in Figure 9. It is immediately apparent in Figure 9a that viscosity decreases at the bottom of the upper mantle and viscosity increases below about 1500 km depth will be most effective in improving the fit to the Y_2^0 nonhydrostatic geoid coefficient. Unfortunately, the other harmonic coefficients of the geoid, in the degree range $l = 2 - 5$ (see Figures 9a and 9b), are also quite sensitive to such adjustments of the viscosity profile. If we focus only on the degree-2 geoid (see Figure 9a), we see that improved fits to the Y_2^0 geoid may be obtained by increasing the viscosity below 1500 km depth and then decreasing the viscosity near the core-mantle boundary (where the Y_2^0 sensitivity is negligible) to correct for the now reduced fit to the Y_2^2 geoid (which is evidently quite sensitive to viscosity perturbations near the core-mantle boundary). By this reasoning we arrive at the viscosity profile shown in Figure 13a, obtained from

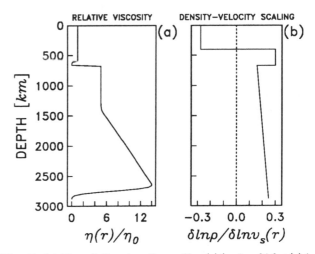

Fig. 13. (a) The relative viscosity profile $\eta(r)/\eta_0$ in which $\eta(r)$ is the actual mantle viscosity and η_0 is a reference mantle viscosity. This viscosity profile is obtained by modifying the lower-mantle portion of the viscosity profile in Figure 10a. (b) The corresponding density-velocity proportionality in the mantle inferred from the geoid data (see text for details).

Figure 10a by further increasing the viscosity in the deep mantle and by inserting a low-viscosity D" layer. The density-velocity scaling that is inferred by least-squares fitting to the geoid data, using this new viscosity profile, is shown in Figure 13b. The agreement between the predicted and observed nonhydrostatic geoid is now a considerable improvement (variance reduction = 79%) over the fit previously provided by the viscosity profile in Figure 10a. We trace this improvement to the greatly increased amplitude of the predicted Y_2^0 geoid (predicted $N_2^0 = -23.0$ metres). We find however that such improvements are occurring at the expense of the (significantly reduced) fits to the other geoid coefficients and hence our measure of fit (i.e. variance reduction) is rather misleading in this case. A satisfactory resolution of this problem requires a detailed consideration of the viscosity Fréchet kernels, in the context of a formal inversion of the geoid data, in which we also account for the uncertainties in the seismic tomographic models. Such work is now in progress.

It is useful at this stage to consider how our analysis of the geoid data, in terms of the mantle flow models, differs from the previous detailed studies by Hager and Richards [1989] and Forte and Peltier [1991b]. It is important to appreciate, at the outset, that the geoid-derived inferences of mantle viscosity and $\partial ln\rho/\partial ln v$ are strongly conditioned by the particular seismic model which is employed to represent the lateral heterogeneity in the mantle. This point has been emphasized in the discussion of Forte and Peltier [1991b] who also point out that the specific physical assumptions employed in the viscous flow theory itself (e.g. including the gravitational effects due to the compressibility of the mantle, the ocean layer, and perturbed outer core) are important factors. We noted above the striking correlation between the $\delta v_S/v_S$-heterogeneity in model $SH8/U4L8$ and the geoid (especially at $l = 2$) in the vicinity of the 670 km seismic discontinuity. This important observation led us to propose the existence of a thin layer of very low viscosity at the base of the upper mantle in order to obtain geoid kernels whose amplitudes peak at this depth. The possibility of low viscosity zones in the upper mantle has been considered by Hager and Clayton [1989] and Hager and Richards [1989] although in these studies the lowest viscosities were assigned to a thick 'asthenospheric' channel in the depth range 100 - 400 km. This channel causes the geoid kernels to peak at about 400 km depth and at 670 km depth the degree-2 geoid kernels have negligible amplitude (see Figure 6 in Forte and Peltier [1991b]). On the basis of the $\delta v_S/v_S$-heterogeneity in model $SH8/U4L8$ we would argue that such a depth variation for the geoid kernels (especially at $l = 2$) is inappropriate.

In the study of Forte and Peltier [1991b] the preferred two-layer viscosity profile consisted of a factor of 9 jump at 1200 km depth. This model, derived on the basis of the early heterogeneity models M84C [Woodhouse and Dziewonski, 1984] and L02.45 [Dziewonski, 1984], obviously lacks the low-viscosity layer at the base of the upper

mantle. It does however agree with the viscosity profiles in Figure 10 to the extent that they all indicate the importance of viscosity increases deep in the lower mantle. The $\partial ln\rho/\partial ln v_S$ values which have been inferred for the transition zone (see Figures 10b and 10d) are somewhat smaller than the value, $\partial ln\rho/\partial ln v_S = +0.5$, effectively employed in Forte and Peltier [1991b] and they are six times larger than those effectively employed by Hager and Richards [1989] (in which $\partial ln\rho/\partial ln v_S = +0.06$ throughout the upper mantle in their model 'WL'). The very small $\partial ln\rho/\partial ln v_S$ value employed in Hager and Richards [1989] is, on the basis the present results, inappropriate. This has important consequences for the value of the absolute mantle viscosity which may be inferred from the observed plate velocities [e.g. Hager, 1990; Forte et al., 1991a] since the vigour of the predicted plate velocities is most sensitive to the amplitude of the density heterogeneities in the transition zone (see the next section).

We conclude this discussion of the geoid by pointing out that, although the viscosity models in Figures 10 and 13 provide good fits to the geoid data, they are by no means the only such models. As is evident in Figure 9, the viscosity Fréchet kernels for the geoid data indicate that the resolving power of this data is quite limited. There exist significant trade-offs between viscosity inferences at different depths and thus any model of mantle viscosity derived from the geoid data is certain to be non-unique. It is also important to remember that an explicit treatment of lateral viscosity variations in the viscous flow models may lead to important trade-offs between the amplitude of such lateral viscosity variations and the radial viscosity profile inferred from the geoid data [Forte and Peltier, 1987b; Ribe, 1992]. For the purposes of this study, the viscosity profile in Figure 10a is considered adequate and it will be employed in the predictions of plate-motions and CMB topography provided below.

3.2 The Predicted Plate Motions

The most obvious manifestation of the mantle convective circulation is of course provided by the observed 'drift' of the continents as the tectonic plates move relative to each other. It was clear from the earliest seismic tomographic models [e.g. Masters et al., 1982; Woodhouse and Dziewonski, 1984] that the large scale movement of the Earth's plates was driven by the deep-seated buoyancy forces associated with the seismically-inferred lateral density variations [Peltier and Forte, 1984; Forte and Peltier, 1987a]. This realization led to the detailed mantle-flow studies by Forte and Peltier [1987b, 1991a] which demonstrated that the surface flow excited by seismically-inferred density perturbations was in good agreement with the large-scale divergence and convergence of plate velocities at ridges and trenches. From a rheological perspective the plate velocities are very important because, unlike the geoid, they are sensitive to the absolute value of mantle viscosity. This sensitivity has been exploited by Forte and Peltier [1987b, 1991b,c] and Forte et al. [1991a] to infer the absolute mantle viscosity by matching the observed plate motions to the surface mantle flow predicted on the basis of the seismic tomographic models of heterogeneity.

The vector nature of the plate-velocity field may be completely characterized in terms of two complimentary scalar functions. In Hager and O'Connell [1981] the plate velocities were summarized in terms of the so-called poloidal and toroidal generating scalars. In Forte and Peltier [1987a,b] the plate velocities $\mathbf{v}(\theta, \varphi)$ were also shown to be completely characterized by their horizontal divergence $\nabla_H \cdot \mathbf{v}$ and by their radial vorticity $\hat{\mathbf{r}} \cdot \nabla \times \mathbf{v}$. Although the horizontal divergence and radial vorticity scalars are directly related to the poloidal and toroidal scalars considered by Hager and O'Connell [1981], the former are convenient because they are amenable to direct physical interpretation. In fact, the field $\nabla_H \cdot \mathbf{v}$ describes the rate of plate divergence at ridges and trenches while the field $\hat{\mathbf{r}} \cdot \nabla \times \mathbf{v}$ describes plate motions at transform plate boundaries [Forte and Peltier, 1987a].

As in the case of the nonhydrostatic geoid, it is possible to derive the kernel functions relating the plate-velocity scalars to the internal density perturbations in the mantle. The horizontal divergence kernels $D_l(r)$ relate the spherical harmonic coefficients of the horizontal divergence of the predicted surface flow , $(\nabla_H \cdot \mathbf{u})_l^m(r = a)$, to the radially-varying spherical harmonic coefficients of the density perturbation $\delta\rho_l^m(r)$ as follows :

$$(\nabla_H \cdot \mathbf{u})_l^m(r = a) = \frac{g_0}{\eta_0} \int_b^a D_l(r)\delta\rho_l^m(r)dr, \qquad (19)$$

in which $g_0 = 10 \ m/s^2$ is the gravitational acceleration in the mantle and η_0 is the reference viscosity which normalizes the viscosity profile $\eta(r)/\eta_0$ used in the geoid modelling (see previous section). In Figure 14 we show a selection of horizontal divergence kernels $D_l(r)$, calculated for the viscosity profile $\eta(r)/\eta_0$ in Figure 10a, for both an incompressible mantle and for a (PREM) compressible mantle. It is quite evident from this figure that the finite compressibility of the mantle has a substantial impact on the longest-wavelength (i.e. $l \leq 5$) surface flow. As in the case of the geoid kernels in Figure 11, the peak amplitude in the longest wavelength horizontal divergence kernels is found in the vicinity of the 670 km seismic discontinuity and this is due to the low-viscosity zone at the base of the upper mantle. It is therefore obvious that density heterogeneities near the bottom of the upper mantle provide the greatest contribution to the predicted surface flow.

The mantle flow predicted by models with a spherically symmetric viscosity is entirely poloidal and therefore cannot account for the strong toroidal flows (or equivalently, the radial vorticity at transform faults) which characterize the actual plate motions. The toroidal component of the plate velocities may be explained in terms of a mantle

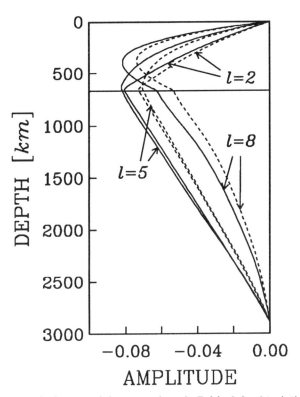

Fig. 14. The horizontal divergence kernels $D_l(r)$, defined in (19), calculated for the viscosity profile shown in Figure 10a. The solid lines represent the divergence kernels for a compressible mantle with a PREM [Dziewonski and Anderson, 1981] radial density profile and the dashed lines represent the divergence kernels for an incompressible mantle. The location of the 670 km seismic discontinuity is indicated by the heavy horizontal line.

possessing lateral viscosity variations [e.g. Ribe, 1992; Forte, 1992]. The tectonic plates themselves are generally regarded as the most extreme manifestation of lateral rheology variations [e.g. Hager and O'Connell, 1981] and this assumption has motivated several recent studies of mantle flow which attempt to explicitly account for the presence of rigid plates. Two approaches have been taken in these studies which may be distinguished by the specific assumptions made regarding the nature of the constraints posed by the presumed rigidity of the tectonic plates. One approach, advocated by Hager and O'Connell [1981], involves matching the average stresses acting on the the plates and this is the technique employed by Ricard and Vigny [1989] and Gable et al. [1991]. The second approach, taken by Forte and Peltier [1991b,c], does not require the evaluation of flow-induced stresses on the plates and instead involves an explicit consideration of the limited class of surface flows which are permitted by the requirement that plate motions may only occur by rigid-body rotations around Euler rotation poles. In this second approach it may be shown [Forte and Peltier, 1991c] that the rigid-plate constraint on predicted surface flows may ultimately be summarized by the following expressions :

$$(\nabla_H \cdot \mathbf{v})_l^m = \sum_{s,t} P_{ls}^{mt}(\nabla_H \cdot \mathbf{u})_s^t, \qquad (20)$$

$$(\hat{\mathbf{r}} \cdot \nabla \times \mathbf{v})_l^m = \sum_{s,t} Q_{ls}^{mt}(\nabla_H \cdot \mathbf{u})_s^t, \qquad (21)$$

in which P_{ls}^{mt} and Q_{ls}^{mt} are matrix operators (which depend only on the observed plate geometries) describing the effects of the surface plates on the buoyancy-induced flow in (19), and $(\nabla_H \cdot \mathbf{v})_l^m$ and $(\hat{\mathbf{r}} \cdot \nabla \times \mathbf{v})_l^m$ are the spherical harmonic coefficients of the predicted plate-like horizontal divergence and radial vorticity, respectively. These predicted plate-like surface flows are sensitive to the absolute mantle viscosity via the reference viscosity η_0 in (19).

It is important to keep in mind that the technique for predicting plate-like surface flows employed here (and by Gable et al. [1991] and Ricard and Vigny [1989]) is basically kinematic because the plates are simply introduced via a surface boundary condition. The nonhydrostatic geoid predictions are very sensitive to amplitude of the flow-induced surface topography and hence to the assumed boundary conditions (e.g. free-slip versus no-slip boundaries). It is therefore unlikely that the simplified treatment of the plates described above is adequate for modelling the nonhydrostatic geoid (and, in fact, is not employed in the geoid calculations presented above). We find, as in Ricard and Vigny [1989], that when we employ such kinematic treatments of the plates, the flow models which improve the fit to the observed plate motions degrade the fit to the geoid data and vice-versa. The resolution of this incompatibility will have to await the development of dynamically consistent (and hopefully realistic) models with a lithosphere possessing lateral rheology variations [e.g. Forte, 1992; Ribe, 1992].

On the basis of expressions (19)-(21) we can now calculate the plate-like surface flows expected from the $\delta v_S/v_S$ heterogeneity in model $SH8/U4L8$. In these calculations we shall employ the $\partial ln\rho/\partial lnv_S$ values listed in (18) and the divergence kernels, shown in Figure 14, obtained with the same viscosity profile used in the geoid modelling (i.e. in Figure 10a). The results of these calculations are shown in Figure 15 in which we present the observed horizontal divergence and radial vorticity fields along with the corresponding predicted fields. Although model $SH8/U4L8$ describes lateral heterogeneity only up to spherical harmonic degree $l = 8$, the matrix elements in (20) and (21) were calculated up to degree $l = 15$ and hence the predictions in Figure 15 also include harmonics up to degree 15. The degree-8 truncation level in model $SH8/U4L8$ is not a serious limitation from our point of view since the lateral heterogeneity in the upper mantle is apparently dominated by long wavelength features corresponding to the degree range $l \leq 7$ [e.g. Su and Dziewonski, 1991; Dziewonski et al., 1992]. Also, as shown by Ricard and Vigny [1989] and Forte and Peltier [1991b,c], the interaction of very-long wavelength mantle

flow with the tectonic plates induces shorter wavelength plate motions which agree closely with the corresponding observed plate motions. The agreement between the predicted and observed divergence fields in Figure 15 is evidently very good (global correlation coefficient for the degree range $l = 1$–15 is $+0.81$ and the variance reduction is 66%) while the agreement between the predicted and observed radial vorticity fields is not as good (global correlation coefficient in the range $l = 1$–15 is $+0.56$ and the variance reduction is 20%). We find that the radial vorticity predictions are especially sensitive to any 'misalignment' between the observed plate boundaries and the corresponding pattern of seismically-inferred density heterogeneities in the mantle.

We have exploited the sensitivity of the predicted plate-motions to the absolute value of mantle viscosity by determining the value of η_0 in (19) which minimizes the least-squares misfit between the observed and predicted plate divergence. In this manner we find $\eta_0 = 1.2 \times 10^{21}$ $Pa\ s$ and we thus infer, on the basis of the $\eta(r)/\eta_0$ profile in Figure 10a, that the upper-mantle viscosity is (with the exception of the thin low-viscosity zone at the bottom of the upper-mantle) 1.2×10^{21} $Pa\ s$. If, instead, we employ the viscosity profile and density-velocity scaling in Figures 10c and 10d we find $\eta_0 = 0.82 \times 10^{21}$ $Pa\ s$. Such inferences for the upper-mantle viscosity are encouraging because they agree well with the "Haskell"-value of 1×10^{21} $Pa\ s$ which has been traditionally inferred from the analysis of glacial isostatic adjustment data [e.g. Haskell, 1935,1936,1937; Cathles, 1971; Peltier and Andrews, 1976; Fjeldskaar and Cathles, 1991].

4. VISCOUS FLOW CALCULATION OF THE DYNAMIC CMB TOPOGRAPHY

The previous section demonstrated that the mantle flow predicted on the basis of the $\delta v_S/v_S$ heterogeneity in model $SH8/U4L8$ provides good matches to both the observed nonhydrostatic geoid and the observed plate motions. In particular we wish to emphasize that the inferences of $\partial ln\rho/\partial lnv_S(r)$ and $\eta(r)$ obtained from the geoid and plate-motion data are in good agreement with independent laboratory data and also the independent viscosity inferences derived from postglacial rebound data. Such a degree of consistency suggests to us that both the geometry and the magnitude of the mantle flow calculated using model $SH8/U4L8$ is plausible and therefore our predictions of flow-induced CMB topography, presented below, should be equally valid.

The CMB kernel functions $B_l(r)$ relate the spherical harmonic coefficients of the flow-induced CMB topography, δb_l^m, to the radially-varying coefficients of the driving density contrasts $\delta\rho_l^m(r)$ as follows :

$$\delta b_l^m = \frac{1}{\Delta\rho_{cm}} \int_b^a B_l(r)\delta\rho_l^m(r)dr, \qquad (22)$$

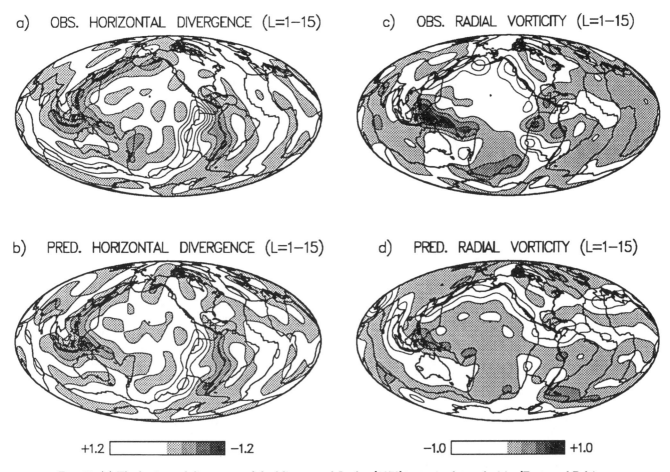

a) OBS. HORIZONTAL DIVERGENCE (L=1–15)

c) OBS. RADIAL VORTICITY (L=1–15)

b) PRED. HORIZONTAL DIVERGENCE (L=1–15)

d) PRED. RADIAL VORTICITY (L=1–15)

+1.2 [   ] −1.2   −1.0 [   ] +1.0

Fig. 15. (a) The horizontal divergence of the Minster and Jordan [1978] tectonic plate velocities [Forte and Peltier, 1987b] in the degree range $l = 1 - 15$. The contour interval is 0.3×10^{-7} rad/yr. (b) The horizontal divergence, in the degree range $l = 1 - 15$, predicted with model $SH8/U4L8$ using the mantle viscosity profile in Figure 10a and the $\partial ln\rho/\partial ln v_S$ values in (18). The contour interval is 0.3×10^{-7} rad/yr. (c) The radial vorticity of the Minster and Jordan [1978] tectonic plate velocities [Forte and Peltier, 1987b] in the degree range $l = 1 - 15$. The contour interval is 0.25×10^{-7} rad/yr. (d) The radial vorticity, in the degree range $l = 1 - 15$, predicted with model $SH8/U4L8$ using the mantle viscosity profile in Figure 10a and the $\partial ln\rho/\partial ln v_S$ values in (18). The contour interval is 0.25×10^{-7} rad/yr. In the calculation of the predicted plate motions in (b) and (d), according to (19)–(21), the reference viscosity $\eta_0 = 1.2 \times 10^{21}$ Pa s was employed (see text for details).

in which $\Delta\rho_{cm} = $ -4.434 Mg/m^3 is the density jump across the CMB. A selection of CMB topography kernels $B_l(r)$, calculated for the radial viscosity profile in Figure 10a, is shown in Figure 16. The predicted CMB topography is, like the geoid, sensitive only to the relative viscosity variations $\eta(r)/\eta_0$ and cannot therefore be used to constrain the absolute mantle viscosity.

The $\partial ln\rho/\partial ln v_S$ values in (18), inferred from the geoid data, and the kernels in Figure 16 will now be employed to calculate the flow-induced CMB topography expected on the basis of the $\delta v_S/v_S$ heterogeneity in model $SH8/U4L8$. In Figure 17 we show the resulting CMB topography prediction and, for comparison, we also show the seismically inferred CMB topography of Morelli and

Dziewonski [1987; hereafter referred to as MD]. Although the MD model only includes spherical harmonics up to degree $l = 4$ it is possible that the least-squares inversion upon which this model is based may have aliased into it the effect of higher degree CMB topography. It is for this reason that we show predicted CMB topography for both the degree range $l = 1$–8 and for the range $l = 1$–4. It is evident that, apart from the mismatch beneath southern Africa and beneath the Indian ocean, the CMB topography prediction is in good agreement with the MD model as both show the well-defined ring of depressed CMB below the circum-Pacific region and the elevated CMB below the central Pacific and Atlantic oceans. It is also rather evident that the major difference between the CMB topography

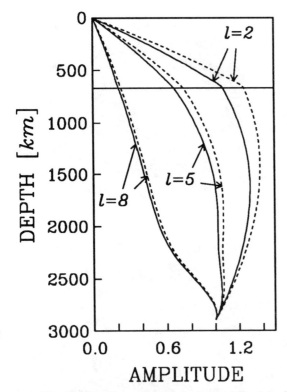

Fig. 16. The CMB topography kernels $B_l(r)$, defined in (22), calculated for the viscosity profile shown in Figure 10a. The solid lines represent the topography kernels for a compressible mantle with a PREM [Dziewonski and Anderson, 1981] radial density profile and the dashed lines represent the topography kernels for an incompressible mantle. The location of the 670 km seismic discontinuity is indicated by the heavy horizontal line.

prediction and the MD model is the approximately factor of 2 difference in the overall amplitude of the CMB undulations.

The flow-induced CMB topography is essentially a representation of the radially-integrated lower-mantle heterogeneity as is evident from the kernels in Figure 16. It is thus clear that if we were to adjust the overall amplitude of the predicted CMB undulations to agree with that in the MD model we should double the lower-mantle $\partial ln\rho/\partial lnv_S$ value in (18). Unfortunately, such an adjustment would degrade the good match obtained between the predicted and observed nonhydrostatic geoid. In fact we find that if we employ the value $\partial ln\rho/\partial lnv_S = +0.2$ for the depth-range 1000-2891 km, and employ the other values in (18) for the depth-interval 0-1000 km, the variance reduction obtained with the predicted geoid is 54% (compared to the original variance reduction of 65% - see section 3.1). This reduced fit is almost entirely due to the increased misfit between the degree-2 components of the predicted and observed geoids. The CMB topography predictions in Figures 17b and 17c are clearly dominated by very long wavelength signal and this may be understood both

in terms of the low-pass filtration effect of the kernels in Figure 16 and, most importantly, in terms of the simple fact that the $\delta v_S/v_S$ heterogeneity in model $SH8/U4L8$ is dominated by degree-2 structure in the bottom half of the lower mantle. Indeed the circum-Pacific ring of depressed CMB evident in Figure 17b is a direct manifestation of the dominant $Y_2^2(\theta, \varphi)$ pattern of lateral heterogeneity in the lower mantle.

To further assess the plausibility of our flow-induced CMB topography prediction we compare it with other published seismic inferences of CMB topography in Figure 18. In this figure we show our CMB topography prediction alongside the MD model, model 'SAF' of Li et al. [1991; hereafter referred to as LGW] retrieved from normal-mode multiplet splitting data, and the 'Model 6' of Doornbos and Hilton [1989; hereafter referred to as DH] retrieved from PKP and $PKKP$ travel-time residuals. There is clearly a very good agreement in the patterns of the CMB topography in Figures 18a–c, all of which display the characteristic circum-Pacific depression. The agreement between the LGW model in Figure 18c and our prediction in Figure 18a is especially good. Since the normal mode splitting data employed in the derivation of the LGW model is only sensitive to even-degree structure we present in Figure 19a a map of our predicted CMB topography which includes only the harmonics for degrees $l = 2$ and $l = 4$ and in Figure 19b we again show the LGW model. It is quite evident from this figure that the agreement between the two models is very good, both in terms of the pattern and in terms of the amplitude of the CMB deflections. Such agreement may perhaps be fortuitous since as Li et al. [1991] point out there may be significant trade-offs between the normal-mode inferences of CMB topography and volumetric mantle heterogeneity.

The predictions of flow-induced CMB topography presented in Forte and Peltier [1989, 1991a] were based only on the earliest available seismic heterogeneity models of the mantle, namely model $M84C$ of Woodhouse and Dziewonski [1984] which describes the S-velocity heterogeneity in the upper mantle and model $L02.56$ (or $L02.45$) of Dziewonski [1984] which describes the P-velocity heterogeneity in the lower mantle. We shall now consider the CMB topography predicted on the basis of models $M84C$ and $L02.56$. As a prerequisite to this prediction we first obtain (as in section 3.1) the optimal $\partial ln\rho/\partial lnv_S$ and $\partial ln\rho/\partial lnv_P$ values inferred from least-squares fitting to the geoid data (using the geoid kernels in Figure 11). For model $M84C$ we thus infer the following conversion factors :

$$\partial ln\rho/\partial lnv_S = \begin{cases} +0.19, & 0\text{–}400 \text{ km}, \\ +0.11, & 400\text{–}670 \text{ km}, \end{cases} \tag{23}$$

and for model $L02.56$ we have :

$$\partial ln\rho/\partial lnv_P = \begin{cases} +0.27, & 670\text{–}1000 \text{ km}, \\ +0.36, & 1000\text{–}2891 \text{ km} \end{cases} \tag{24}$$

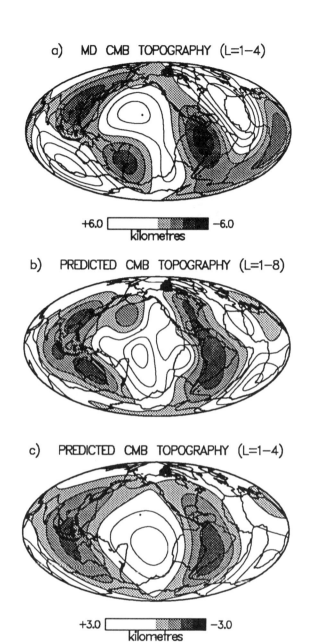

a) MD CMB TOPOGRAPHY (L=1-4)

+6.0 ☐▨■ -6.0
kilometres

b) PREDICTED CMB TOPOGRAPHY (L=1-8)

c) PREDICTED CMB TOPOGRAPHY (L=1-4)

+3.0 ☐▨■ -3.0
kilometres

Fig. 17. (a) The seismically-inferred CMB topography model of Morelli and Dziewonski [1987] in the degree range $l = 1 - 4$. The contour interval is 1.5 kilometres. (b) The flow-induced CMB topography, in the degree range $l = 1 - 8$, predicted with model $SH8/U4L8$ using the viscosity profile in Figure 10a and the $\partial ln\rho/\partial lnv_S$ values in (18). The contour interval is 0.75 kilometres. (c) The flow-induced CMB topography in (b) shown for the truncated degree range $l = 1 - 4$. The contour interval is 0.75 kilometres.

If we now compare the $\partial ln\rho/\partial lnv_S$ inferences in (23) with those in (18) it is clear that such inferences are quite sensitive to the particular choice of lateral heterogeneity models used in the mantle-flow and geoid modelling. We also find that the results in (23) are strongly conditioned by the particular choice of model $L02.56$ to represent lower-

mantle structure and there is indeed a strong statistical correlation between the uncertainties in the inferences in (23) and (24). A comparison of $\partial ln\rho/\partial lnv_P$ in (24) with $\partial ln\rho/\partial lnv_S$ in (18) is interesting because it suggests that $\partial lnv_S/\partial lnv_P = 3.6$ in most of the lower mantle. Although the precise value of this result is dependent on our assumed viscosity profile and on our choice of models $L02.56$ and $SH8/U4L8$, it agrees with previous studies showing that the temperature sensitivity of shear velocity is much greater than that of compressional velocity in the lower mantle [e.g. Anderson, 1987; Giardini et al., 1987; Dziewonski and Woodhouse, 1987; Yeganeh-Haeri et al., 1989; Li, 1990; Agnon and Bukowinski, 1990].

The flow-induced CMB topography obtained using models $M84C$ and $L02.56$, and using the kernels in Figure 16 along with the conversion factors in (23) and (24), is shown in Figure 20a alongside the prediction for model $SH8/U4L8$ in Figure 20b. The agreement between these two maps of flow-induced CMB topography is clearly very good (global correlation coefficient for the degree range $l = 1-8$ is +0.64 and for the range $l = 1-4$ it is +0.74). We view this striking agreement as an important confirmation of the plausibility, and indeed the validity, of either of the two fields shown in Figure 20. This agreement is rather remarkable when we consider that the map in Figure 20a is a radial integral of a P-velocity heterogeneity model for the lower mantle while the map in Figure 20b is a radial integral of an S-velocity model. This agreement suggests that both compressional- and shear-velocity heterogeneity in most, if not all, of the mantle is due to the same field of lateral temperature variation maintained by the thermal convection process. Such a conclusion is also supported by independent seismic modelling by Woodward and Masters [1991b] who show that when model $L02.56$ is scaled appropriately it provides a rather good fit to the observed pattern of global $ScS - S$ differential travel-time residuals.

Finally we turn to a consideration of the effect of a low-viscosity layer, at the base of the mantle, on the amplitude of the flow-induced CMB topography. The presence of a thermal boundary layer at the base of the mantle would be expected if the mantle convective circulation were partially driven by a heat flux out of the core [e.g. Turcotte and Oxburgh, 1967; Jeanloz and Richter, 1979; Yuen and Peltier, 1980; Peltier, 1981; Jarvis and Peltier, 1982; Stacey and Loper, 1983; Solheim and Peltier, 1990]. In the PREM radial Earth model [Dziewonski and Anderson, 1981] the lowermost 150 km of the mantle, referred to as the D" layer [Bullen, 1950], is characterized by reduced gradients in the depth variation of P- and S-wave velocities. Although the actual value of these gradients, as given in PREM, is somewhat uncertain it indicates a reduction in the pressure derivative of the bulk modulus, dK/dP, in D". This reduction may be caused by compositional gradients and/or increased temperatures. Stacey and Loper [1983] have interpreted the reduced dK/dP in D" as

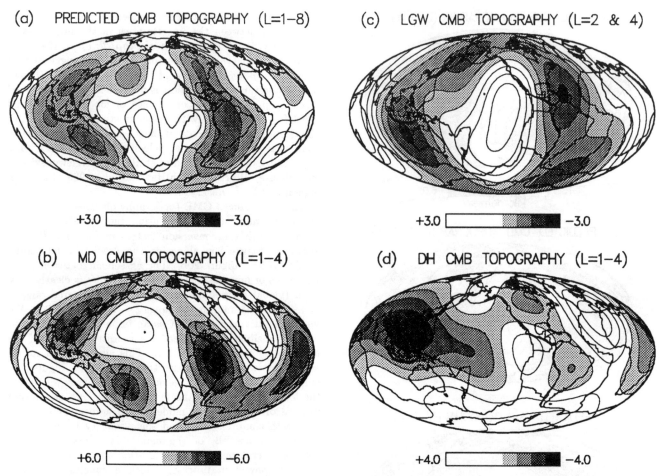

(a)   PREDICTED CMB TOPOGRAPHY (L=1–8)

+3.0 ▭ −3.0

(b)   MD CMB TOPOGRAPHY (L=1–4)

+6.0 ▭ −6.0

(c)   LGW CMB TOPOGRAPHY (L=2 & 4)

+3.0 ▭ −3.0

(d)   DH CMB TOPOGRAPHY (L=1–4)

+4.0 ▭ −4.0

Fig. 18. (a) The predicted flow-induced CMB topography, in the degree range $l = 1 - 8$, shown previously in Figure 17b. The contour interval is 0.75 kilometres. (b) The seismically-inferred CMB topography model of Morelli and Dziewonski [1987] in the degree range $l = 1 - 4$. The contour interval is 1.5 kilometres. (c) The seismically-inferred CMB topography model "SAF" of Li et al. [1991] for harmonic degrees $l = 2$ and 4. The contour interval is 0.75 kilometres. (d) The seismically-inferred CMB topography model "Model 6" of Doornbos and Hilton [1989] in the degree range $l = 1 - 4$. The contour interval is 1 kilometre.

a manifestation of a thermal boundary layer and they have formulated a detailed dynamical and thermal model of D" in which they determined that the presence of a thermal boundary layer would result in a radial viscosity profile in D" which takes the form

$$\eta(r)/\eta_0 = (\eta_{CMB}/\eta_0) \exp\left[\frac{r-b}{h}\right], \qquad (25)$$

in which η_{CMB}/η_0 is the relative mantle viscosity at the CMB, $r = b$ is the radius of the CMB, and h is the scale height for radial viscosity variations in D". Expression (25) reflects the fact that, owing to the temperature sensitivity of mantle viscosity, the viscosity in D" will be lowest immediately adjacent to the CMB where the superadiabatic temperature is greatest. A softening of the D" layer will decrease the flow-induced normal stress acting on the CMB and should therefore reduce the amplitude of the dynamic CMB topography.

The effect of a low-viscosity layer at the base of the mantle has been previously considered by Hager and Richards [1989] and Forte and Peltier [1991a,b]. In the study by Hager and Richards [1989] the low-viscosity zone was modelled with a 300 km - thick layer at the base of the mantle in which the viscosity was 100 times smaller than in the overlying mantle. It appears that their choice of such a thick layer was dictated by the need to reduce the very large Y_2^0 component of their predicted CMB topography. As we show below, the effect of a low-viscosity zone at the bottom of the mantle is such as to reduce the high-harmonic degree components of the predicted CMB more strongly than the low-degree components. As the thickness of this layer is increased the reduction of the low-degree components is also increased. It seems, therefore, that Hager and Richards [1989] have invoked a thick, constant-viscosity, low-viscosity D" layer in order to reduce the excessively large Y_2^0-CMB topography arising from the anomalously

(a) PREDICTED CMB TOPOGRAPHY (L=2 & 4)

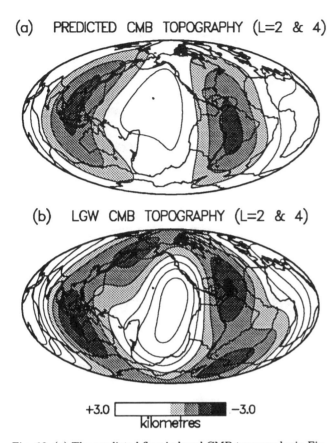

(b) LGW CMB TOPOGRAPHY (L=2 & 4)

+3.0 ▭ −3.0
kilometres

Fig. 19. (a) The predicted flow-induced CMB topography in Figure 17b shown only for harmonic degrees $l = 2$ and 4. (b) The seismically-inferred CMB topography model "SAF" of Li et al. [1991] for harmonic degrees $l = 2$ and 4. In each case the contour interval is 0.75 kilometres.

large lower-mantle Y_2^0-heterogeneity in the Clayton and Comer [1983] P-velocity model (see section 3.1 above for a discussion of this problem). In the study by Forte and Peltier [1991b] the seismologically-defined D" layer was modelled with the viscosity profile given in (25) and it was found that, even with very steep decreases of viscosity across D", the degree-2 CMB topography (which dominates all other components) was not appreciably affected.

We now wish to consider the extent to which the flow-induced CMB topography in Figure 17b may be reduced by introducing a low-viscosity D" layer as described in (25). In Figure 21a we show the viscosity profile obtained by inserting a low-viscosity D" layer in the viscosity profile shown in Figure 10a. The parameters η_{CMB}/η_0 and h in (25) are chosen so that the mantle viscosity at the CMB is 1000 times smaller than the viscosity at the top of D". This viscosity decrease is somewhat smaller than the viscosity decrease estimated by Stacey and Loper [1983] who assume that the temperature dependence of viscosity is given by $\eta = A \exp(gT_m/T)$ in which T/T_m is the homologous temperature and g is an empirical constant.

For silicates a reasonable estimate for g seems to be 30 (such a value has also been estimated by Beauchesne and Poirier [1989] in their creep experiments on a perovskite analogue) in which case Stacey and Loper estimate that the viscosity at the bottom of D" will be about 6000 times smaller than the viscosity at the top of D". We have found that the perturbations to the predicted CMB topography arising from perturbations in the radial viscosity profile of the mantle are scaled according to the logarithm of the viscosity perturbations (i.e. according to $\delta ln[\eta(r)]$) and such a scaling has also been determined for the predicted geoid and plate motions [Forte et al., 1991a]. It is therefore straightforward to understand why the dynamical effect of a factor of 6000 viscosity decrease across D" is similar to a factor of 1000 decrease. In view of the significant uncertainties regarding the precise rheological behaviour and the actual geotherm of the deep mantle we consider our choice of a factor of 1000 viscosity decrease as merely representative of the actual possible decrease across D". In Figure 21b we now show the impact of such a low-viscosity

(a) PREDICTED CMB TOPOGRAPHY (L=1−8)

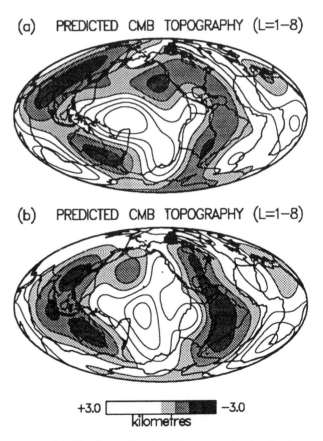

(b) PREDICTED CMB TOPOGRAPHY (L=1−8)

+3.0 ▭ −3.0
kilometres

Fig. 20. (a) The flow-induced CMB topography, in the degree range $l = 1 - 8$, predicted with models $M84C$ and $L02.56$ using the mantle viscosity profile in Figure 10a and the $\partial ln\rho/\partial lnv$ conversion factors in (23) and (24). (b) The flow-induced CMB topography in Figure 17b. In each case the contour interval is 0.75 kilometres.

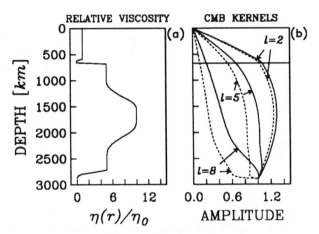

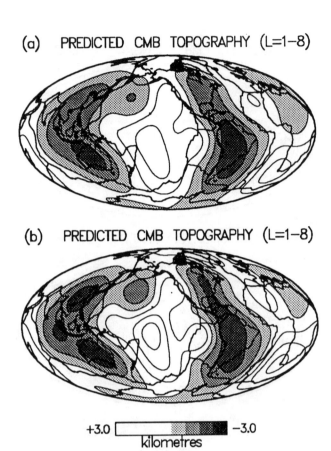

Fig. 21. (a) The relative viscosity profile $\eta(r)/\eta_0$ of Figure 10a which has been modified in the lowermost 150 km according to expression (25) in the text (in which h and η_{CMB}/η_0 are chosen so that the viscosity at the CMB is 1000 times smaller than the viscosity at the top of D"). (b) CMB topography kernels $B_l(r)$ defined in (22). The solid lines represent the kernels calculated for the original viscosity profile on Figure 10a while the dashed lines represent the corresponding kernels calculated for the viscosity profile shown in (a). The location of the 670 km seismic discontinuity is indicated by the heavy horizontal line.

Fig. 22. (a) The flow-induced CMB topography, in the degree range $l = 1 - 8$, predicted with model $SH8/U4L8$ using the viscosity profile in Figure 21a and the $\partial ln\rho/\partial lnv_S$ values in (26). (b) The flow-induced CMB topography, shown previously in Figure 17b, calculated for the viscosity profile in Figure 10a. In each case the contour interval is 0.75 kilometres.

zone on the behaviour of the CMB topography kernels. It is quite apparent from this figure that the low-degree harmonic coefficients (e.g. $l = 2$) of the CMB topography are not as strongly affected as the higher-degree (e.g. $l = 8$) coefficients.

Prior to calculating the flow-induced CMB topography expected on the basis of the viscosity profile shown in Figure 21a we shall first determine the $\partial ln\rho/\partial lnv_S$ values which provide the best least-squares fit between the predicted and observed nonhydrostatic geoid. This procedure, which has been described above in section 3.1, provides us with the following result :

$$\partial ln\rho/\partial lnv_S = \begin{cases} +0.14, & 0\text{--}400 \text{ km}, \\ +0.35, & 400\text{--}670 \text{ km}, \\ +0.29, & 670\text{--}1000 \text{ km}, \\ +0.08, & 1000\text{--}2891 \text{ km} \end{cases} \quad (26)$$

A comparison of (26) with (18) shows that the inference of $\partial ln\rho/\partial lnv_S$ in the top 400 km of the mantle is rather sensitive to the details of the assumed radial viscosity model. On the basis of the geoid kernels (not shown here) calculated for the viscosity model in Figure 21a and using (26) we find that the geoid predicted with $SH8/U4L8$ accounts for 60% of the variance of the observed geoid. The major source of misfit between predicted and observed geoid is again due to the small amplitude of the predicted Y_2^0-geoid coefficient and, were it not for this Y_2^0-misfit, the variance reduction would increase substantially to 83%.

The flow-induced CMB topography calculated on the basis of the CMB topography kernels for the viscosity

profile in Figure 21a, the conversion factors in (26), and the $\delta v_S/v_S$ heterogeneity in model $SH8/U4L8$ is now shown in Figure 22a. In Figure 22b we show, for the purpose of comparison, the CMB topography (shown previously in Figure 17b) calculated without a low-viscosity D" layer. It is quite obvious from Figure 22 that the impact of the low-viscosity D" layer is not appreciable. The root-mean-square amplitude of the predicted CMB topography in Figure 22b is 1.01 km while in Figure 22a it is reduced to 0.86 km. It is evident from a more detailed examination of these maps that this reduction in overall amplitude is due mainly to the reduction of elevated topography below the central Pacific ocean and below southern Africa (i.e. below the 'mega-plumes' shown in Figure 5a). We find that the topography maxima in these regions is reduced from 2.7 km in Figure 22b to 2.0 km in Figure 22a.

5. SUMMARY

In this study we have shown, on the basis of simple viscous flow models (with a spherically symmetric viscosity structure), that the most recent seismic inferences of seismic velocity heterogeneities in the mantle are capable of providing a very good fit to the principal convection-related surface observables, namely the nonhydrostatic geoid and the tectonic plate motions. An important outcome of this study was the inference of mantle viscosity profiles, shown in Figure 10, which are characterized by a pronounced low-viscosity zone at the bottom of the upper mantle and a region of substantially increased stiffness within the lower mantle. On the basis of this new relative viscosity profile we have inferred, using the plate-motion data, an absolute upper-mantle viscosity which agrees well with that generally inferred in independent analyses of glacial-isostatic adjustment data.

The very good agreement between the two maps of flow-induced CMB topography in Figure 20 is rather interesting when we consider that the mantle-heterogeneity models which were employed in each map are separated by nearly a decade of rapid progress in both seismic imaging techniques and the acquisition of high-quality global seismic data. It is therefore natural to enquire, from a geodynamic perspective, whether any real progress has been made with the most recent mantle-heterogeneity model $SH8/U4L8$ employed in this study. Our answer is that, indeed, this model provides a marked improvement in the quality of the predicted geodynamic surface observables. We can illustrate this improvement very simply by pointing out that the geoid predicted with the early heterogeneity models $M84C$ and $L02.56$, using the geoid kernels in Figure 11 and the conversions in (23) and (24), can only account for 38% of the variance (in the degree range $l = 2-8$) of the GEM-T2 nonhydrostatic geoid. When we compare this to the 65–80% variance reductions delivered by model $SH8/U4L8$ (see section 3.1) it is evident that seismic tomography has indeed made substantial progress. The remaining mismatch between the predicted and observed convection-related observables indicates that there is of course still room for improvement in both the seismic tomographic models and the theoretical flow models of the mantle [e.g. Forte, 1992; Ribe,1992]. Nonetheless, the agreement between the CMB topography predictions in Figure 20 suggests that this improvement should not alter significantly our present image of the topography on the CMB generated by the mantle-convective circulation.

Acknowledgments. We thank We-jia Su for the use of his 'GE-OMAP' program which we employed to plot the maps and cross-sections shown above. We also thank B. Romanowicz and an anonymous reviewer for useful comments. This work has been supported by a Canadian NSERC postdoctoral fellowship award-ed to A.M.F. and by grant EAR90-05013 from the National Science Foundation.

REFERENCES

Agnew, D. C., J. Berger, R. Buland, W. Farrell, and F. Gilbert, International deployment of accelerometers: A network for very long period seismology, *EOS Trans. AGU*, **57**, 180–188, 1976.

Agnon, A., and M. S. T. Bukowinski, δ_S at high pressure and $dlnV_S/dlnV_P$ in the lower mantle, *Geophys. Res. Lett.*, **17**, 1149–1152, 1990.

Anderson, D. L., A seismic equation of state II. Shear properties and thermodynamics of the lower mantle, *Phys. Earth Planet. Inter.*, **45**, 307–323, 1987a.

Anderson, D. L., Thermally induced phase changes, lateral heterogeneity of the mantle, continental roots, and deep slab anomalies, *J. Geophys. Res.*, **92**, 13,968–13,980, 1987b.

Anderson, O. L., E. Schreiber, R. C. Lieberman, and N. Soga, Some elastic constant data on minerals relevant to geophysics, *Rev. Geophys. Space Phys.*, **6**, 491–524, 1968.

Anderson, O. L., A. Chopelas, and R. Boehler, Thermal expansion vs. pressure at constant temperature: a re-examination, *Geophys. Res. Lett.*, **17**, 685–688, 1990.

Beauchesne, S., and J.-P. Poirier, Creep of barium titanate perovskite: a contribution to a systematic approach to the viscosity of the lower mantle, *Phys. Earth Planet. Inter.*, **55**, 187–199, 1989.

Bloxham, J., Length-of-day variations, topographic core-mantle coupling, and the steady and time-dependent components of core-flow, *EOS Trans. AGU*, **72**, 451, 1991.

Borch, R. S., and H. W. Green II, Dependence of creep in olivine on homologous temperature and its implications for flow in the mantle, *Nature*, **330**, 345–348, 1987.

Bullen, K. E., A suggested new 'seismological' latitude, *Mon. Not. R. Astr. Soc., Geophys. Suppl.*, **4**, 158–164, 1937.

Bullen, K. E., An Earth model based on a compressibility-pressure hypothesis, *Mon. Not. R. Astr. Soc., Geophys. Suppl.*, **6**, 50–59, 1950.

Cathles, L. M., *The Viscosity of the Earth's Mantle*, Ph.D. dissertation, Princeton University, Princeton, N.J., 1971.

Chopelas, A., New accurate sound velocity measurements of lower mantle materials at very high pressures, *EOS Trans. AGU*, **69**, 1460, 1988.

Chopelas, A., Thermal expansion, heat capacity, and entropy of MgO at mantle pressures, *Phys. Chem. Minerals*, **17**, 249–257, 1990.

Chopelas, A., and R. Boehler, Thermal expansion measurements at very high pressure, systematics, and a case for a chemically homogeneous mantle, *Geophys. Res. Lett.*, **16**, 1347–1350, 1989.

Christensen, U., Convection with pressure- and temperature-dependent non-Newtonian rheology, *Geophys. J. R. Astr. Soc.*, **77**, 343–384, 1984.

Clayton, R. W., and R. P. Comer, A tomographic analysis of mantle heterogeneities from body wave travel time data, *EOS Trans. AGU*, **64**, 776, 1983.

Doornbos, D. J., and T. Hilton, Models of the core-mantle boundary and the travel times of internally reflected core phases, *J. Geophys. Res.*, **94**, 15,741–15,751, 1989.

Duffy, T. S., and T. J. Ahrens, Sound velocities at high pressure and temperature and their geophysical implications, *J. Geophys. Res.*, **97**, 4503–4520, 1992.

Durek, J. J., M. H. Ritzwoller, and J. H. Woodhouse, Estimating aspherical Q in the upper mantle using surface wave amplitude data, *EOS Trans. AGU*, **70**, 1212, 1989.

Dziewonski, A. M., Resolution of large scale velocity anomalies in the mantle, *EOS Trans. AGU*, **56**, 395, 1975.

Dziewonski, A. M., Mapping the lower mantle, *EOS Trans. AGU*, **63**, 1035, 1982.

Dziewonski, A. M., Mapping the lower mantle: determination of lateral heterogeneity in P velocity up to degree and order 6, *J. Geophys. Res.*, **89**, 5929–5952, 1984.

Dziewonski, A. M., and F. Gilbert, The effect of small, aspherical perturbations on travel times and a re-examination of the corrections for ellipticity, *Geophys. J. R. Astr. Soc.*, **44**, 7–18, 1976.

Dziewonski, A. M., B. H. Hager, and R. J. O'Connell, Large-scale heterogeneities in the lower mantle, *J. Geophys. Res.*, **82**, 239–255, 1977.

Dziewonski, A. M., and D. L. Anderson, Preliminary reference Earth model, *Phys. Earth Planet. Inter.*, **25**, 297–356, 1981.

Dziewonski, A. M., and J. Steim, Dispersion and attenuation of mantle waves through wave-form inversion, *Geophys. J. R. Astr. Soc.*, **70**, 503–527, 1982.

Dziewonski, A. M., and J. H. Woodhouse, Global images of the earth's interior, *Science*, **236**, 37–48, 1987.

Dziewonski, A. M. and R. L. Woodward, Acoustic imaging at the planetary scale, *Acoustical Imaging*, **19**, 785–797, 1992.

Dziewonski, A. M., W.-J. Su, and R. L. Woodward, A corner at 3,000 km in the spectrum of heterogeneity, *EOS Trans. AGU*, **73**, 200, 1992.

Fjeldskaar, W., and L. M. Cathles, Rheology of mantle and lithosphere inferred from post-glacial uplift in Fennoscandia, in *Glacial Isostasy, Sea-Level and Mantle Rheology*, edited by R. Sabadini, K. Lambeck, and E. Boschi, pp. 1–19, NATO ASI Series *334*, Kluwer, Boston, 1991.

Forte, A. M., *Mantle Convection and Global Geophysical Observables*, Ph.D. dissertation, University of Toronto, Toronto, 1989.

Forte, A. M., and W. R. Peltier, Surface plate kinematics and mantle convection, in *The Composition, Structure, and Dynamics of the Lithosphere-Asthenosphere System*, edited by K. Fuchs and C. Froidevaux, pp. 125–136, AGU Geodynamic Series *16*, Washington D.C., 1987a.

Forte, A. M., and W. R. Peltier, Plate tectonics and aspherical Earth structure: The importance of poloidal-toroidal coupling, *J. Geophys. Res.*, **92**, 3645–3679, 1987b.

Forte, A. M., and W. R. Peltier, Core-mantle boundary topography and whole-mantle convection, *Geophys. Res. Lett.*, **16**, 621–624, 1989.

Forte, A. M., and W. R. Peltier, Mantle convection and core-mantle boundary topography: Explanations and implications, *Tectonophysics*, **187**, 91–116, 1991a.

Forte, A. M., and W. R. Peltier, Viscous flow models of global geophysical observables. I. Forward problems, *J. Geophys. Res.*, **96**, 20,131–20,159, 1991b.

Forte, A. M., and W. R. Peltier, Gross Earth data and mantle convection: New inferences of mantle viscosity, in *Glacial Isostasy, Sea-Level and Mantle Rheology*, edited by R. Sabadini, K. Lambeck, and E. Boschi, pp. 425–444, NATO ASI Series *334*, Kluwer, Boston, 1991c.

Forte, A. M., W. R. Peltier, and A. M. Dziewonski, Inferences of mantle viscosity from tectonic plate velocities, *Geophys. Res. Lett.*, **18**, 1747–1750, 1991a.

Forte, A. M., A. M. Dziewonski, and R. L. Woodward, Geodynamic implications of seismically inferred heterogeneity, *EOS Trans. AGU*, **72**, 451, 1991b.

Forte, A. M., The kinematics and dynamics of poloidal-toroidal coupling of mantle flow, *EOS Trans. AGU*, **73**, 273, 1992.

Gable, C. W., R. J. O'Connell, and B. J. Travis, Convection in three dimensions with surface plates, *J. Geophys. Res.*, **96**, 8391–8405, 1991.

Giardini, D., X.-D. Li, and J. H. Woodhouse, Three dimensional structure of the Earth from splitting in free oscillation spectra, *Nature*, **325**, 405–411, 1987.

Giardini, D., X.-D. Li, and J. H. Woodhouse, The splitting functions of long period normal modes of the Earth, *J. Geophys. Res.*, **93**, 13,716–13,742, 1988.

Goldreich, P., and A. Toomre, Some remarks on polar wandering, *J. Geophys. Res.*, **74**, 2555–2567, 1969.

Gudmundsson, O., *Some Problems in Global Tomography: Modeling the Core-Mantle Boundary and Statistical Analysis of Travel-Time Data*, Ph.D. dissertation, California Institute of Technology, Pasadena, 1989.

Hager, B. H., The viscosity profile of the mantle: a comparison of models on postglacial and convection time scales, *EOS Trans. AGU*, **71**, 1567, 1990.

Hager, B. H., and R. J. O'Connell, A simple global model of plate dynamics and mantle convection, *J. Geophys. Res.*, **86**, 4843–4867, 1981.

Hager, B. H., R. W. Clayton, M. A. Richards, R. P. Comer, and A. M. Dziewonski, Lower mantle heterogeneity, dynamic topography and the geoid, *Nature*, **313**, 541–545, 1985.

Hager, B. H., and R. W. Clayton, Constraints on the structure of mantle convection using seismic observations, flow models, and the geoid, in *Mantle Convection*, edited by W. R. Peltier, pp. 657–763, Gordon and Breach Science Publishers, 1989.

Hager, B. H., and M. A. Richards, Long-wavelength variations in the Earth's geoid: physical models and dynamical implications, *Phil. Trans. R. Soc. Lond. A*, **328**, 309–327, 1989.

Haskell, N. A., The motion of a viscous fluid under a surface load, *Physics*, **6**, 265–269, 1935.

Haskell, N. A., The motion of a viscous fluid under a surface load, Part II., *Physics*, **7**, 56–61, 1936.

Haskell, N. A., The viscosity of the asthenosphere, *Amer. J. Sci.*, **33**, 22–28, 1937.

Hide, R., Interaction between the Earth's liquid core and solid mantle, *Nature*, **222**, 1055–1056, 1969.

Hide, R., Fluctuations in the Earth's rotation and the topography of the core-mantle interface, *Phil. Trans. R. Soc. Lond. A*, **328**, 351–363, 1989.

Hinderer, J., H. Legros, D. Jault, and J.-L. Le Mouël, Core-mantle topographic torque: a spherical harmonic approach and implications for the excitation of the Earth's rotation by core motions, *Phys. Earth Planet. Inter.*, **59**, 329–341, 1990.

Inoue, H., Y. Fukao, K. Tanabe, and Y. Ogata, Whole mantle P-wave travel time tomography, *Phys. Earth Planet. Inter.*, **59**, 294–328, 1990.

Isaak, D. G., O. L. Anderson, T. Goto, and I. Suzuki, Elasticity of single-crystal forsterite measured up to 1700 K, *J. Geophys. Res.*, **94**, 10,637–10,646, 1989.

Jarvis, G. T., and W. R. Peltier, Mantle convection as a boundary layer phenomenon, *Geophys. J. R. Astr. Soc.*, **68**, 385–424, 1982.

Jarvis, G. T., and W. R. Peltier, Lateral heterogeneity in the convecting mantle, *J. Geophys. Res.*, **91**, 435–451, 1986.

Jarvis, G. T., and W. R. Peltier, Long wavelength features of mantle convection, in *Mathematical Geophysics*, edited by N. J. Vlaar, G. Nolet, M. J. R. Wortel, and S. A. P. L. Cloetingh, pp. 209–226, D. Reidel Publishing Co., 1988.

Jault, D., and J.-L. Le Mouël, Core-mantle boundary shape: constraints inferred from the pressure torque acting between the core and mantle, *Geophys. J. Int.*, **101**, 233–241, 1990.

Jeanloz, R., and F. M. Richter, Convection, composition, and thermal state of the lower mantle, *J. Geophys. Res.*, **84**, 5497–5504, 1979.

Jeffreys, H., On the hydrostatic theory of the figure of the Earth, *Geophys. J. R. Astr. Soc.*, **8**, 196–202, 1963.

Jordan, T. H., The continental tectosphere, *Rev. Geophys. Space Phys.*, **13**, 1–12, 1975.

Jordan, T. H., Mineralogies, densities, and seismic velocities of garnet lherzolites and their geophysical implications, in *The Mantle Sample: Inclusions in Kimberlites and Other Volcanics*, edited by F. R. Boyd and H. O. A. Meyer, pp. 1–14, AGU Publications, 1979.

Karato, S.-I., M. S. Paterson, and J. D. Fitzgerald, Rheology of synthetic olivine aggregates: Influence of grain size and water, *J. Geophys. Res.*, **91**, 8151–8176, 1986.

Kohlstedt, D. L., and P. Hornack, Effect of oxygen partial pressure on the creep of olivine, in *Anelasticity in the Earth*, edited by F. D. Stacey, M. S. Paterson, and A. Nicolas, pp. 101–107, AGU Geodynamic Series *4*, 1981.

Li, X.-D., *The Asphericity of the Earth from Free Oscillations*, Ph.D. dissertation, Harvard University, Cambridge, Massachusetts, 1990.

Li, X.-D., D. Giardini, and J. H. Woodhouse, Large-scale three-dimensional even-degree structure of the Earth from splitting of long-period normal modes, *J. Geophys. Res.*, **96**, 551–577, 1991.

MacDonald, G. J. F., The figure and long term mechanical properties of the Earth, in *Advances in Earth Sciences*, edited by P. M. Hurley, pp. 199–245, MIT Press, 1965.

Marsh, J. G., F. J. Lerch, B. H. Putney, T. L. Felsentreger, B. V. Sanchez, S. M. Klosko, G. B. Patel, J. W. Robbins, R. G. Williamson, T. L. Engelis, W. F. Eddy, N. L. Chandler, D. S. Chinn, S. Kapoor, K. E. Rachlin, L. E. Braatz, and E. C. Pavlis, The GEM-T2 gravitational model, *J. Geophys. Res.*, **95**, 22,043–22,071, 1990.

Masters, G., and F. Gilbert, Structure of the inner core inferred from observations of its spheroidal shear modes, *Geophys. Res. Lett.*, **8**, 569–571, 1981.

Masters, G., T. H. Jordan, P. G. Silver, and F. Gilbert, Aspherical earth structure from fundamental spheroidal-mode data, *Nature*, **298**, 609–613, 1982.

McKenzie, D. P., The viscosity of the lower mantle, *J. Geophys. Res.*, **71**, 3995–4010, 1966.

Minster, J. B., and T. H. Jordan, Present-day plate motions, *J. Geophys. Res.*, **83**, 5331–5354, 1978.

Montagner, J.-P., and T. Tanimoto, Global upper mantle tomography of seismic velocities and anisotropies, *J. Geophys. Res.*, **96**, 20,337–20,351, 1991.

Morelli, A., and A.M. Dziewonski, 3D structure of the Earth's core inferred from travel time residuals, *EOS Trans. AGU*, **67**, 311, 1986.

Morelli, A., and A. M. Dziewonski, Topography of the core-mantle boundary and lateral homogeneity of the liquid core, *Nature*, **325**, 678–683, 1987.

Morelli, A., and A. M. Dziewonski, Joint determination of lateral heterogeneity and earthquake location, in *Glacial Isostasy, Sea-Level and Mantle Rheology*, edited by R. Sabadini, K. Lambeck, and E. Boschi, pp. 515–534, NATO ASI Series, Kluwer, Boston, 1991.

Munk, W. H., and G. J. F. MacDonald, *The Rotation of the Earth*, Cambridge University Press, 1960.

Nakanishi, I., and D. L. Anderson, Worldwide distribution of group velocity of mantle Rayleigh waves as determined by spherical harmonic inversion, *Bull. Seism. Soc. Am.*, **72**, 1185–1194, 1982.

Nakanishi, I., and D. L. Anderson, Measurement of mantle wave velocities and inversion for lateral heterogeneity and anisotropy, I. Analysis of great circle phase velocities, *J. Geophys. Res.*, **88**, 10,267–10,283, 1983.

Nakanishi, I., and D. L. Anderson, Measurement of mantle wave velocities and inversion for lateral heterogeneity and anisotropy, II. Analysis by the single station method, *Geophys. J. R. Astron. Soc.*, **78**, 573–618, 1984.

Nakiboglu, S. M., Hydrostatic theory of the Earth and its mechanical implications, *Phys. Earth Planet. Int.*, **28**, 302–311, 1982.

Nataf, H.-C., I. Nakanishi, and D. L. Anderson, Anisotropy and shear-velocity heterogeneities in the upper mantle, *Geophys. Res. Lett.*, **11**, 109–112, 1984.

Nataf, H.-C., I. Nakanishi, and D. L. Anderson, Measurements of mantle wave velocities and inversion for lateral heterogeneities and anisotropy, 3. Inversion, *J. Geophys. Res.*, **91**, 7261–7307, 1986.

Nicolas, A., and J.-P. Poirier, *Crystalline Plasticity and Solid State Flow in Metamorphic Rocks*, John Wiley and Sons, 1976.

O'Connell, R. J., Pleistocene glaciation and the viscosity of the lower mantle, *Geophys. J. R. Astr. Soc.*, **23**, 299–327, 1971.

Pekeris, C. L., Thermal convection in the interior of the Earth, *Mon. Not. R. Astr. Soc., Geophys. Suppl.*, **3**, 343–367, 1935.

Peltier, W. R., Surface plates and thermal plumes: separate scales of the mantle convective circulation, in *The Thermal and Chemical Evolution of the Mantle*, edited by R. J. O'Connell and W. S. Fyfe, pp. 59–77, AGU Geodyn. Series, **6**, 1981.

Peltier, W. R., and J. T. Andrews, Glacial-isostatic adjustment-I. The forward problem, *Geophys. J. R. Astr. Soc.*, **46**, 605–646, 1976.

Peltier, W. R., and A. M. Forte, The gravitational signature of plate tectonics, *Terra Cognita*, **4**, 251, 1984.

Peterson, J., H. M. Butler, L. G. Holcomb, and C. R. Hutt, The seismic research observatory, *Bull. Seismol. Soc. Am.*, **66**, 2049–2068, 1976.

Reynard, B., and G. D. Price, Thermal expansion of mantle minerals at high pressure - a theoretical study, *Geophys. Res. Lett.*, **17**, 689–692, 1990.

Ribe, N. M., A thin-shell model for the interaction of surface plates and convection, *EOS Trans. AGU*, **73**, 273, 1992.

Ricard, Y., L. Fleitout, and C. Froidevaux, Geoid heights and lithospheric stresses for a dynamic Earth, *Ann. Geophys.*, **2**, 267–286, 1984.

Ricard, Y., C. Froidevaux, and L. Fleitout, Global plate motion and the geoid: a physical model, *Geophys. J. Int.*, **93**, 477–484, 1988.

Ricard, Y., and C. Vigny, Mantle dynamics with induced plate tectonics, *J. Geophys. Res.*, **94**, 17,543–17,559, 1989.

Ricard, Y., C. Doglioni, and R. Sabadini, Differential rotation between lithosphere and mantle: A consequence of lateral mantle viscosity variations, *J. Geophys. Res.*, **96**, 8407–8415, 1991.

Richards, M. A., and B. H. Hager, Geoid anomalies in a dynamic Earth, *J. Geophys. Res.*, **89**, 5987–6002, 1984.

Richards, M. A., and B. H. Hager, Effects of lateral viscosity variations on long-wavelength geoid anomalies and topography, *J. Geophys. Res.*, **94**, 10,299–10,313, 1989.

Ricoult, D. L., and D. L. Kohlstedt, Experimental evidence for the effect of chemical environment upon the creep rate of olivine, in *Point Defects in Minerals*, edited by R. N. Shock, pp. 171–184, AGU Geophys. Monogr. Ser. *31*, 1985.

Ritzwoller, M., G. Masters, and F. Gilbert, Observations of anomalous splitting and their interpretation in terms of aspherical structure, *J. Geophys. Res.*, **91**, 10,203–10,228, 1986.

Ritzwoller, M., G. Masters, and F. Gilbert, Constraining aspherical Earth structure with low-degree interaction coefficients: application to uncoupled multiplets, *J. Geophys. Res.*, **93**, 6369–6396, 1988.

Roberts, P. H., On topographic core-mantle coupling, *Geophys. Astrophys. Fluid Dynamics*, **44**, 181–187, 1988.

Romanowicz, B. A., The upper mantle degree 2: constraints and inferences from global mantle wave attenuation measurements, *J. Geophys. Res.*, **95**, 11051–11071, 1990.

Romanowicz, B., Seismic Tomography of the Earth's Mantle, *Annu. Rev. Earth Planet. Sci.*, **19**, 77–99, 1991.

Ryerson, F. J., W. B. Durham, D. J. Cherniak, and W. A. Lanford, Oxygen diffusion in olivine: Effect of oxygen fugacity and implications for creep, *J. Geophys. Res.*, **94**, 4105–4118, 1989.

Solheim, L. P., and W. R. Peltier, Heat transfer and the onset of chaos in a spherical, axisymmetric anelastic model of whole mantle convection, *Geophys. Astrophys. Fluid Dyn.*, **53**, 205–255, 1990.

Speith, M. A., R. Hide, R. W. Clayton, B. H. Hager, and C. V. Voorhies, Topographic coupling of core and mantle, and changes in length of day, *EOS Trans. AGU*, **67**, 908, 1986.

Stacey, F. D., and D. E. Loper, The thermal boundary-layer interpretation of D" and its role as a plume source, *Phys. Earth Planet. Int.*, **33**, 45–55, 1983.

Stewart, C. A., Thermal convection in the Earth's mantle: Mode coupling induced by temperature-dependent viscosity in a 3-dimensional spherical shell, *Geophys. Res. Lett.*, **19**, 337–340, 1992.

Stocker, R. L., and M. F. Ashby, On the rheology of the upper mantle, *Rev. Geophys. Space Phys.*, **11**, 391–426, 1973.

Su, W.-J., and A. M. Dziewonski, Predominance of long-wavelength heterogeneity in the mantle, *Nature*, **352**, 121–126, 1991.

Tanimoto, T., The three dimensional shear wave structure in the mantle by overtone waveform inversion I. Radial seismogram inversion, *Geophys. J. R. Astr. Soc.*, **89**, 713–740, 1987.

Tanimoto, T., Long-wavelength S-wave velocity structure throughout the mantle, *Geophys. J. Int.*, **100**, 327–336, 1990.

Turcotte, D. L., and E. R. Oxburgh, Finite amplitude convective cells and continental drift, *J. Fluid Mech.*, **28**, 29–42, 1967.

Turcotte, D. L., and G. Schubert, Structure of the olivine-spinel phase boundary in the descending lithosphere, *J. Geophys. Res.*, **76**, 7980–7987, 1971.

Verhoogen, J., Phase changes and convection in the Earth's mantle, *Phil. Trans. R. Soc. Lond. A*, **258**, 276–283, 1965.

Wahr, J. M., The effects of the atmosphere and oceans on the Earth's wobble-I. Theory, *Geophys. J. R. Astron. Soc.*, **70**, 349–372, 1982.

Weertman, J., Creep laws for the mantle of the Earth, *Phil. Trans. R. Soc. Lond. A*, **288**, 9–26, 1978.

Weertman, J., and J. R. Weertman, High temperature creep of rock and mantle viscosity, *Annual Rev. Earth Planet. Sci.*, **3**, 293–315, 1975.

Wong, Y. K., *Upper Mantle Heterogeneity from Phase and Amplitude Data of Mantle Waves*, Ph.D. Dissertation, Harvard University, Cambridge, Massachusetts, 1989.

Woodhouse, J. H., and A. M. Dziewonski, Mapping the upper mantle: three-dimensional modeling of earth structure by inversion of seismic waveforms, *J. Geophys. Res.*, **89**, 5953–5986, 1984.

Woodhouse, J. H., and D. Giardini, Inversion for the splitting function of isolated low order normal mode multiplets, *EOS Trans. AGU*, **66**, 301, 1985.

Woodhouse, J. H., and A. M. Dziewonski, Three dimensional mantle models based on mantle wave and long period body wave data, *EOS Trans. AGU*, **67**, 307, 1986.

Woodhouse, J. H., and A. M. Dziewonski, Seismic modeling of the Earth's large-scale three-dimensional structure, *Phil. Trans. R. Soc. Lond. A*, **328**, 291–308, 1989.

Woodward, R. L., and G. Masters, Global upper mantle structure from long-period differential travel times, *J. Geophys. Res.*, **96**, 6351–6377, 1991a.

Woodward, R. L., and G. Masters, Lower mantle structure from $ScS - S$ differential travel times, *Nature*, **352**, 231–233, 1991b.

Woodward, R. L., and G. Masters, Upper mantle structure from long-period differential traveltimes and free oscillation data, *Geophys. J. Int.*, **109**, 275–293, 1992.

Woodward, R. L., A. M. Forte, W.-J. Su, and A. M. Dziewonski, Constraints on the large-scale structure of the Earth's mantle, in *Chemical Evolution of the Earth and Planets*, edited by Takahashi *et al.*, pp. , J. Geophys. Res., in press, 1992.

Yeganeh-Haeri, A., D. J. Weidner, and E. Ito, Elasticity of $MgSiO_3$ in the perovskite structure, *Science*, **243**, 787–789, 1989.

Yuen, D. A., and W. R. Peltier, Mantle plumes and the thermal stability of the D'' layer, *Geophys. Res. Lett.*, **7**, 625–628, 1980.

Yuen, D. A., A. M. Leitch, and U. Hansen, Dynamical influences of pressure-dependent thermal expansiity on mantle convection, in *Glacial Isostasy, Sea-Level, and Mantle Rheology*, edited by R. Sabadini, K. Lambeck, and E. Boschi, pp. 663–701, NATO ASI Series, Kluwer, Boston, 1991.

A. M. Dziewonski, A. M. Forte, and R. L. Woodward, Department of Earth and Planetary Sciences, Harvard University, 20 Oxford St., Cambridge, MA, 02138, USA.

Constraints on Mantle Structure from Seismological and Convection Results

PHILIPPE MACHETEL

U.P.R. 234, C.N.R.S., Groupe de Recherche de Geodesie Spatiale, Toulouse, France

An endothermic phase change at the 670 km depth discontinuity is able to modify the structure of convection. Depending upon the value of the Clapeyron slope, mantle convection can be organized in one, two or one-and-an-half layers. In the course of this work, an endeavor is made to compare the theoretical results given by mantle convection models with seismological data. The spherical axisymmetrical numerical models assume an anelastic compressible Earth where the 670 km depth phase change is taken into account. These models predict chaotic behaviors for the mixing rate of upper and lower mantles which is found to strongly depend on the value of the Clapeyron slope. The chaotic regimes are associated with short-wavelength thermal anomalies which spectra display Kolmogorov slopes (-5/3). The comparison of these slopes with those obtained in recent seismological models from surface waves let to think that the structure of the mantle is layered or almost layered. In spite of filtering effects due to a weak seismic resolution, the depth dependence of the thermal anomalies seems hardly compatible with a one-layer convection (whole mantle convection). The propagation of P-waves in the short-wavelength thermal anomalies due to chaotic convection has been studied but the interpretation of results is strongly depending on local behaviors and do not allow to argue for or against layered convection in the mantle. However, our results are clues to indicate that the mantle is likely layered or at least intermittently layered.

INTRODUCTION

The existence of an endothermic phase change at the 670 km discontinuity is able to alter the global structure of mantle convection [Christensen and Yuen, 1984; 1985; Machetel and Weber, 1991]. The debate about this global structure is not definitely settled but several geophysical results indicate that the mantle is layered or weakly layered [i.e; Galer and Goldstein, 1991]. Considering that the purpose of this paper is not to compile the geophysical data arguing for or against mantle layering we would like to invite the reader to refer to review works [i.e. Zindler and Hart, 1986; Silver et al, 1988; Nataf, 1989; Olson et al, 1990]. According to the value of the Clapeyron slope, three kinds of regime have been proposed by Machetel and Weber [1991] : 1, a one-layer pattern with sinking currents going down from the top to the bottom of the shell; 2, an intermittent regime for a Clapeyron slope $\gamma = -2$ 10^6 Pa/K, with a layered structure almost everywhere and intermittent penetrations of upper mantle material deep into the lower mantle; 3, a perfectly layered structure, persistent with time, with no mass transfer between the upper and lower mantles reservoirs. These results were found in a restricted but realistic range of mantle parameters. The present work offers a broader exploration of geophysical parameters such as the internal heating rate or the bottom mantle temperature, and an endeavor to study the effects of these structures on seismic wave propagation. The thermal fields obtained during this study result of direct approaches of mantle properties [i.e. computational dimensionless parameters are not arbitrary imposed but calculated, for each case, from geophysical values, given in Table 1). Thus, the Rayleigh numbers are much higher than the ones considered in previous studies during which low Rayleigh number chaotic behaviors were established as consequences of compressibility and of internal heating [Machetel and Yuen, 1989]. Short-wavelength thermal heterogeneities arise from the chaotic fluid motion. In this work, the temperature field will be considered as a rough first estimate of seismic velocity anomalies [Honda, 1987; Machetel, 1990]. The numerical and mathematical difficulties of the problem impose simplifying assumptions such as constant viscosity or constant thermal expansion; however, the numerical processing takes into account a spherical, axisymmetrical geometry, mantle compressibility, realistic rates of radiogenic heating and a phase change at the 670 km discontinuity.

EQUATIONS OF CONVECTION WITH PHASE CHANGE

The temperature, pressure, and phase change variations of density have been introduced in the equations of compressible mantle convection through the equation of state (1).

Dynamics of Earth's Deep Interior and Earth Rotation
Geophysical Monograph 72, IUGG Volume 12

TABLE 1. Geophysical Values used in this Study

Description	Variable	Units	Value	Ref
Acceleration of gravity	g	m/s^2	10	
Thermal expansion	α	K^{-1}	1.4 10^{-5}	8
Sub-plate temperature	T_o	K	1400	2
CMB temperature	T_i	K	3-4 10^3	3
Sub-plate radius	r_o	m	6.291 10^6	1
CMB radius	r_i	m	3.480 10^6	1
Heat capacity	C_p	J/(kg K)	1.25 10^3	2
Thermal diffusivity	k	m^2/s	0.8 10^{-6}	4
Dynamic viscosity	η	kg/(m s)	10^{22}	5
Chondritic internal heating	H	J/(kg s)	5.3 10^{-12}	6
Sub-plate density	ρ_o	kg/m^3	3.37 10^3	1
Mid-depth density	ρ_m	kg/m^3	4.85 10^3	1
Bulk compressibility	K_s	Pa	4.3 10^{11}	1
Density jump (670 km)	$\delta\rho_{670}$	kg/m^3	390	1
Mean density (670 km)	ρ_{670}	kg/m^3	4.18 10^3	1
670 km pressure	P_{670}	Pa	2.383 10^{10}	1
670 km temperature	T_{670}	K	2000	7
Thickness of transition	e	m	10^4	

Table 1. Geophysical values used in this paper. The subscript o refers to the upper surface values, the subscript 670 to the 670 km discontinuity and the subscript i to the bottom value. Reference are (1) : Dziewonski and Anderson (1981); (2) : Stacey (1977); (3) : Jeanloz and Morris (1986); (4) : Anderson (1989); (5) : Hager and Richards (1989); (6) : Stacey and Loper (1988); (7) : Boehler and Chopelas (1990) and (8) : Wang et al (1988).

$$\rho(T,P,X) = \rho_{aw}(r)\left(1 - \alpha(T - T_s) + K_T^{-1}(P - P_H)\right) + \delta\rho_{670}\Gamma(X) \quad (1)$$

$$X = \frac{P - P_{670} - \gamma(T - T_{670})}{g\,\rho_{670}\,e} \quad (2)$$

All the notations of the paper are given in Table 2. The density profile $\rho_{aw}(r)$ in Equation (1) has been obtained with a classical Adams-Williamson compression from the sub-lithospheric density [Jarvis and McKenzie, 1980; Machetel and Yuen, 1989]. The function G(X) varies abruptly but continuously from zero to one in a narrow range of temperatures and pressures around the 670 km depth phase change.

$$\Gamma(X) = \frac{1}{2}\big(1 + \tanh(X)\big) \quad (3)$$

To define a reference density profile which takes into account the mean density jump of the phase change, $\Gamma(X)$ has been expanded around the value X_H as $\Gamma(X_H + h)$.

$$X_H = \frac{P_H - P_{670}}{g\,\rho_{670}\,e}\;;\; h = \frac{P_d - \gamma(T - T_{670})}{g\,\rho_{670}\,e} \quad (4)$$

$$\Gamma(X_H + h) = \Gamma(X_H)\big(1 + 2h\,(1 - \Gamma(X_H))\big) \quad (5)$$

Thus, it is possible to establish an equation of state taking into account the thermal and dynamical perturbations around this reference.

$$\rho(P,T) = \rho_{ref}(r)\left(1 - \alpha_\Phi(T - T_s) + K_{T\Phi}^{-1}(P - P_H) + \gamma_\Phi\,h\right) \quad (6)$$

In Equation (6) ρ_{ref}, α_F, $K_{T\Phi}^{-1}$ and γ are equal to :

$$\rho_{ref}(r) = \rho_{aw}(r) + \delta\rho_{670}\,\Gamma(X_H) \quad (7)$$

$$\alpha_\Phi = \alpha\frac{\rho_{aw}(r)}{\rho_{ref}(r)} \quad (8)$$

$$K_{T\Phi}^{-1} = K_T^{-1}\frac{\rho_{aw}(r)}{\rho_{ref}(r)} \quad (9)$$

$$\gamma_\Phi = 2\,\Gamma(X_H)(1 - \Gamma(X_H))\frac{\delta\rho_{670}}{\rho_{ref}(r)} \quad (10)$$

The reference density, given by Equation (7) has been borne in the classical conservation equation (11) in which sonic waves have been neglected.

$$\text{div}(\rho_{ref}\mathbf{v}) = 0 \quad (11)$$

TABLE 2. Notations used in this Paper

Variable	Description	Units
ρ	Density	kg/m^3
r	Radius	m
r_o	Top radius	m
r_i	Bottom radius	m
Θ	Colatitude	radian
α	Thermal expansion	K^{-1}
T	Temperature	K
T_o	Top temperature	K
T_i	Bottom temperature	K
T_s	Adiabatic temperature	K
K_T	Bulk compressibility	Pa^{-1}
P	Pressure	Pa
P_H	Hydrostatic pressure	Pa
P_{670}	Hydrostatic pressure at 670 km	Pa
T_{670}	Temperature at 670 km	K
ρ_{670}	Mean density at 670 km	kg/m^3
$\delta\rho_{670}$	670 km density jump	kg/m^3
γ	Clapeyron slope	Pa/K
g	Acceleration of gravity	m/s^2
e	Thickness of transition	m
η	Dynamic viscosity	kg/(m s)
k	Thermal diffusivity	m^2/s
H	Internal heating rate	J/(kg s)

Table 2. Notations used in this paper. The subscript o refers to the upper surface values, the subscript 670 to the 670 km depth discontinuity and the subscript i to the CMB.

The equation of state (6) has been borne in the motion equation (12) in which inertial terms have been neglected (infinite Prandtl number approximation).

$$\rho(P,T)\ \mathbf{g} - \mathrm{grad}(P) + \mathrm{div}(\tau) = 0 \qquad (12)$$

The expression of the deviatoric stress tensor τ is given by :

$$\tau_{ij} = \eta\ (\mathrm{grad}(\mathbf{v}) + \mathrm{grad}(\mathbf{v})^{T}) - \frac{2}{3}\ \eta\ \mathrm{div}(\mathbf{v})\ \delta_{ij} \qquad (13)$$

The latent heat release Q_l of the phase change is related to the Clapeyron slope by the relationship (14).

$$Q_l = \frac{T\ \gamma\ \delta\rho_{670}}{\rho_{670}^{2}} \qquad (14)$$

This latent heat has been included in the energy balance equation (15).

$$\rho_{ref}\ C_p\ \frac{d(T - T_s)}{dt} - \rho_{ref}\ Q_l\ \frac{d\Gamma}{dt} = F + \rho_o\ C_p\ k\ \nabla^2 T + \rho_{ref}\ H \qquad (15)$$

Where F represents the viscous dissipation.

$$\Phi = \tau : \nabla \mathbf{v} \qquad (16)$$

The resulting equations (11), (12), (13), (15) and (16) have been expressed in axisymmetrical spherical geometry following the conventions described by Machetel and Yuen [1989].

COMPUTATIONAL APPROACH

As mentioned above, the numerical calculations have been performed in terms of dimensionless variables which will appear only in this section. All the results of this paper will be given in SI (or directly derived) Units The scaling factors used for distances, times, velocities, temperatures and pressures; and the related dimensionless numbers are given in Table 3. But, in the following, the prime convention for dimensionless variables will be omitted.

The dimensionless temperature equation becomes :

$$\rho_{ref}\left(\frac{\partial T}{\partial t} + \mathbf{v}.\mathrm{grad}(T) + D(T+\tau_0)\left(v_r + \frac{\partial\Gamma}{\partial X}\frac{\bar{\gamma}\ \Delta\rho_{670}}{\mathrm{Ra}\ e\ \rho_{670}^{3}}\left(\frac{\bar{\gamma}}{\mathrm{Ra}}\mu\frac{dT}{dt} + \rho_{ref}v_r\right)\right)\right)$$
$$= \frac{D}{\mathrm{Ra}}\ \Phi + \nabla^2 T + \rho_{ref}\ R \qquad (17)$$

Where $\bar{\gamma}$ is the dimensionless Clapeyron slope and $\mu = \alpha\ \Delta T$. Ra, the Rayleigh number, D, the dissipation number, and R the dimensionless internal heating are defined in Table 3. The dimensionless continuity equation becomes :

$$\mathrm{div}(\rho_{ref}\ \mathbf{v}) = 0 \qquad (18)$$

TABLE 3. Scaling Conventions and dimensionless Numbers

Notation	Description	Expression
	Distances	$r = d\ r'\ ;\ d = r_o - r_i$
	Times	$t = \dfrac{d^2}{k}\ t'$
	Velocities	$v = \dfrac{k}{v}\ v'$
	Pressures	$P = \dfrac{k\ \eta}{d^2}\ P'$
	Temperatures	$T = \Delta T\ T' + T_o\ ;\ \Delta T = T_i - T_o$
D	Dissipation number	$\dfrac{d\ g\ \alpha}{C_p}$
Ra	Rayleigh number	$\dfrac{g\ \alpha\ \Delta T\ d^3\ \rho_o}{k\ \eta}$
$\bar{\gamma}$	Clapeyron slope	$\dfrac{g\ d^2\ \Delta T}{k\ \eta}$
R	Internal heating	$\dfrac{d^2 H}{C_p\ k\ \Delta T}$

Table 3. Scaling conventions used for the numerical computations and resulting dimensionless parameters.

Then, the velocity can be obtained from the spatial derivatives of the stream-function ψ :

$$v_r = \frac{1}{\rho_{ref}\ r^2\ \sin\theta}\frac{\partial\psi}{\partial\theta},\ v_\theta = \frac{-1}{\rho_{ref}\ r\ \sin\theta}\frac{\partial\psi}{\partial r} \qquad (19)$$

and the one-component vorticity vector is equivalent to a scalar quantity ω:

$$\omega = r\ \sin\theta\ \mathrm{curl}(\mathbf{v}) \qquad (20)$$

Combination of Equations (19) and (20) gives :

$$D^2(\psi) = -\rho_{ref}\ \omega \qquad (21)$$

Where is the derivative operator :

$$D^2 = \frac{\partial^2}{\partial r^2} + \frac{1}{r^2}\frac{\partial^2}{\partial\theta^2} - \frac{\cot\theta}{r^2}\frac{\partial}{\partial\theta} - \frac{1}{\rho_{ref}}\frac{d\rho_{ref}}{dr}\frac{\partial}{\partial r} \qquad (22)$$

The vorticity equation (23) is obtained by taking the curl of the dimensionless form of Equation (12) in which the latitudinal derivative of pressure has been calculated by projecting out the θ component [Machetel and Yuen, 1989].

$$D^2(\omega) = \sin\theta\left(\mathrm{Ra}\ \rho_{aw} + \frac{\Delta\rho_{670}\ \bar{\gamma}}{\rho_{670}\ e}\left(\frac{\partial\Gamma}{\partial X}\right)_{x_H}\right)\frac{\partial T}{\partial\theta} - \frac{4}{3}\left(\frac{1}{\rho_{ref}}\frac{\partial\rho_{ref}}{\partial r}\right)^2\frac{\partial v_r}{\partial\theta} \qquad (23)$$

Isothermal, zero stream-function and free-slip boundary conditions have been assumed at the top and at the bottom of the shell. Zero stream-function, free-slip, and no thermal flux

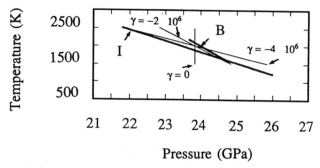

Fig. 1. Upper and lower mantle phase boundary diagram. The thin lines show the hypothesized phases boundaries in our models for $\gamma = 0$, $-2 \ 10^6$, and $-4 \ 10^6$ Pa/K when the thick lines give the phase boundaries as they can be extracted from experimental results (B: by Boehler and Chopelas (1990); I: by Ito and Katsura (1989)).

conditions have been assumed along the axisymmetrical axis. The equations (17), (21) and (23) have been solved with second-order, alternate direction, implicit finite difference methods. The convergence of the coupled system of Poisson's equations (21) and (23) has been checked each time step. The temporal behavior of the temperature equation (17) has been described thanks to a half-implicit scheme : the laplacian operator has been solved explicitly while a half-implicit scheme has been employed to include the viscous dissipation and the advection terms. The mesh includes 41553 points with 81 radial and 513 latitudinal levels.

The dimensionless parameters of Table 3 have been calculated for each case from the values given in Table 1.

The experimental and theoretical range of uncertainty about the value of the Clapeyron slope has been surrounded by setting it to 0, $-2 \ 10^6$ and $-4 \ 10^6$ Pa/K. With no thermal or dynamical perturbation, the phase change occurs at a pressure of $2.383 \ 10^{10}$ Pa for a temperature of 2000 K. The thick lines in Figure 1 give the phase boundaries as they can be extracted from experimental results by Boehler and Chopelas [1990] and by Ito and Katsura [1989]. The thin lines delimit the phase

boundaries obtained with our assumptions. Setting the Clapeyron slope to $\gamma = -2 \ 10^6$ Pa/K, the phase boundaries hypothesized in the models are quite close to experimental but also to theoretical results [Price et al, 1989].

INFLUENCE OF THE CLAPEYRON SLOPE

As emphasized by Christensen and Yuen [1984 and 1985] and, in a recent work, by Machetel and Weber [1991] an endothermic phase change at the 670 km discontinuity is able to modify the pattern of convection. In this last paper, it was shown that the global structure (one, two or one-and-a-half layers) was depending on the value of the Clapeyron slope. This series of computations were run with a half-chondritic internal heating rate and a bottom temperature of 3000 degrees Kelvin. Now we have extended these results to a higher bottom temperature and to a larger range of internal heating rates. Figure 2 displays the stream-functions and the isocontours of the thermal fields for a bottom temperature of 4000 K. The main properties of the flow structures that have been observed in our previous work [Machetel and Weber, 1991] are still valid. The layering of mantle convection occurs for a Clapeyron slope $\gamma = -4 \ 10^6$ Pa/K when no typical feature of the phase change can be detected for $\gamma = 0$. With an intermediate value of γ ($-2 \ 10^6$ Pa/K) the "one-and-a-half" structure appears again with large areas of layering and intermittent penetration of upper mantle into the lower mantle. The time-dependent rate of mixing mix(t) can be defined as :

$$\text{mix}(t) = 100 \ \frac{\text{mass flux through the discontinuity}}{\text{mass of the lower mantle}} \ 3.16 \ 10^{16} \qquad (24)$$

The temporal behavior of the mixing rate, recorded during the computations of the cases of Figure 2 are given in Figure 3. The very low rate of mixing of curve (c) tallies with a two-layer mantle when the one-layer structure involves a very efficient mixing (curve a). Now, with $\gamma = -2 \ 10^6$ Pa/K (curve b), the intermittences of layering and mixing induce broader amplitude variations but a lower time-averaged value. As a consequence of the chaotic nature of the high Rayleigh number solutions,

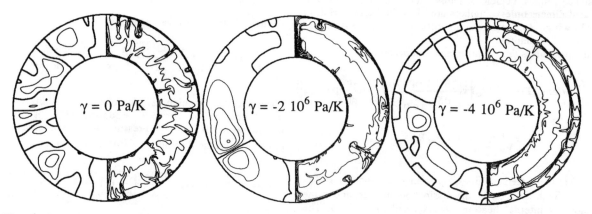

Fig. 2. Stream lines (left hemispheres) and temperature fields (right hemispheres) obtained with a bottom temperature of 4000 K and a half-chondritic internal heating rate for $\gamma = 0$ (left), $-2 \ 10^6$ (middle) and $-4 \ 10^6$ Pa/K (right). Thick lines give the cells boundaries. Isocontour level for temperature is 300 K.

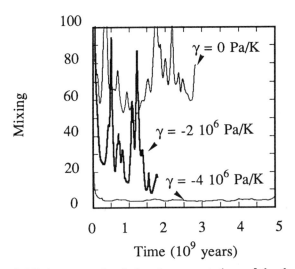

Fig. 3. Mixing properties during the computations of the three cases of Figure 2. mix(t) gives the fraction of lower mantle crossing the discontinuity by second. It has been multiplied by $3.15567 \cdot 10^{16}$ (number of seconds in one billion years) and by 100 (to obtain a percentage). With these conventions a rate of 100 constant over one billion years would mean that a quantity of matter equivalent to the lower mantle has crossed the discontinuity.

these quantities oscillate chaotically with time. However, their time-averaged values allow to better distinguish the long-term trends. For $\gamma = 0$, the rate oscillates chaotically between 50 and 100 % per billion years with a time-averaged value of 70 %. This value corresponds to a fairly high mixing since it would mean that more than half of the lower mantle crosses the discontinuity each billion years. With a more endothermic Clapeyron slope $\gamma = -2 \cdot 10^6$ Pa/K, mix(t) peaks up to 90 % (if we except the very short transient period) but lows to minima less than 10 % over short lapses of time. The brutal events of mixing, following cold material accumulations above the 670 km discontinuity, are accountable for this sudden temporal behavior. These intermittences of mixing occur roughly every half billion years, a period which is comparable to the estimated one for large scale tectonic events like the breaking of Pangea [Le Pichon and Huchon, 1984]. The time-averaged rate of mixing is now of the order of 20 % per billion years. If we could retain this rate for the mantle, that would mean that a mass equivalent to the lower mantle has crossed the 670 km discontinuity since the origin of the Earth. The third case, with $\gamma = -4 \cdot 10^6$ Pa/K, corresponds to a stratified mantle with almost no mixing between the upper and the lower parts. If this scenario was correct, the lower mantle could have retain the same composition it had when the Earth formed 4.5 billion years ago. The comparison of these results with those of Machetel and Weber [1991] shows that an increase of the bottom temperature from 3000 to 4000 K does not change the structure of mantle convection. The number of layers in the mantle, one, one-and-a-half or two depends on the value of the Clapeyron slope. This result is not depending on the rate of internal heating. The top panels of Figure 4 display the stream-functions and the isotherms obtained with a Clapeyron slope γ

= -2 10^6 Pa/K but with various rates of internal heating. From left to right, the rates of internal heating are zero, half-chondritic and chondritic. In all cases, persistent one-and-a-half layer structures were reached, as it is proved by the temporal evolutions of the mixing rates during the computations (Figure 4, bottom panels). The large peak to trough amplitudes of the mixing rates match the events of penetration of upper mantle into the lower one. Each computer run has been started from the same initial condition : the final thermal state of the case $\gamma = -2 \cdot 10^6$ Pa/K obtained by Machetel and Weber [1991]. This is true for all the runs of the paper. That explains the similarity of the temporal behavior just at the beginning of the computation. But, after a short time, this is ruled over by the chaotic nature of the flow. However, in spite of these divarications, the global structure remains the same with intermittent mixing sequences.

For the values tested here, the structure of convection does not depend on the bottom temperature or on the internal heating rate. Recently, Solheim and Peltier [1991] found that the level of layering was depending on the Rayleigh number. This result is not contradictory to ours since the Rayleigh number is also a function, among other physical values, of the coefficient of thermal expansion. The choice of the value of alpha at the discontinuity -the place where buoyancy acts to control the structure- is therefore crucial. Unfortunately they do not give this information in the paper. Finally, it can be emphasized that the final structure (one, one-and-a-half or two layers) has been reached very soon during the computations and was persistent after that. This result let to think that the structure of mantle has been acquired very early in its history and has probably remained the same through geological times.

SHORT WAVELENGTH THERMAL ANOMALIES

The chaotic nature of mantle convection induces short-wavelength temperature fluctuations [Machetel and Yuen, 1989; Machetel, 1990]. The spherical harmonic expansions of the temperature fields are given by :

$$T(r,\theta) = \sum_{l=0}^{\infty} T_l(r) \, P_l(\cos\theta) \tag{25}$$

The $l = 0$ term of theses expansions characterize the mean value of the temperature field over the spherical surface. The root mean square $RMS_l(r)$ of the spherical harmonic expansions have been computed for each degree l (except zero) as :

$$RMS_l(r) = \sqrt{\frac{1}{2l+1} T_l(r)^2} \tag{26}$$

and have been represented in the log-log diagrams of Figures 5, 6 and 7. It is reasonable to assume that, at least to a rough first order, the thermal anomalies are equivalent to seismic velocity anomalies. Then, the spectra of Figures 5, 6 and 7 are directly comparable to the RMS of Love wave phase velocity given by Tanimoto in a recent paper [1991]. For a period of 100 seconds, the maximum of sensivity of such data is reached for depth around one hundred kilometers below the Earth's surface that is to say, inside the thermal boundary layers just below the lithospheric plates. Tanimoto emphasized that a

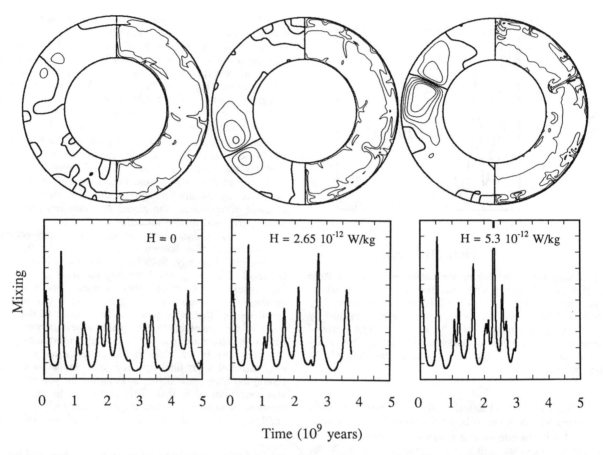

Fig. 4. Effects of the internal heating on the layering of mantle convection. Top panels display the stream-functions and the temperature fields at the end of computations. The temporal behaviors of mixing rates are given in the bottom panels The internal heating ranges from 0 W/kg (left column) to a half-chondritic rate (middle column) to a chondritic rate (right column).

slope greater than one can be found for this spectrum. Indeed, if the three first values of the expansion are excluded, a slope very close to -1.7 can be measured for the Figure 4 of Tanimoto's paper. Thus this spectrum can be considered like a turbulent temperature anomaly spectrum. Until degree 6, the harmonics characterize the large scale properties of the top of the mantle. As shown by Souriau and Souriau [1983], degrees 2 and 3 can be related to the geographical distributions of the oceans, degree 4 to the distribution of trenches and subduction zones and degrees 5 and 6 to the shields. This part of the spectrum may have no significance for the turbulent dynamics at the top of the upper mantle. The next part of the Tanimoto's spectrum corresponds to the inertial-convective range whose slope is -5/3 [Lesieur, 1987]. The viscous-convective range does not appear because the expansion of the seismic velocity anomalies is not known to sufficiently high harmonics. The present number of points used during our computations allows us to compute spectral expansions of the temperature anomalies until degree 256. Figures 5, 6 and 7 give the spectra of the RMS of thermal anomalies, obtained from our theoretical models, for various radial depths (a : 2855 km, b : 2188 km, c : 1485 km, d : 817 km, e : 115 km and f : 677 km). Thus panels a

and e are located in the thermal boundary layers just above the CMB and just below the lithospheric plates. Panels c, d and e sample the lower mantle and panels f are close the 670 km discontinuity. Convection is responsible for the advection of thermal anomalies. Therefore, the wavelengths of these anomalies are closely related to the horizontal wavelength of the radial velocity. Indeed, as a direct consequence of axisymmetrical geometry, the flows are purely poloidal and the only vortices, necessary for the cascading of energy in the spectra are due to horizontal distortions of vertical motions. With a Clapeyron slope $\gamma = -4 \cdot 10^6$ Pa/K (Figure 5), the structure of convection is layered and the typical horizontal wavelength of the cells is set by the thickness of the upper mantle. Isotherms of the right panel of Figure 2 show that 8 or 9 rising currents can be counted in the upper mantle from one pole to the other. These rising structures constitute the large scale flow pattern in the upper mantle and correspond to the non-inertial part (the low degree harmonics before the inertial-convective part) of the spectrum (Figure 5, e) which ends approximately for $l = 8$. For higher degrees of the thermal anomaly expansion, the slope of the spectrum follows the Kolmogorov rule (-5/3). With a layered mantle structure a thermal boundary layer exists

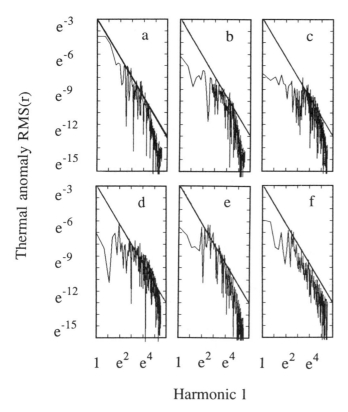

Harmonic 1

Fig. 5. RMS spectra (see text) of the thermal anomalies at various depths (a : 2855 km, b : 2188 km, c : 1485 km, d : 817 km, e : 115 km and f : 677 km). Levels a and e are located inside of the bottom and top thermal boundary layers; levels b, c and d sample the lower mantle and level f gives the spectrum close to the 670 km discontinuity. The straight line gives the Kolmogorov slope (-5/3). The spectra correspond to the case γ = -4 10^6 Pa/K of Figure 2.

at the level of the 670 km discontinuity. It is why the spectrum of panel f gives the same trend to follow the Kolmogorov rule. Now, we are going to consider what happens farther from the thermal boundary layers, inside of the lower mantle (Figure 5, panels a,b,c and d). The maximum of vortex is obtained when the sinking fluid encounters the bottom surface (Figure 5, panel a). Then the global thermal structure of the lower mantle (Figure 2, right panel) disappears and almost all the wavelengths participate to the creation of whirl. Therefore, the inertial-convective part of the spectrum starts for low degree harmonics. This is no more the case in the mid-lower mantle. Now, the large scale sinking currents do not contribute to create vortex and the slopes of the spectra are broken into two parts (Figure 5, panels b,c and d) : a non-inertial part which reaches higher degrees when the distance from the boundary layers increases and an inertial-convective part with Kolmogorov slopes (-5/3). The point is that the beginning of the inertial-convective part of the slope for the top of the upper mantle (panel e) is intimately related to layering of the mantle through its effects on the wavelength of the upper mantle cells. The same trend is reached when convection is not perfectly

layered but almost layered with intermittent penetrations of upper mantle into the lower one (γ = -2 10^6 Pa/K). The same behaviors are obtained within the thermal boundary layers (Figure 6, panels a, e and f) and inside of the lower mantle. If we consider now a mantle structure with only one layer, obtained with γ = 0, then the absence of a thermal boundary layer at the 670 km depth discontinuity produces a broad non-inertial part for the spectrum at this level (Figure 7, panel f). This result was expected from the previous spectra (Figure 5 and 6) outside of the thermal boundary layers. However, the typical wavelength of the sinking currents at the top of the upper mantle is now controlled by the thickness of the whole mantle. In this case, the inertial-convective part of the spectrum (Figure 6, panel e) starts for higher degree harmonics (around 13). The beginning of the inertial-convective part part of the Tanimoto's spectrum to l = 8 proves that the statistical sizes of the upper mantle cells are not compatible with a one-layer turbulent convective structure in the mantle. However, this seismological results of Love phase velocity can be explained as the natural expression of turbulent thermal anomalies at the top of the mantle within a two-layered or a one-and-a-half layered flow pattern.

FILTERING OF SEISMIC TOMOGRAPHY

It is tempting to consider the seismic velocity anomalies given by seismic tomography as pictures of the temperature field in the mantle. Indeed as a rough first approximation [Honda, 1987], the thermal anomalies can be converted following the rule :

$$\delta V_p(r,\theta) = A\, V_p(r)\, \delta T(r,\theta) \qquad (27)$$

where the coefficient A may be chosen equal to -50 10-6 K^{-1} from the results of Chung [1981], and is the P-wave velocity given by the PREM model [Dziewonski and Anderson, 1981]. This law is only an approximation since the lateral variations of P-wave velocity depend also on the sensitivity of the bulk modulus to density [Agnon and Bukowinski, 1990]. Furthermore, this approximation does not take into account the anisotropy which can result from convective motion (i.e. Tanimoto and Anderson, 1984]. We have to keep these cautions in mind to examine the results of the next sections. However, since the aim of the paper is to study the global structure of the mantle, we will more focus on the shape of the anomalies than on their amplitudes and, within this frame, the results should remain qualitatively valid. Following Machetel [1990], the filtering effects have been studied by operating a double radial and latitudinal Legendre transform with 256 latitudinal and 40 radial harmonics and a truncated inverse Legendre transform. The velocity anomalies computed from Equation (27) have been expressed as :

$$dV_p(u,\theta) = \sum_{k=0}^{40} \sum_{l=0}^{256} C_{lk}\, p_k(u)\, p_l(\cos\theta) \qquad (28)$$

where and are the normalized Legendre polynomial verifying :

$$\int_{-1}^{1} p_i(u)\, p_j(u)\, du = \delta_{ij} \qquad (29)$$

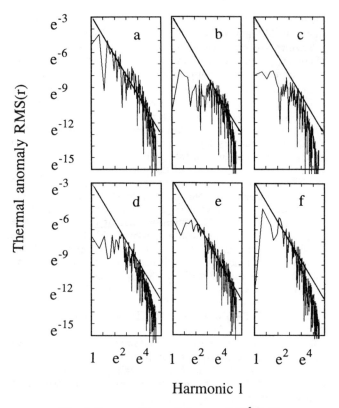

Fig. 6. Same as Figure 5, for $\gamma = -2 \ 10^6$ Pa/K.

The coefficients are calculated from the velocity anomalies.

$$C_{lk} = - \int_{-1}^{1} \int_{0}^{\pi} dV_p(u,\theta) \ p_l(\cos\theta) \ \sin\theta \ d\theta \ p_k(u) \ du \qquad (30)$$

Left hemispheres of the left column of Figure 8 show the velocity anomalies computed with Equation (27) for the three values of the Clapeyron slope γ. The right part of the first column, and the second and third columns show the effect of truncature. With (100 x 40) harmonics both the amplitudes and the shapes of the anomalies are almost perfectly retrieved for the three values of the Clapeyron slope. With a truncature of (50 x 20) (second column), the global structures of the solutions are conserved but the amplitudes of these anomalies suffer. The locations and the depth of penetration of the large sinking plumes for $\gamma = 0$ Pa/K or the layered structure for $\gamma = -4$ 10^6 Pa/K can be successfully compared with those of the non-filtered solution. It seems possible to decipher what kind of coupling is prevailing at the 670 km depth discontinuity. At least, it is possible to distinguish a one-layer structure for mantle convection from two-layer or from one-and-a-half layer structures. This is no more the case for the (20 x 10) filtering. Now the anomalies are spread over large areas. In all the cases, $\gamma = 0$, $-2 \ 10^6$ and $-4 \ 10^6$ Pa/K, radial anomalies extend continuously from the surface until the CMB. Indeed, with $\gamma = -2 \ 10^6$ or $-4 \ 10^6$ Pa/K, the spreading of neighboring narrow

plumes into large anomalies destroys the signature of the coupling at the 670 km discontinuity. When they are filtered at the accuracy of seismic tomography, the seismic velocity anomalies do not allow to characterize the structure of the mantle.

RMS OF THE TEMPERATURE ANOMALIES

The structure of convection influences the thermal state of the mantle. Keeping all the other parameters unchanged, an increase of the endothermic character of the Clapeyron slope increases the trend toward layering that results into a strong insulating effect. Then, the mean mantle temperature rises but the amplitude of the thermal anomalies do not change except in the neighboring of the phase change where the layering creates a new thermal boundary layer [Machetel and Weber, 1991]. The Root Mean Square of the thermal anomalies has been computed as a function of the depth with Equation (31).

$$RMS(r) = \sqrt{\frac{1}{2} \int_{0}^{\pi} (\delta T)^2 \sin\theta \ d\theta} \qquad (31)$$

As described in a previous work [Machetel, 1990], the amplitude of the signal decreases with truncation. In a whole mantle convection, $\gamma = 0$ Pa/K, the RMS of the filtered thermal anomalies is roughly constant outside of the top and bottom thermal boundary layer (Figure 9, left). Now, for a partial

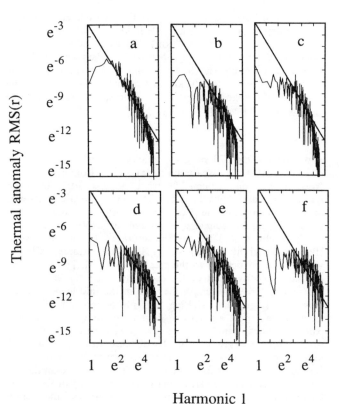

Fig. 7. Same as Figure 5, for $\gamma = 0$ Pa/K.

$\gamma = 0$ Pa/K

$\gamma = -2\ 10^6$ Pa/K

$\gamma = -4\ 10^6$ Pa/K

Fig. 8. Effects of filtering on the shape and on the amplitude of velocity anomalies in the mantle. Isocontours range from -1% to 1% by steps of .25%. The thick lines give the level 0. From top row to bottom row, $\gamma = 0$, -2 10^6 and -4 10^6 Pa/K. Left hemispheres of left column display the anomalies without filtering. The same anomalies have been filtered using radial and latitudinal Legendre transform with $(lmax_\theta, lmax_r)$ harmonics. From left to right the levels of filtering are : $(100,40)$ in the right hemispheres of left column, $(50,20)$ in the second column and $(20,10)$ in the third column.

layering of the mantle, $\gamma = -2 \ 10^6$ Pa/K, or for layered mantle convection, $\gamma = -4 \ 10^6$ Pa/K, the amplitudes of the upper mantle anomalies remain much stronger even after filtering than those of the lower mantle. Middle and right panels of Figure 9 show that this result does not significantly depend on the level of filtering. Thus, the increase of the RMS anomaly given by tomographic models at the top of lower mantle (see Tanimoto [1990] or Gudmunsson [1989] for synthetic representations) is more compatible with a layering or a partial layering of mantle convection. Furthermore, the ratio of the amplitude of RMS between the top of the lower and the mid-lower mantle ranges approximately between 3 to 5, that is roughly the rate given by layered and one-and-a-half layered convection models. This last result is independent of the uncertainties on the A coefficient of Equation (27).

P-WAVE PROPAGATION

The method proposed by Jacob [1970] has been used to compute the propagation of P-wave in a heterogeneous medium with short-wavelength velocity anomalies. This method allows to perform ray-tracing through an arbitrary velocity structure in a spherical Earth. According to Fermat's principle a seismic ray will choose the specific path S that takes the least travel time T from points A to B.

$$T = \int_A^B \sqrt{U(l_i l_i)} \ ds = \text{Extremum} \tag{32}$$

where $U(r,\theta)$ is the slowness and l_i is the tangent vector at any point along the ray : $l_i = (r', r\theta')$. The quantities r' and θ' are the derivatives of the coordinates along the ray increment ds. The complete equations, developed by Jacob [1970] for 3-D spherical geometry have been simplified to take the axisymmetrical geometry into account. Then the ray propagates from point to point through an arbitrary velocity structure with heterogeneities. A first test has been conducted by comparing the hodochrone P-wave obtained from the classical ray parameter method for a spherically stratified Earth

[Bullen, 1963] with those obtained with our program without lateral heterogeneities. In this case, the velocity is not constant but increasing with depth according to the PREM model. The ray propagation has been calculated considering this radial dependence as a heterogeneity. Figure 10 (panel a) shows the rays propagating through a spherically stratified mantle assuming a PREM distribution of radial velocity. Panel (b) gives the corresponding hodochrones and panel (c) shows those obtained from the ray parameter method. The agreement between both methods is satisfactory except for the triplication zone of the waves. This is due to our method of computation of the wave propagation which is based on only 81 radial levels. This number of level is not sufficient to accurately describe the triplication. However, for the higher incidences angles the agreement is almost perfect and the ray parameter hodochrones will serve as reference to compute the delays due to the propagation through the short-wavelength heterogeneities. The propagation of P-waves through the thermal anomalies is shown in Figure 11 for $\gamma = 0$, $-2 \ 10^6$ and $-4 \ 10^6$ Pa/K. In right panels of figure 11, the thin line gives the ray parameter hodochrones, the thick lines show the hodochrones of waves which have travelled through the anomalies and the difference between the two arrival times is given by the middle curve (with an amplified scale). The trajectories of the rays are modified by the presence of velocity gradients. Following Fermat's principle, the rays converge toward the cold anomalies and diverge from the hot ones. When the source of P-waves is located in a region of weak velocity gradient (as it is the case at the top of Figure 11 where it has been placed just between two sinking currents), the incidence angle at the starting point of the rays are not really affected by the local anomalies. The resulting hodochrone remains approximately flat for the small angular distances (until 60 degrees). However, the spreading of cold anomalies at the bottom draw the ray path toward core and induce delays for the large angular distances. Now, if the P-wave source is inside the narrow sinking areas (Figure 11, middle panels), the wave are trapped in the high velocity zone. If the cold zone is more horizontal than the local propagation direction of the ray, this will result in a shortening of the ray path and in a propagation

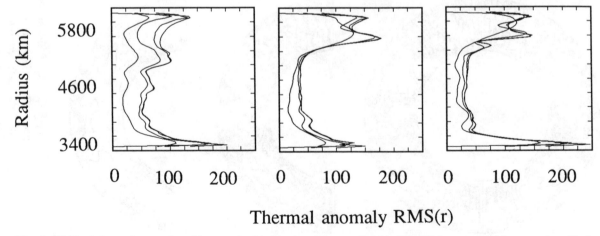

Fig. 9. RMS of thermal anomaly without and with filtering. The amplitude of RMS decreases with truncation. Each panel contains four curves corresponding to : no filtering, (100,40), (50,20) and (20,10) harmonics. From left to right the Clapeyron slopes are 0, -2 10^6 and -4 10^6 Pa/K.

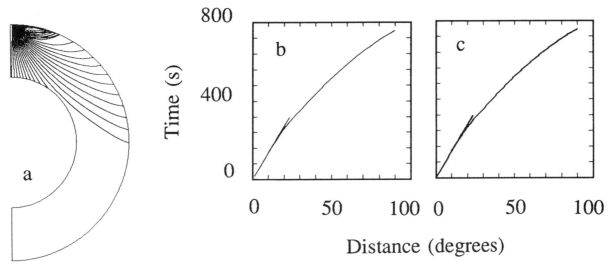

Fig. 10. P-wave propagation in a radially stratified Earth with PREM velocity. Panel (a) shows the rays path obtained in a structure with no thermal (velocity) anomaly. The corresponding hodochrone is given in panel (b) and can be compared with this given by the ray-parameter method (Bullen, 1963) (c).

in a faster medium. In this case, the rays will reach the surface sooner (Figure 11, middle). A third case is observed when the source is located in a hot anomaly just beside a strong negative anomaly (Figure 11, bottom). Then the horizontal velocity gradients strongly distort the beginning of the path by increasing the incidence angle. The rays are lengthened and the travel times are increased leading to positive hodochrone anomalies . This shows the influence of the cold anomalies which act as a ray attractor. However similar hodochrones can be obtained, by moving the starting point of rays, from the three kinds of structures. This is because the ray propagation and the hodochrone give path-averaged properties. At this level of our study we cannot use it to constrain the large scale structure of mantle convection. The distortions of the discontinuity by convection produce lateral velocity anomalies which are not modeled in the present work. Their effects will be taken into account in a following study.

SUMMARY AND DISCUSSION

The numerical models presented in this paper show that the trend to get mantle layering when an endothermic phase change is assumed at the 670 km depth discontinuity is persistent in a large range of bottom mantle temperatures (at least from 3000 to 4000 K). Three structures are possible depending on the value of the Clapeyron slope. For $\gamma = 0$, a one-layer structure exists with currents going from the upper boundary to the lower one. With a more endothermic Clapeyron slope ($\gamma = -2\ 10^6$ Pa/K), the chaotic regime is characterized by accumulations of upper mantle material at the discontinuity level. Intermittent and brutal penetrations of upper mantle into the lower one occur after lapses of time commensurate to those of major tectonic events. The local character of these mixing events which occur in a surrounding layered structure allows to speak of partly layered convection. finally, with $\gamma = -4\ 10^6$ Pa/K, the structure is organized in two layers without significant mixing. These properties are verified independently of the internal

heating rates over the whole possible range of radiogenic heat production (from zero to chondritic rate). The transient time to go from one kind of regime to an other is very short, of the order of 0.1 billion years. As it does not seem to depend on the CMB temperature neither on the internal heating rate , but mainly on the Clapeyron slope of the transition, the structure of mantle convection may have been acquired very early in its history and has probably remained the same after in spite of chaotic time evolution.

The spectra of the present seismic velocity anomalies in the upper mantle displays a turbulent-like spectrum. The Tanimoto's spectrum of Love phase velocity can be split into two parts : for degrees lower than 6-7, and for higher degrees. The first part is characteristic of the long-wavelength features in the upper mantle. The second follows the Kolmogorov rule (the slope in log-log representation is -5/3). This result means that the wavelengths which participate to the inertial-convective part of the turbulent spectrum start approximately from degrees 6-7. The convective models show that such wavelengths are not compatible with a whole mantle convection but only with a layered or a partly layered convection in the mantle.

It is tempting to try to deduce the large scale structure of mantle convection directly from the results of global tomographic studies. However, the filtering of the numerical models of mantle convection at the accuracy of tomographic results leads to fuzzy images where the characteristics of layering or of whole mantle convection are no more decipherable. However, the ratio of the RMS of the velocity anomalies in the upper mantle to the lower mantle, when it is compared to the numerical modelling gives an other clue to a layered or to a one-and-a-half layered convective structure in the mantle.

The short-wavelength thermal anomalies influence the propagation of P-waves but the results are difficult to explain in terms of mantle layering. Similar hodochrones can be found for the three kinds of structures depending on the starting points of the rays.

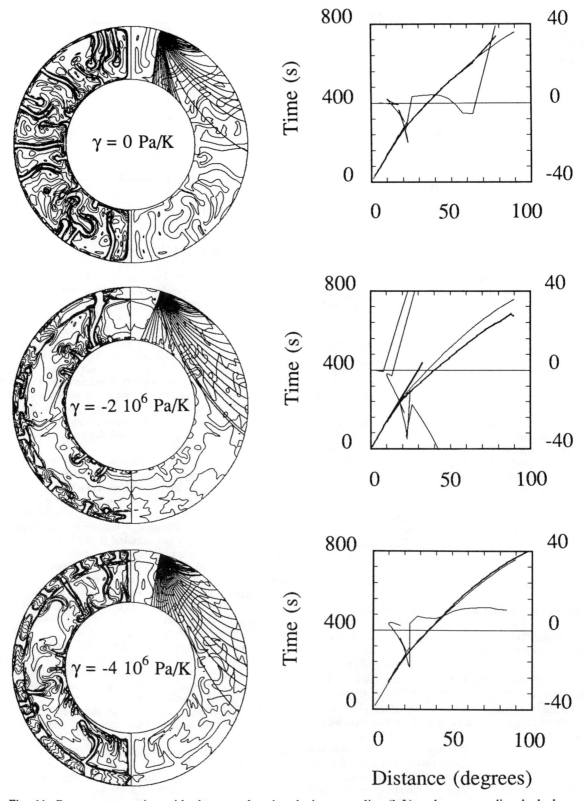

Fig. 11. P-wave propagation with short-wavelength velocity anomalies (left) and corresponding hodochrones (heavy lines in right panels). In right panels, the thin lines give the reference hodochrones obtained with the PREM model (ray-parameter method). The middle curves give the delays due to heterogeneities (with an amplified scale).

The balance of these results shows that a two-layer or a one-and-a-half convective pattern represent the most probable structure for the Earth's mantle. Numerical calculations show that these two structures exist over wide ranges of geophysical parameters. As these structures match well the properties obtained from seismological results, we would like to propose that they represent the pattern of convection in the Earth's mantle.

The dependences of the viscosity on the temperature, the pressure and the phase change has to be explored in a next work. This will help to check the compatibility of such mantle structure with observed geoid anomalies.

Acknowledgments. The authors thanks Annie Souriau for scientific discussions and her help in confronting theoretical and computed hodochrones, Henri Claude Nataf and Claude Jaupart for critical reading of the manuscript. This work is the CNRS/INSU/DBT contribution N° 452.

REFERENCES

Agnon, A., Bukowinski, M.S.T., deltas at high pressure and dln Vs/dln Vp in the lower mantle, *Geophys. Res. Lett.*, 13, 1149-1152, 1990.

Anderson, D.L., *Theory of the Earth*, Blackwell Scientific Publications, Boston, pp366, 1989.

Boehler, R., Chopelas, A., Major phase transitions and constraints on the temperature distribution in the Earth, *Proceedings of the Royal Astronomical Society*, Geophysical Survey, 1990.

Bullen, K.E., *An introduction to the theory of seismology*, Cambridge University press, London, pp 381, 1963.

Christensen, U.R., Yuen, D.A., The interaction of a subducting lithospheric slab with a chemical or phase boundary, *J. Geophys. Res.*, 89, 4389-4402, 1984.

Christensen, U.R., Yuen, D.A., Layered convection induced by phase transitions, *J. Geophys. Res.*, 90, 10291-10300, 1985.

Chung, D.H., Elasticity and Equation of state of olivines in the Mg2SiO4-Fe2SiO4 system, *Geophys. J. Roy. Astron. Soc.*, 25, 511-538, 1971.

Dziewonski, A.M., Anderson, D.L., Preliminary Reference Earth Model, *Phys. Earth Planet. Int.*, 25, 297-356, 1981.

Galer, S.J.G., Goldstein, S.L., Early mantle differentiation and its thermal consequences, *Geochim. Cosmochim. Acta*, 55, 227-239, 1991.

Gudmunsson, O., Some problems in global tomography, *PhD Thesis*, California Institute of Technology, 1989.

Hager, B.H., Richards, M.A., Long wavelength variations in the Earth's geoid : physical models and dynamical implications, *Phil. Trans. R. Soc. Lond., A*, 328, 309-327, 1989.

Honda, S., The RMS residual temperature in the convecting mantle and seismic heterogeneities, *J. Phys. Earth*, 35, 195-207, 1987.

Inoue, H., Fukao, Y., Tanabe, K., Ogata, Y., Whole mantle P-wave travel time tomography, *Phys. Earth Planet. Int.*, 59, 294-328, 1990.

Ito, E., Katsura, T., A temperature profile of the mantle transition zone, *Geophys. Res. Lett.*, 16, 425-428, 1989.

Jacob, K.H., Three-dimensional ray tracing in a laterally heterogeneous spherical Earth, *J. Geophys. Res.*, 75, 6675-6689, 1970.

Jarvis, G.T., McKenzie, D.P., Convection in a compressible fluid with infinite Prandtl number, *J. Fluid Mech.*, 96, 515-583, 1980.

Jeanloz, R., Morris, S., Temperature distribution in the crust and mantle, *Ann. Rev. Earth Planet. Sci.*, 14, 377-415, 1986.

Le Pichon, X., Huchon, P., Geoid Pangea and convection, *Earth Planet. Sci. Lett.*, 67, 123-135, 1984.

Lesieur, M., *Turbulence in fluids*, Martinus Nijhoff Publishers, Dordrecht, Boston, Lancaster, pp 286, 1987.

Machetel, P., Short-wavelength lower mantle seismic velocity anomalies, *Geophys. Res. Lett.*, 17, 1145-1148, 1990.

Machetel, P., Weber, P., Intermittent layered convection in a model mantle with an endothermic phase change at 670 km, *Nature*, 350, 55-57, 1991.

Machetel, P., Yuen, D.A., Penetrative convective flows induced by internal heating and mantle compressibility, *J. Geophys. Res.*, 94, 10609-10626, 1989.

Nataf, H.C., One-and-a-half layer convection, in *Crust and mantle recycling at convergence zones*, edited by S.R. Hart and L. Gulen, Kluwer Academic Publishers, 197-200, 1989.

Olson, P., Silver, P.G., Carlson, R.W., The large-scale structure of mantle convection in the Earth's mantle, *Nature*, 344, 209-215, 1990.

Price, G.D., Wall, A., Parker, S.C., The properties and behavior of mantle minerals: a computer simulation approach, *Phil. Trans. R. Soc. Lond., A*, 328, 391-407, 1989.

Silver, P.G., Carlson, R.W., Olson, P., Deep slabs, geochemical heterogeneity and the large structure of mantle convection, *Ann. Rev. Earth Planet. Sci.*, 16, 477-541, 1988.

Solheim, L.P., Peltier, W.R., Mantle phase transitions and layered convection, *Canadian Journal of Earth Sciences*, submitted, 1991.

Souriau, M., Souriau, A., Global tectonics and the geoid, *Phys. Earth Planet. Int.*, 33, 126-136, 1983.

Stacey, F.D., A thermal model of the Earth, *Phys. Earth Planet. Int.*, 15, 341-348, 1977.

Stacey, F.D., Loper, D.E., Thermal history of the Earth a corollary concerning non-linear mantle rheology, *Phys. Earth Planet. Int.*, 53, 167-174, 1988.

Tanimoto, T., Predominance of large scale heterogeneity and the shift of velocity anomalies between the upper and lower mantle, *J. Phys. Earth*, 38, 493-509, 1990.

Tanimoto, T., Anderson, D.L., Mapping convection in the mantle, *Geophys. Res. Lett.*, 11, 287-290, 1984.

Wang, Y., D.j., Weidner, R.C., Lieberman, X., Liu, J., Ko, M.T., Vaughan, Y., Zhao, A., Yeganeh-Haeri, R.E.G., Pacalo, Phase transition and thermal expansion of MgSiO3 perovskite, *Sciences*, 251, 410-413, 1991.

Zindler, A., Hart, S., Chemical geodynamics, *Ann. Rev. Earth Planet. Sci.*, 14, 493-571, 1986.

P. Machetel, U.P.R.234/C.N.R.S./G.R.G.S. 18 avenue Edouard Belin, 31055 Toulouse cedex, France.

Constraints on the Temperature and Composition of the Base of the Mantle

MICHAEL E. WYSESSION,[1] CRAIG R. BINA AND EMILE A. OKAL

Department of Geological Sciences, Northwestern University, Evanston, Illinois

P and *S* waves diffracted around the core-mantle boundary (CMB) are examined to obtain measurements of long-wavelength average velocities in D″, the base of the mantle. Observations are made of profiles of diffracted waves (*Sd* and *Pd*) from WWSSN and Canadian stations, and are compared to synthetic seismograms generated with the reflectivity method. The apparent ray parameters of the data and synthetic profiles serve as the basis of comparisons, which suggest significant lateral heterogeneity on the order of about 4% for both *P* and *S* velocities at the base of the mantle. While most of the D″ average velocity anomalies are on the order of ± 1% relative to PREM, our range in seismic heterogeneity is largely the result of a 3% *S* and *P* velocity low in D″ beneath Indonesia, which is made even more unusual by the fact that it is adjacent to a regional fast velocity anomaly beneath Southeast Asia. This velocity low is over a major rising plume in the outer core, as calculated by *Voorhies* [1986], and if this plume has been held in place over time through core-mantle coupling then the low velocity would be expected due to the increased mantle influx of heat and iron. We undertake a calculation of the variations in D″ seismic velocities due to changes in temperature and composition using a third-order Birch-Murnaghan equation of state with current available thermoelastic data on perovskite and magnesiowüstite. Using this model, small seismic velocity anomalies in D″ could be the result of temperature variations, though small fluctuations in relative amounts of magnesium and iron would have a greater effect on the velocities. For example, the Indonesian anomaly cannot be explained by only a thermal anomaly, but requires only a 20% increase in the Fe/(Mg+Fe) ratio (and even less if accompanied by a raise in temperature). In some regions of D″ the *P* and *S* velocities do not vary in tandem, as under Northern North America, where shear velocities are fast but *P* velocities are slightly slow. The implied lateral change in Poisson ratio could be the result of variations in the relative amounts of silicates and oxides, exacerbated by the high thermal gradients that are expected to exist in D″.

INTRODUCTION

In examining the structure of D″ (the very base of the mantle) the combined use of both *P* and *S* velocities supplies a greater constraint than only one or the other, and this is true as well with waves that diffract around the CMB. In *Wysession and Okal* [1988, 1989] we looked at the diffracted *Sd* and *Pd* waves separately, beginning to map out lateral heterogeneities in D″. What we present here is a joint examination of profiles of both *Pd* and *Sd* waves, facilitated with the use of reflectivity synthetic seismograms, which can model both arrivals at the same frequency levels. These diffracted wave profiles give us long wavelength information about the velocities in D″, and combined with the results from other disciplines (geomagnetics, in particular) we can begin to sug-

gest and identify dynamic processes occurring in particular regions at the base of the mantle.

Mineral physics, especially the breakthroughs of recent experimental work, has begun to give us a thermochemical description of the materials (i.e., perovskite, magnesiowüstite) that comprise the lower mantle. It is therefore possible to begin comparing the seismic velocity anomalies observed with those predicted from thermoelastic principles. In this light, we will also attempt to compare the seismic velocity variations we find with those calculated from Birch-Murnaghan equations of state, using as initial parameters the recent results of mineral physics experiments.

Diffracted waves have long been identified as being as important tools for examining the CMB, as they can travel for thousands of km along the base of the mantle (shown in Figure 1). A careful examination of diffracted arrivals travelling along a similar azimuth to several stations from a single earthquake gives a good indication of the average velocity of that strip of D″ within which all the waves travel, once corrections are made so that we have removed all other effects, and a comparison is made with synthetic seismograms modeling identical earthquake-station geometries. These velocities, when compiled as a map for D″, will serve as the

---

[1]Now at Department of Earth and Planetary Sciences, Washington University, St. Louis, Missouri.

Dynamics of Earth's Deep Interior and Earth Rotation
Geophysical Monograph 72, IUGG Volume 12

Geometrical Waves

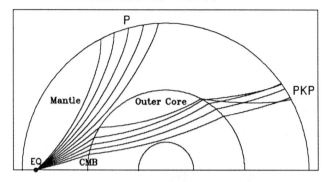

CMB Diffracted Waves

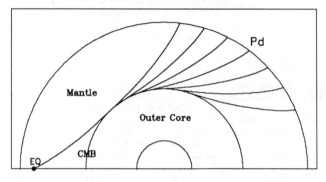

Figure 1. A demonstration of the difference between geometrical waves (obeying Snell's Law) and diffracted waves. *Top*: Ray tracing for P waves within the Earth, showing the shadow zone that exists between the P and PKP arrivals. *Bottom*: Diffracted P waves (Pd) that travel along the core-mantle boundary and leak back to the surface, arriving within the shadow zone and beyond.

basis for further determinations about the temperatures, compositions and core/mantle dynamics of the CMB region.

Pd and *Sd* waves have been observed since the start of the century (see *Sacks* [1966] and *Cleary* [1974] for discussions), though with the recognition of the complexity of of the CMB diffracted waveforms it has been through simulations with synthetic seismograms that recent studies have been able to model mantle properties. The results presented here were attained through comparison of the *Sd* and *Pd* arrivals with synthetic arrivals calculated with the reflectivity method, though our earlier studies used synthetics from normal mode summation [*Wysession and Okal*, 1988, 1989]. The reflectivity method was previously used by *Mula and Müller* [1980] and *Mula* [1981] in their studies of CMB diffracted waves.

Diffracted wave studies can involve investigations of both travel times and amplitudes, though we will concentrate here on the diffracted wave ray parameters as determined from the arrival times. The ray parameter (slowness) of a linear profile of diffracted arrivals is given by $p = dT/d\Delta = R_{CMB}/V_{CMB}$ (where T is the travel time, Δ the epicentral distance, R_{CMB} the radius of the CMB, and V_{CMB} the apparent velocity at the CMB) and is a direct indicator of the seismic velocities at the base of the mantle. However, p is only an *apparent* slowness and is a complicated function of D″ structure and particular earthquake-station geometries [*Chapman and Phinney*, 1972; *Mula and Müller*, 1980; *Wysession*, 1989, 1991], and therefore actual velocities at the base of the mantle cannot be simply taken from the ray parameter but must be inferred through comparisons with synthetic profiles. There are other precautions and corrections to be taken and made before D″ velocities can be inferred.

PROCEDURE

For our profiles we used the diffracted P and S arrivals at WWSSN and Canadian stations from 21 large earthquakes. This gave us 20 azimuthally independent Pd profiles and 12 Sd profiles, which each contain at least 4 stations (and as many as 10) along a similar azimuth that span from the start of the shadow zone to as great as 160°. Details are given in *Wysession* [1991].

The diffracted profiles are constrained to narrow azimuthal windows (a maximum of about 20°) so that the velocities of a particular strip through D″ are examined, and so that the downswing paths of all arrivals are essentially the same. This means that since we are measuring the slowness between arrivals and not absolute arrival times, we do not need to worry about source effects and mislocations, mantle heterogeneities (on the downswings), or slab diffraction [*Cormier*, 1989]. The Sd and Pd ray parameters, once given their necessary corrections, should be entirely a function of the velocities within D″.

In an attempt to have the data and synthetics as comparable as possible, we correct the arrival times of the data for ellipticity using the relationships of *Jeffreys and Bullen*, 1970, and for mantle upswing path heterogeneities. The latter are calculated by ray tracing our waves through the full mantle 3-D tomographic velocity model of *Woodhouse and Dziewonski* [1987] and summing the velocity heterogeneities along the paths. The procedure is explained in detail in *Wysession and Okal* [1988, 1989].

The synthetic seismograms are generated using the reflectivity method, which gives easier access to the high frequency portions of the diffracted arrivals [*Wysession*, 1991], and are based on the algorithms of *Kennett* [1983]. The synthetic seismograms are generated for the radially symmetric PREM structure of *Dziewonski and Anderson* [1981], and are generated using the same focal mechanisms, path geometries and instrument responses as the data. Examples of the data and synthetic diffracted waves are given in Figure 2, which shows the Sd profile from Loyalty Island (Oct 7, 1966) to the Mid-East.

The data and synthetics are compared, and D″ velocities determined, on the basis of their apparent ray parameters (apparent slownesses), and we emphasize the necessity of this. The value of the measured ray parameter will be biased by the distance covered along the CMB, the geometry of stations used, the instrument frequency responses and the method used to determine it. One example of *Mula and Müller* [1980] used reflectivity synthetics to show that a particular apparent P-wave ray parameter of $p_{app} = 4.54$ s/deg, which would suggest an apparent velocity of $\alpha_{app} = 13.38$

km/s, actually was generated for an average D″ P velocity of $\alpha_{ave} =$ 13.74 km/s.

The slownesses represent the linear slope of the arrivals with increasing epicentral distance, as can be seen in Figure 2. For both the data and synthetics we determine the slownesses using two different robust techniques: peak maxima picks and multi-waveform cross-correlation. Though each adds bias into the data slownesses, this is recreated in the reflectivity synthetics. Details are explained in *Wysession and Okal* [1989]. The effects of the profile geometries, instrument responses and determination techniques can be seen in the ranges of slowness values for the synthetics. The ranges for the synthetic Pd slownesses were 4.51-4.59 s/deg (cross-correlation) and 4.49-4.55 s/deg (peak maxima), and for the Sd slownesses were 8.44-8.50 s/deg (cross-correlation) and 8.43-8.45 s/deg (peak maxima).

Once the data slownesses are determined relative to the synthetics, we can average the results for profiles that travel the same CMB paths. What we have is a measure of the percentage difference of the diffracted wave ray parameter in different parts of D″ relative to a PREM D″. What we would like is a measure of the percentage difference of the average velocities in different parts of D″ relative to the average velocity in PREM's D″. The apparent slownesses can be easily converted into apparent velocities by definition, but the translation of these apparent velocities into actual D″ velocities is very complicated. A determination of the radial structure of D″ is beyond the scope of this study, though eventually broadband arrays may give enough frequency amplitude information to invert for D″ radial velocities, much as is done in surface wave inversions.

It is possible, however, using the results of *Mula and Müller* [1980], to translate the apparent velocities into average velocities if a specific depth is determined for D″. Using 12 different velocity models that varied essentially only in the bottom 190 km of the mantle, *Mula and Müller* [1980] generated reflectivity synthetic diffracted Pd waves and determined the apparent ray parameters, which gave them the apparent D″ velocities. They found a striking linear correlation between these apparent velocities α_{app} and the P velocity averaged over the bottom 190 km of the mantle (α_{190}). Their conclusion was that if D″ was assumed to be 190 km thick, then $\alpha_{190} \propto 0.83\alpha_{app}$ (km/s). A similar result for shear waves yielded $\beta_{190} = \propto 0.65\beta_{app}$. These relationships are determined at a period of T = 20 sec, the approximate response peak of long period WWSSN instruments. In our discussions of average D″ velocities we will use these relationships to convert apparent velocities into D″ average velocities with the arbitrary assumption that D″ is 190 km thick, though it should be understood that the average velocities must increase if D″ is thinner, and must decrease if D″ is thicker.

RESULTS AND DISCUSSION

The variations in averaged velocity, determined by differences between the data slownesses and PREM synthetics, is shown in Table 1. The actual range in individual apparent ray parameters was large: 4.44-4.80 s/deg for Pd, and 8.27-9.01 s/deg for Sd. When compared with the synthetics and averaged by region, the lateral variations were more moderate, but still amounted to sev-

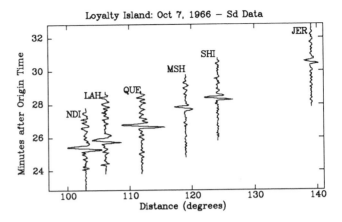

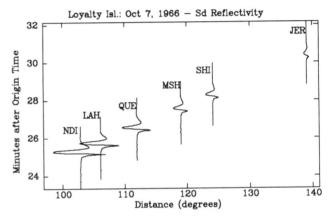

Figure 2. An example of the reflectivity synthetic seismograms, shown here modeling the diffracted S arrivals from the Loyalty Island 10/7/66 earthquake. Data are shown at the *top* and synthetics at the *bottom*.

eral percent for both P and S velocities. As can be seen in Table 1, PREM does a fairly good job of serving as a reference model for the diffracted data, though there are roughly twice as many CMB regions that are slightly slower than PREM than are faster than PREM. It is not uncommon for a region to be faster or slower by about 1%, with the significant exception being under Indonesia. The regions we examined, including the path through Indonesia, displayed 4.0% and 3.9% lateral variation in α_{190} and β_{190}, respectively, implying that the level of heterogeneity for P and S is approximately the same. In our regional discussion it is important to bear in mind that for the 1 sec PREM velocities, α_{190} and β_{190} (over the 190 km at the base of the mantle) are 13.690 km/s and 7.264 km/s.

Certainly the most unusual region of the CMB that we found was underneath Northern Indonesia and Southeast Asia, as sampled from the diffracted wave profiles from the Tonga/Kermadec region across the Mid-East to the Mediterranean. There was very good coverage along this profile, with as many as eight stations well separated with a total distance range of up to 49°, and this allowed us to examine the first and second halves of the path separately, both retaining high quality profiles of several stations. The path profiles, when examined whole, showed no unusual velocities. Both α_{190} and β_{190} were nearly identical to those of PREM, with each being slightly slow (see Table 1). When separated, however, the second part showed slightly fast P and S velocities, but the first half had extremely slow velocities. We attempted to split up other profiles that had long distance ranges, and though in none of those cases did we find significant differences between the halves, their cross-correlograms were not good enough to include in the study.

Along the first half of the Tonga-to-Mid-East path, sampling CMB under Northern Indonesia, the apparent slowness was 3.8% slower than PREM for P waves and 4.7% slower for S waves. Again, using the assumptions above for a 190 km thick D'', this would imply the velocity anomalies to be $\Delta\alpha_{190}$ = -3.2% and $\Delta\beta_{190}$ = -3.1% (α_{190} = 13.25 km/s and β_{190} = 7.04 km/s). These are by far the lowest average D'' velocities that we have yet found. The results are very robust, found nearly identically in profiles from four earthquakes for Pd and three earthquakes for Sd. What makes this even more unusual is that the second half of these profiles, under Southeast Asia, is unusually fast, relative to PREM. The implied velocity anomalies along this segment were $\Delta\alpha_{190}$ = +0.7% and $\Delta\beta_{190}$ = +0.8%. This juxtaposition of slow and fast velocities also appears in tomographic studies that use non-diffracted arrivals, such as *Woodhouse and Dziewonski* [1987] (Model V3), *Tanimoto* [1987], and *Inoue et al.* [1990].

A possible geodynamic explanation for this may involve coupling with core flow. There is a strong correlation between our D'' velocities and the geomagnetically determined core flow model of *Voorhies* [1986] (and to a lesser extent, though still evident, *Bloxham* [1989]). Our slow velocity region beneath Indonesia sits right over one of the largest regions of CMB core upwelling in the *Voorhies* [1986] models, and the adjacent fast velocities are above the largest *Voorhies* [1986] region of downwelling. Thermochemically, a reduction in D'' seismic velocity would most likely be the result of increased temperature or iron content, and both of these would be expected above a core plume. If iron-silicate reactions [*Knittle and Jeanloz*, 1989b; *Jeanloz*, 1990] are occurring at the CMB, and liquid iron is seeping into the mantle through capillary action [*Stevenson*, 1986], then we would expect there to be both an increase in heat flux into the mantle and an increase in the amount of denser (and seismically slower) iron oxides in D'' above a region of vigorous core upwelling.

There is a difficulty in understanding why a correlation should exist between mantle and outer core features, when the core flow patterns are transitory in comparison to the longer times scales of mantle dynamics. Even though these core features under Indonesia and Southeast Asia have changed little over the last century and a half [*Bloxham and Jackson*, 1989], over much longer times scales we would require a dynamic coupling between the mantle and core. However, it is possible that a mantle anomaly may give rise to a preferential core flow, and cause a positive feed-back that will reinforce the mantle anomaly. Many of the studies of secular variation of core flow patterns do suggest the necessity of mantle-core coupling [*Bloxham and Gubbins*, 1987]. This could take the form of either gravitational coupling between density inhomogeneities, or topographic coupling, due to the pressure gradients in the core near CMB topography, both of which are discussed in *Jault and LeMouël* [1989, 1990] and *Bloxham and Jackson* [1991].

Electromagnetic coupling between the mantle and core was suggested by *Jeanloz* [1990] due to lateral variations in D'' electrical

TABLE 1. D'' Velocities Relative to PREM Averaged over 190 km and Inferred From the Apparent Slownesses of Diffracted P and S Wave Profiles

Number of Profiles	Path Description	CMB Region Sampled	ΔV_{190}
	Diffracted P		
1	Taiwan to the Eastern Americas	Arctic Ocean / northern Canada	+0.8%
3	Taiwan/Korea to the Western Americas	northwest North America	-1.0%
3	Indonesia to North America	North Pacific Rim	-0.9%
3	Tonga to North America	east central Pacific	-0.1%
1	Sandwich Islands to North America	northeast South America	-0.6%
3	South America to Europe/Asia	northeast Atlantic / Mediterranean	-1.0%
1	Tonga to Europe/Asia	northern Pacific	-0.3%
1	Indonesia to Europe	north central Asia	+0.6%
4	Tonga to Mid-East (whole)	Indonesia / Southeast Asia	-0.3%
4	Tonga to Mid-East (first part)	Indonesia	-3.2%
4	Tonga to Mid-East (second part)	Southeast Asia	+0.7%
	Diffracted S		
1	Burma to North America	Artic Ocean	-0.1%
1	Indonesia to Europe	north central Asia	-0.4%
3	Japan/Kuriles to Americas	northwest North America	+1.0%
3	Tonga to North America	east central Pacific	+0.5%
4	Tonga to Mid-East (whole)	Indonesia / Southeast Asia	-0.2%
3	Tonga to Mid-East (first part)	Indonesia	-3.1%
3	Tonga to Mid-East (second part)	Southeast Asia	+0.8%

conductivity of more than 11 orders of magnitude. Metal-rich heterogeneities in D″ would pin the magnetic field lines from the core, either distorting the image of flow patterns or controlling core flow near the CMB. The metal-rich D″ rock needed to maintain this electromagnetic coupling, FeO and FeSi created from iron-silicate reactions and locally aggregated through intra-D″ convective sweeping [*Davies and Gurnis*, 1986; *Zhang and Yuen*, 1988; *Hansen and Yuen*, 1989; *Sleep*, 1988], would have significantly slower velocities than perovskite, and only small additional amounts would be required to give us the slow D″ velocities we see under Indonesia. So while the correlation between seismic and geomagnetic images may be coincidental, it is not unlikely that this is an indication of significant coupling between the mantle and core.

It is interesting to note that the *P* and *S* velocities do not always differ from PREM in the same way. In *Wysession and Okal* [1988, 1989] we found that the CMB region underneath Alaska and Canada, along the northern rim of the Pacific, had relatively fast *S* velocities and relatively slow *P* velocities, and in quantifying this with the reflectivity synthetics we still found this to be true. The three *Sd* profiles from Japan and the Kurile Trench to the Americas had a velocity anomaly of $\Delta\beta_{190} = +1.0\%$, whereas for three similar profiles from Taiwan/Korea to the Americas (as well as three from Indonesia to North America) the *P* anomaly was $\Delta\alpha_{ave} = -1.0\%$. The fast shear velocities occur in the same region where *ScS*-precursor studies like *Lay and Helmberger* [1983] and *Young and Lay* [1990] have found evidence of a very high *S*-velocity zone, and are also seen in the tomographic shear velocity models of *Tanimoto* [1987] and *Grand* [pers. comm., 1991]. The same high velocity zone has not been seen there from *PcP* precursors, and in fact tomographic *P* velocity models [*Morelli and Dziewonski*, 1987; *Inoue et al.*, 1990] also find slightly slower anomalies. The occurrence of fast shear velocities at the base of the mantle beneath the rim of the Northern Pacific would not be surprising under the geodynamic circumstances. Subduction has been occurring for a long time there, and because the absolute plate motion of the North America/Pacific trench is very slow - on the order of 1 cm/yr [*Gripp and Gordon*, 1990] - there has been a lot of cold material that has been put into the mantle above where our diffracted waves sample the CMB. If the slabs penetrated into the lower mantle, or if convection limited to the upper mantle was thermally coupled to the lower mantle, then we might expect to see an accumulation of the mantle dregs there at the CMB [*Ringwood*, 1975; *Hofman and White*, 1982].

Poisson Ratio

The fact that *P* velocities here are slow, however, suggests that the shear and bulk elastic moduli here are behaving in a manner different from D″ rock elsewhere, again suggesting a different chemical signature. We can quantify this behavior with the Poisson ratio, ν, which equals 0.5 for a liquid but decreases with more pronounced rigidity. As we will show, variations in temperature and pressure can give rise to changes in ν. For the Japan-Americas path the reversal of velocity variations leads to a Poisson ratio of $\nu_{190} = 0.292$. This is 4% less than for PREM, which yields $\nu_{190} = 0.304$. These values are listed in Table 2, along with those for the

TABLE 2. Velocity Results for Dual *P/S* Coverage, Including Poisson Ratios

Path Description	α_{190}, km/s	β_{190}, km/s	ν
PREM	13.69	7.26	0.304
Taiwan/Japan to Western Americas	13.55	7.34	0.292
Indonesia to Europe	13.77	7.23	0.309
Tonga to North America	13.68	7.30	0.301
Tonga to Mid-East (first part)	13.25	7.04	0.303
Tonga to Mid-East (second part)	13.79	7.32	0.304

other CMB regions for which we had both *Pd* and *Sd* coverage. While the paths from Tonga, to North America and both halves to the Mid-East, do not display any variation in ν, the Indonesia to Europe path, sampling D″ beneath Northern Siberia, shows an increase in ν because the shear velocity is slow but the *P* velocity is fast. While the 5.5% variation seen in the Poisson ratios of our profiles does not seem large given the possible errors in P and S velocities (1% errors for V_p and V_s would suggest approximately 3.5% errors in ν), it is significant because contaminating factors (source and mantle heterogeneities, conversions from slownesses to average velocities) will effect V_p and V_s in similar manners.

Thermal gradient

One explanation for the 5.5% range we see in the D″ Poisson ratio could be a lateral variation in the thermal expansivity due to slight compositional changes, in connection with a rapidly increasing thermal gradient. D″ probably sees a departure from a mid-mantle adiabat of around 0.5°C/km to as much as 20°C/km by the time the core is reached, much like the Earth's surface. And the top of the lithosphere, the other major thermal boundary layer in the Earth, also displays strong variations in ν [*Clarke and Silver*, 1991]. Extending the derivations of *Stacey and Loper* [1983] and *van Loenen* [1988] to include the shear modulus terms, if we expand the vertical *P* and *S* velocity gradients into their fundamental thermoelastic constituents, we go from

$$\frac{dv_s}{dz} = \left[\frac{\partial v_s}{\partial T}\right]_P \frac{dT}{dz} + \left[\frac{\partial v_s}{\partial P}\right]_T \frac{dP}{dz} \qquad (1)$$

and

$$\frac{dv_p}{dz} = \left[\frac{\partial v_p}{\partial P}\right]_T \frac{dP}{dz} + \left[\frac{\partial v_p}{\partial T}\right]_P \frac{dT}{dz} \qquad (2)$$

to

$$\frac{dv_s}{dz} = \left\{\frac{1}{2v_s\rho}\left[\frac{\partial\mu}{\partial T}\right]_P + \frac{\alpha v_s}{2}\right\}\frac{dT}{dz} + \qquad (3)$$

$$+ \frac{g}{2v_s}\left\{\left[\frac{\partial\mu}{\partial P}\right]_T - \frac{\rho v_s^2}{K_s}(1 + \gamma_{th}\alpha T)\right\}$$

and

$$\left[\frac{dv_p}{dz}\right] = \frac{1}{2v_p\,\rho}\left\{\left[\frac{\partial K_S}{\partial T}\right]_P + \frac{4}{3}\left[\frac{\partial \mu}{\partial T}\right]_P + v_p^2\,\rho\alpha\right\}\frac{dT}{dz} + \quad (4)$$

$$+ \frac{g}{2v_p}\left\{\left[\frac{\partial K_S}{\partial P}\right]_T + \frac{4}{3}\left[\frac{\partial \mu}{\partial P}\right]_T - \frac{\rho v_p^2}{K_S}(1 + \gamma_{th}\,\alpha T)\right\}$$

where v_S is the S velocity, v_P the P velocity, z the depth, P the pressure, T the temperature, μ the shear modulus, K_S the adiabatic incompressibility, ρ the density, g the acceleration of gravity, γ_{th} the thermal Grüneisen ratio, and α the coefficient of thermal expansion. An interesting result occurs when reasonable lower mantle values are inserted, shown in Figures 3 and 4 [values were taken from *Dziewonski and Anderson*, 1981; *Lay*, 1989; *Isaak et al.*, 1989; *Knittle at al.*, 1986; *Hemley et al.*, 1987; *Yeganeh-Haeri et al.*, 1989]. The only difference between the two is that in Figure 4 the thermal expansivity is increased from 1.3×10^{-5} K^{-1} to 4.0×10^{-5} K^{-1}, and $\partial K_S/\partial T$ from -0.015 GPa/K to -0.035 GPa/K. Such a change is not unreasonable, as there have been a wide range of experimental values put forward, and the temperature derivatives of elastic moduli at 3000 K and 136 GPa are certainly not known to within a factor of 2. These changes have reversed the order in which the P and S velocity gradients become negative with increasing thermal gradients. In the case of Figure 3 the S velocity gradient becomes negative first, causing there to be a region along the thermal gradient axis where the P velocities would still be increasing but the S velocities would be decreasing. Though this linear approximation is really only meaningful at a particular point in temperature/pressure space, it can serve to demonstrate that

since the thermal gradient will be increasing with depth in D″ there would be a physical layer within D″, corresponding to this region along the thermal gradient axis, where the P and S velocity gradients would be reversed. In Figure 4 just the opposite occurs - there will be a layer within D″ where the S velocities will still be increasing but the P velocities will be decreasing. Beneath these layers both velocities will decrease as the much hotter iron core draws near.

The implication here is that if we were to travel laterally along the CMB between regions whose materials had different physical properties, such as differing amounts of perovskite and magnesiowüstite, we might expect the P and S velocities to vary in different ways. If there really is as much variation in the thermal expansivity in D″ as there is in laboratories at the surface, then we could expect lateral variations of the increase in the thermal gradient to be driving the differences between P and S velocities.

D″ Equations of State

One avenue of modeling seismic velocities in D″ is through the use of equations of state of mineral phases that we know are stable at CMB conditions and presume comprise a significant part of the lower mantle. This method uses the standard temperature and pressure (STP) elastic moduli and their temperature and pressure derivatives of the iron and magnesium end members of perovskite and magnesiowüstite, and calculates the elastic moduli and density at CMB conditions. This allows us to see what kinds of thermochemical variations are necessary in order to explain the seismic lateral heterogeneity, such as the anomalously slow D″ velocities underneath Indonesia and the adjacent region of fast velocities.

Our investigations have used a third-order Birch-Murnaghan equation of state in order to model our D″ velocity anomalies [*Birch*, 1952; *Bina and Helffrich*, 1991].CMB velocities are cal-

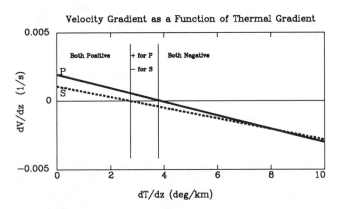

Figure 3. An example of the P and S velocity gradients as functions of the thermal gradient at the base of the mantle. The thermoelastic parameters used for the calculation are K_S = 685 GPa; μ = 291 GPa; $\alpha = 1.3 \times 10^{-5}$ K^{-1}; $\partial K_S/\partial T$ = -0.015 GPa/K; $\partial \mu/\partial T$ = -0.035 GPa/K; K_S' = 4.0; μ' = 1.9; T = 3000 K (sources are given in the text). Note that the shear velocity gradient becomes negative before the P velocity gradient, creating a layer in D″ in which P velocities are still increasing but S velocities are decreasing.

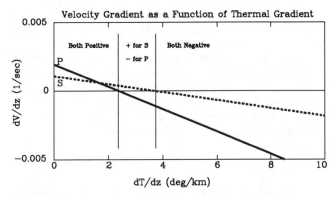

Figure 4. An example of the P and S velocity gradients as functions of the thermal gradient at the base of the mantle. The thermoelastic parameters used for the calculation are K_S = 685 GPa; μ = 291 GPa; $\alpha = 4.0 \times 10^{-5}$ K^{-1}; $\partial K_S/\partial T$ = -0.035 GPa/K; $\partial \mu/\partial T$ = -0.035 GPa/K; K_S' = 4.0; μ' = 1.9; T = 3000 K (sources are given in the text).Note that in this case the P velocity gradient becomes negative before the S velocity gradient, creating a layer in D″ in which S velocities are still increasing but P velocities are decreasing. This is the opposite of the case represented in Figure 3.

culated by starting with the elastic moduli and their derivatives for the iron and magnesium end members of perovskite and magnesiowüstite, listed in Table 3, and then making the independent temperature and pressure corrections as demonstrated in *Bina and Silver* [1990]. While these initial values are difficult to obtain experimentally and are therefore subject to change with future research, they will at least give us an order of magnitude understanding of the sensitivity of D″ velocities to changes in temperature and composition. For any combination of minerals, the resulting velocities calculated for each are combined according to the molar proportions desired. We make the assumption that bulk material velocities vary linearly with the volume proportions of the minerals include, and while this assumption may not be perfectly accurate, it is much less of a worry that the assumptions we make for the starting STP parameters of (Mg, Fe)SiO₃ perovskite and (Mg, Fe)O magnesiowüstite.

Because of uncertainties in the thermoelastic parameters we do not present absolute values but rather the percentage variations in velocities due to changes in temperature and composition. The results of the computations are shown in Tables 4-6, assuming an Fe-Mg partitioning coefficient between perovskite (Pv) and magnesiowüstite (Mw) of 0.1 [*Bell et al.*, 1979; *Ito and Yamada*, 1982] and a D″ pressure of 135 GPa. We used an initial model of pyrolitic composition (Pv/(Pv+Mw) = 2/3) with a magnesium/metal ratio of 0.9 at a temperature of 3500 K and varied these three parameters. Given the particular set of thermoelastic parameters we used, the seismic velocities were sensitive to both changes in temperatures and iron/magnesium ratios, though much less so for Pv/Mw deviations.

In Table 4 we see that a 1% variation in seismic velocities could be explained by lateral variations of approximately 200° C for *P*

TABLE 3. Thermoelastic Parameters Used for the Equation of State Velocity Calculations

Parameter	MgSiO₃	FeSiO₃	MgO	FeO
V_0	24.46[a]	25.49[a]	11.25[a]	12.25[a]
K_{S0}, GPa	268[b]	268[b]	163[a]	180[a]
K'_{S0}	4.0[b,c]	4.0[b,c]	4.1[a]	3.6[a]
δ_S	2.7[d]	2.7[d]	2.8[e]	3.0[f]
α_0[g] ($\times 10^{-5}$), K⁻¹	4.9[c]	4.9[c]	4.7[e]	5.9[e,a]
$d\alpha/dT$[h] ($\times 10^{-8}$), K⁻²	1.7[c]	1.7[c]	1.04[e]	2.2[e,a]
$d\alpha/dT$[i] ($\times 10^{-8}$), K⁻²	1.6[c]	1.6[c]	0.8[e]	1.7[e,a]
μ, GPa	185[j]	185[j]	132[e]	118[k]
μ'	1.9[j]	1.9[j]	2.5[l]	2.5[l]
$d\mu/dT$ ($\times 10^{-4}$), GPa/K	-3.3[j]	-3.3[j]	-2.5[e]	-2.5[e]

[a]*Jeanloz and Thompson* [1983].
[b]*Knittle and Jeanloz* [1987].
[c]*Mao et al.* [1991].
[d]*Bukowinski and Wolf* [1990].
[e]*Isaak et al.* [1989].
[f]*Sumino and Anderson* [1984].
[g]determined at 1300 K.
[h]linear value for *T* < 1300 K.
[i]linear value for *T* > 1300 K.
[j]*Yeganehi-Haeri et al.* [1989].
[k]*Jeanloz* [1990].
[l]*Agnon and Bukowinski* [1988].

TABLE 4. Velocity Variations Due to Changes in Temperature

Temperature, K	ΔV_P	ΔV_S	Δv
3100	+1.6%	+1.1%	+1.6%
3300	+0.8%	+0.6%	+0.9%
3500	-----	-----	-----
3700	-0.9%	-0.6%	-1.0%
3900	-1.7%	-1.2%	-2.0%

and 300° C for *S*. The effect of changes in temperature on seismic velocities is most likely significantly less in D″ than at the surface, because the temperature derivative of the thermal expansivity is much smaller [*Mao et al.*, 1991] and perovskite and magnesiowüstite seem stable and far from their solidi [*Knittle and Jeanloz*, 1989a, 1991; *Vassiliou and Ahrens*, 1982] under D″ conditions. Nonetheless, most of the seismic variations from PREM for our profiles are on the order of approximately 1%, and if lateral variations in temperature over the top 200 km of the earth are any indication, then temperature could be a dominant factor driving the seismic heterogeneities. Excluding the region under Indonesia, the ranges of anomalies from our averaged profiles correspond here to $\Delta T \approx 400°$ C for *P* and $\Delta T \approx 500°$ C for *S*. These seem slightly larger than one might reasonably allow, but probably not by more than a factor of two.

However, the D″ velocity low under Indonesia, more than 3% slow for both *P* and *S* velocities, cannot be explained just as a thermal anomaly, but is well modeled by an increase in iron, as is shown in Table 5. The commonly accepted Mg/(Mg+Fe) ratio for the lower mantle is approximately 0.9, but a value of 0.7 would satisfy the Indonesian low. In actuality we would not even need quite this much iron, because regions of high iron content would be areas where the products of mantle-core reactions were swept together, and these regions would experience a thermal anomaly from the inclusion of so much core material. Both an increase in heat flux from the core and iron percolation into the mantle would be expected to increase where core liquid was flowing up and against the CMB, if it could be held in one place for a significant amount of time. So the scenario for the D″ area under Indonesia sitting over an upwelling core plume [*Voorhies*, 1986], held in

TABLE 5. Velocity Variations Due to Mg/Fe Variations

Mg/(Mg+Fe)	ΔV_P	ΔV_S	Δv
1.0	+1.9%	+1.8%	+0.3%
0.9	-----	-----	-----
0.8	-1.8%	-1.7%	-0.3%
0.7	-3.5%	-3.3%	-0.5%

TABLE 6. Velocity Variations Due to Oxide/Silicate Ratio Changes

Model	ΔV_P	ΔV_S	Δv
Pv	+0.2%	-0.9%	+3.6%
2Pv + Mw	-----	-----	-----
Pv + 2Mw	-0.1%	+1.6%	-5.8%

place by electromagnetic coupling due to the pinning of magnetic field lines by conductive iron-rich rock [*Jeanloz*, 1990], would be compatible with the *P* and *S* seismic anomaly we see there.

Changes in the relative amounts of perovskite and magnesiowüstite were not as significant as with the Mg/Fe ratio, as is seen in Table 6. Modest changes in shear velocity require very large Pv/Mw variations, and the *P* velocities are insensitive to it, given our initial parameters. Even though the shear velocity of (Mg, Fe)O is much less than for (Mg, Fe)SiO$_3$ at the surface, in our calculations this difference becomes less pronounced at great depths because recent experimental results suggest that the temperature and pressure derivatives of the shear moduli are more favorable for faster Mw than Pv [*Agnon and Bukowinski*, 1988; *Isaak et al.*, 1989; *Yeganeh-Haeri et al.*, 1989]. It is interesting, however, that for the case of Pv/Mw variations the Poisson ratio varies significantly, suggesting that in areas like the CMB under Northern North America, where *S* velocities are fast but *P* velocities slightly slow, this kind of variation may play a role. It is interesting to compare this experiment with the variations in temperature and Mg/Fe ratio - all three affect the *P* and *S* velocities at very different relative rates.

While the comparisons drawn here between seismic anomalies and thermochemical variations are highly speculative, and the correlations will change greatly as future experimental work is done, they represent the direction that CMB research will be taking in the future. As seismic (as well as geomagnetic and geodynamic) models become more refined, and as experimental mineral physics continues to advance methods to simulate deep earth conditions, we will eventually be able to map out the thermal and chemical variations at the base of the mantle and develop a full understanding of the coupling between the earth's core and mantle. There are certainly many other factors involved of which we have not taken account: additional phases such as *SiO*$_2$ stishovite may be present in significant amounts, anisotropy may be affecting our seismic velocities, etc. But CMB research is on the verge of having the insight to know what to look for and the resolution with which to see it

Acknowledgments. We would like to thank Andrea Morelli, Toshiro Tanimoto and John Woodhouse for the use of their respective tomographic mantle models, and Tim Clarke for introducing the reflectivity method. We also thank two anonymous reviewers for their comments. This research was partially supported by NSF grant EAR-84-05040.

REFERENCES

Agnon, A., and M. S. T. Bukowinski, High pressure shear moduli - a many-body model for oxides, *Geophys. Res. Lett.*, *15*, 209-212, 1988.

Bell, P. M., T. Yagi, and H.-K. Mao, Iron-magnesium distribution coefficients between spinel [(Mg,Fe)$_2$SiO$_4$], magnesiowüstite [(Mg,Fe)O], and perovskite [(Mg,Fe)SiO$_3$], Vol. 78, Year Book Carnegie Inst. Washington, 618-621, 1979.

Bina, C. R., and G. R. Helffrich, Calculation of elastic properties from thermodynamic equation of state principles, *Ann. Rev. Earth Planet. Sci.*, in prep., 1991.

Bina, C. R., and P. G. Silver, Constraints on lower mantle composition and temperature from density and bulk sound velocity profiles, *Geophys. Res. Lett.*, *17*, 1153-1156, 1990.

Birch, F., Elasticity and constitution of the Earth's interior, *J. Geophys. Res.*, *57*, 227-286, 1952.

Bloxham, J., Simple models of fluid flow at the core surface derived from geomagnetic field models, *Geophys. J. Int.*, *99*, 173-182, 1989.

Bloxham, J., and D. Gubbins, Thermal core-mantle interactions, *Nature*, *325*, 511-513, 1987.

Bloxham, J., and A. Jackson, Simultaneous stochastic inversion for geomagnetic main field and secular variation, 2, 1820-1980, *J. Geophys. Res.*, *94*, 15,753-15,769, 1989.

Bloxham, J., and A. Jackson, Fluid flow near the surface of Earth's outer core, *Rev. of Geophys.*, *29*, 97-120, 1991.

Bukowinski, M. S. T., and G. H. Wolf, Thermodynamically consistent decompression: Implications for lower mantle composition, *J. Geophys. Res.*, *95*, 12,583-12,593, 1990.

Chapman, C. H., and R. A. Phinney, Diffracted seismic signals and their numerical solution, *Methods in Computational Physics, 12*, 165-230, 1972.

Chopelas, A., and R. Boehler, Thermal expansion measurements at very high pressure, systematics, and a case for a chemically homogeneous mantle, *Geophys. Res. Lett.*, *16*, 1347-1350, 1989.

Clarke, T. J., and P. G. Silver, A procedure for the systematic interpretation of body wave seismograms - I. Application to Moho depth and crustal properties, *Geophys. J. Int.*, *104*, 41-72, 1991.

Cleary, J. R., The D'' region, *Phys. Earth Planet. Inter.*, *30*, 13-27, 1974.

Cohen, R. E., Elasticity and equation of state of MgSiO$_3$ perovskite. *Geophys. Res. Lett.*, *14*, 1053-1056, 1987.

Cormier, V. F., Slab diffraction of *S* waves, *J. Geophys. Res.*, *94*, 3006-3024, 1989.

Davies, G. F., and M. Gurnis, Interaction of mantle dregs with convection: lateral heterogeneity at the core-mantle boundary, *Geophys. Res. Lett.*, *13*, 1517-1520, 1986.

Dziewonski, A. M., and D. L. Anderson, Preliminary reference earth model, *Phys. Earth Planet. Inter.*, *25*, 297-356, 1981.

Fei, Y., H.-K. Mao, and B. O. Mysen, Experimental determination of element partitioning and calculation of phase relations in the MgO-FeO-SiO$_2$ system at high pressure a high temperature, *J. Geophys. Res.*, *96*, 2157-2170, 1991.

Gripp, A. E., and R. G. Gordon, Current plate velocities relative to the hotspots incorporating the NUVEL-1 global plate motion model, *Geophys. Res. Lett.*, *17*, 1109-1112, 1990.

Hansen, U., and D. A. Yuen, Dynamical influences from thermal-chemical instabilities at the core-mantle boundary, *Geophys. Res. Lett.*, *16*, 629-632, 1989.

Hemley, R. J., M. D. Jackson, and R. G. Gordon, Theoretical study of the structure, lattice dynamics, and equations of state of perovskite-type MgSiO$_3$ and CaSiO$_3$, *Phys. Chem. Minerals*, *14*, 2-12, 1987.

Hofman, A. W., and W. M. White, Mantle plumes from ancient oceanic crust, *Earth Planet. Sci. Lett.*, *57*, 421-436, 1982.

Inoue, H., Y. Fukao, K. Tanabe, and Y. Ogata, Whole mantle *P*-wave travel time tomography, *Phys. Earth Planet. Int.*, *59*, 294-328, 1990.

Isaak, D. G., O. L. Anderson, and T. Goto, Measured elastic moduli of single-crystal MgO up to 1800 K, *Phys. Chem. Minerals, 16*, 704-713, 1989.

Ito, E., and H. Yamada, *Stability relations of silicate spinels, ilmenites and perovskites*, In High-Pressure Research in Geophysics, ed. S. Akimoto and M. H. Manghnani, 405-419, 1982.

Jault, D., and J.-L. LeMouël, The topographic torque associated with a tangentially geostrophic motion at the core surface and inferences on the flow inside the core, *Geophys. Astrophys. Fluid Dyn., 48*, 273-296, 1989.

Jault, D., and J.-L. LeMouël, Core-mantle boundary shape: Constraints inferred from the pressure torque acting between the core and mantle, *Geophys. J. Int., 101*, 233-241, 1990.

Jeanloz, R., The nature of the Earth's core, *Annu. Rev. Earth Planet. Sci., 18*, 357-386, 1990.

Jeanloz, R., and A. B. Thompson, Phase transitions and mantle discontinuities, *Rev. Geophys. Space Phys., 21*, 51-74, 1983.

Jeffreys, H., and K. E. Bullen, *Seismological tables*, 50 pp., Brit. Assoc. Adv. Science, London, 1970.

Kennett, B. L. N., *Seismic Wave Propagation in Stratified Media*, Cambridge University Press, Cambridge, MA, 1983.

Knittle, E., and R. Jeanloz, Synthesis and equation of state of (Mg, Fe)SiO_3 perovskite to over 100 GPa, *Science, 235*, 668-670, 1987.

Knittle, E., and R. Jeanloz, Melting curve of (Mg,Fe)SiO_3 perovskite to 96 GPa: evidence for a structural transition in lower mantle melts, *Geophys. Res. Lett., 16*, 421-424, 1989a.

Knittle, E., and R. Jeanloz, Simulating the core-mantle boundary: an experimental study of high-pressure reactions between silicates and liquid iron, *Geophys. Res. Lett., 16*, 609-612, 1989b.

Knittle, E., and R. Jeanloz, The high pressure phase diagram of $Fe_{0.94}O$: A possible constituent of the Earth's core, *J. Geophys. Res., 96*, 16169-16180, 1991.

Knittle, E., R. Jeanloz, and G. L. Smith, Thermal expansion of silicate perovskite and stratification of the earth's mantle, *Nature, 319*, 214-216, 1986.

Lay, T., Structure of the core-mantle transition zone: a chemical and thermal boundary layer, *Eos Trans. AGU, 70*, 49, 1989.

Lay, T., and D. V. Helmberger, A lower mantle S-wave triplication and the velocity structure of D″, *Geophys. J. R. astron. Soc., 75*, 799-837, 1983.

Mao, H. K., R. J. Hemley, Y. Fei, J. F. Shu, L. C. Chen, A. P. Jephcoat, Y. Wu, and W. A. Bassett, Effect of pressure, temperature and composition on lattice parameters and density of (Fe,Mg)SiO_3-perovskites to 30 GPa, *J. Geophys. Res., 96*, 8069-8079, 1991.

Mondt, J. C., *SH* waves: Theory and observations for epicentral distances greater than 90 degrees, *Phys. Earth Planet. Inter., 15*, 46-59, 1977.

Morelli, A., and A. M. Dziewonski, Topography of the core-mantle boundary and lateral heterogeneity of the liquid core, *Nature, 325*, 678, 1987.

Mula, A. H., Amplitudes of diffracted long-period *P* and *S* waves and the velocities and Q structure at the base of the mantle, *J. Geophys. Res., 86*, 4999-5011, 1981.

Mula, A. H., and G. Müller, Ray parameters of diffracted long period *P* and *S* waves and the velocities at the base of the mantle, *Pure Appl. Geophys., 118*, 1270-1290, 1980.

Ringwood, A., *Composition and petrology of the Earth's mantle*, McGraw-Hill, New York, 1975.

Sacks, S., Diffracted waves studies of the earth's core: 1. Amplitudes, core size, and rigidity, *J. Geophys. Res., 71*, 1173-1181, 1966.

Sleep, N. H., Gradual entrainment of a chemical layer at the base of the mantle by overlying convection, *Geophys. J. R. astron. Soc., 95*, 437-447, 1988.

Stacey, F. D., and D. E. Loper, The thermal boundary-layer interpretation of D″ and its role as a plume source, *Phys. Earth Planet. Interiors, 33*, 45-55, 1983.

Stevenson, D. J., On the role of surface tension in the migration of melts and fluids, *Geophys. Res. Lett., 13*, 1149-1152, 1986.

Sumino, Y., and O. L. Anderson, Elastic constants of minerals, In *CRC Handbook of Physical Properties of Rocks, Volume III*, ed. R. S. Carmichael, Boca Raton, Florida: CRC Press, Inc., 39-138, 1984.

Tanimoto, T., The three-dimensional shear wave structure in the mantle by overtone waveform inversion - I. Radial seismogram inversion, *Geophys. J. R. astron. Soc., 89*, 713-740, 1987.

van Loenen, P. M, S velocity at the base of the mantle from diffracted SH waves recorded by the NARS array, M.Sc. Thesis, Department of Theoretical Geophysics, Utrecht, 1988.

Vassiliou, M. S., and T. J. Ahrens, The equation of state of $Mg_{0.6}Fe_{0.4}O$ to 200 GPa, *Geophys. Res. Lett., 9*, 127-130, 1982.

Voorhies, C. V., Steady flows at the top of Earth's core derived from geomagnetic field models, *J. Geophys. Res., 91*, 12,444-12,466, 1986.

Woodhouse, J. H., and A. M. Dziewonski, Models of the upper and lower mantle from waveforms of mantle waves and body waves, *Eos Trans. AGU, 68*, 356-357, 1987.

Wysession, M. E., Diffracted seismic waves and the dynamics of the core-mantle boundary, Ph.D. Thesis, Northwestern University, Evanston, Illinois, 190 pp., 1991.

Wysession, M. E., and E. A. Okal, Evidence for lateral heterogeneity at the core-mantle boundary from the slowness of diffracted *S* profiles, *AGU Monog., 46*, 55-63, 1988.

Wysession, M. E, and E. A. Okal, Regional analysis of D″ velocities from the ray parameters of diffracted *P* profiles, *Geophys. Res. Lett., 16*, 1417-1420, 1989.

Yeganeh-Haeri, A., D. J. Weidner, and E. Ito, Elasticity of $MgSiO_3$ in the perovskite structure, *Science, 243*, 787-789, 1989.

Young, C. J., and T. Lay, The core mantle boundary, *Ann. Rev. Earth Planet. Sci., 15*, 25-46, 1987.

Young, C. J., and T. Lay, Multiple phase analysis of the shear velocity structure in the D″ region beneath Alaska, *J. Geophys. Res., 95*, 17,385-17,402, 1990.

Zhang, S., and D. A. Yuen, Dynamical effects on the core-mantle boundary from depth-dependent thermodynamical properties of the lower mantle, *Geophys. Res. Lett., 15*, 451-454, 1988.

C. R. Bina and E. A. Okal, Department of Geological Sciences, Northwestern University, Evanston, IL 60208.

M. E. Wysession, Department of Earth and Planetary Sciences, Washington University, Campus Box 1169, One Brookings Dr., St. Louis, MO 63130-4899.